D0874302

A Guide to Facilities Planning

ROBERT W. JAMES
Business Department
DeVRY, Inc.
Atlanta, Georgia

PAUL A. ALCORN
General Education Department
DeVRY, Inc.
Atlanta, Georgia

PRENTICE HALL, Englewood Cliffs, New Jersey 07632

Library of Congress Cataloging-in-Publication Data

James, Robert W. (date)
A guide to facilities planning / Robert W. James, Paul A. Alcorn.
p. cm.
Includes bibliographical references and index.
ISBN 0-13-299256-6
1. Factories—Planning. 2. Facility management. 3. Factories—Design and construction. I. Alcorn. Paul. II. Title.
TS177.J36 1991
658.2—dc20

90-21741
CIP

Editorial/production supervision and
interior design: David Ershun
Cover design: Karen Stephens
Manufacturing buyer: Lori Bulwin

A Division of Simon & Schuster
Englewood Cliffs, New Jersey 07632

Printed in the United States of America

10 9 8 7 6 5 4 3 2 1

ISBN 0-13-299256-6

PRENTICE-HALL INTERNATIONAL (UK) LIMITED, *London*
PRENTICE-HALL OF AUSTRALIA PTY. LIMITED, *Sydney*
PRENTICE-HALL CANADA INC., *Toronto*
PRENTICE-HALL HISPANOAMERICANA, S.A., *Mexico*
PRENTICE-HALL OF INDIA PRIVATE LIMITED, *New Delhi*
PRENTICE-HALL OF JAPAN, INC., *Tokyo*
SIMON & SCHUSTER ASIA PTE. LTD., *Singapore*
EDITORA PRENTICE-HALL DO BRASIL, LTDA., *Rio de Janeiro*

This book is dedicated to our families,
those who are near and those who are far,
whose love has helped us become what we are.

R.W.J.

For Lisa with love
because some things are forever

P.A.A.

Contents

Preface and Acknowledgments

The subject of facilities planning spans a wide range of disciplines: real estate, financing, insurance, ergonomics, architecture, law, and management. In fact, the actual breadth of information needed to successfully carry out a major facilities planning and design project would require the expertise of not one but many experts in a host of disparate fields.

Inherent in the development process is a need to plan for financing, site selection, insurance, land procurement, building development, coordination of functions, handling of legal considerations, personnel development, and many other considerations. This makes the creation of any single text on the subject a difficult task indeed.

Yet there is a great need for an overview of the subject, particularly in light of the dramatic increase in small company development now taking place in the United States. These small companies especially need the concepts of just-in-time facilities if they are to compete in the world community.

Large firms have the capacity to devote labor and material to development efforts. They can hire experts in the field and utilize the services of specialized consultants. But the small manufacturer or service organization, as well as the independent entrepreneur just beginning the process of development, contend with a serious lack of knowledge, advice, or possibly even understanding of what is entailed in facilities planning.

For these reasons, and in response to the educational need for a textbook specifically dealing with the facilities planning process, this book has been developed.

This is an introductory text, designed to serve a twofold purpose. First, it provides an excellent overview of the content of facilities planning, offering the student an opportunity to get a general grounding in the field through readings, detailed examples, and a case study designed to offer the practical experience of going through the planning process.

Additionally, it functions as a handbook for those embarking on the facilities planning and design process for the first time. It presents a logical progression of steps from conception of the facility idea to execution of the rough-cut sketches that would be given to construction firms for use in developing the facility.

This will ensure the smooth execution of such a project with a minimum number of surprises. It is our desire and intention that this book be of service to those interested in the topic, for whatever reason, and that it present enough data either to serve the reader's purpose or to point the way to more specialized information where it is needed.

Acknowledgments

Most books have an acknowledgments section, and after coauthoring this book, I understand why. Writing a book is like being a member of a movie production company. The stars get all the glory, but there are countless people behind the scenes who also deserve mention. I hope no one will be overlooked in this list of credits.

It is hard to pinpoint the moment when this book started. The desire to write it was placed in me by God, I am sure, because He was able to clear every obstacle that confronted me, and He also gave me some special people to help in the process.

First, there is my very patient and literate wife, Judy, who acted as our primary grammatical editor throughout the entire five-year process. Second, there is Paul, my coauthor, who has a rare gift; he was able to take my phrases and turn them into sentences. Third, there is my extended family, who encouraged me to keep after it.

Others were involved, too. David Ershun and Judy Ashkenaz worked out all the typesetting details. The production team at Prentice-Hall, some of whom I know only over the phone and some of whom I have never known, gave us the extra effort that brought this book together. Then there were the members of my professional societies, the American Production and Inventory Control Society (APICS) and the American Institute of Building Design (AIBD), whom we called on to answer questions. The administration of DeVRY gave us the chance to help design the course for which this textbook is written. The faculty there supported us as it was being written. We hope this book is up to their high standards of education.

Finally, the many clients who have become friends and the many friends who have become clients will probably see a little bit of their influence in this book. Thank you to all.

R.W.J.

I echo Rob's acknowledgments of that legion of individuals who contributed to the development and production of this book. Special thanks should also be extended to my daughters, Jessica and Meagan Alcorn, for their courage in the face of numerous tirades and for their understanding during those times when my attention was consumed by writing this book rather than attending to their needs.

In particular, I would like to thank the 1988 and 1989 members of the Facilities Planning classes for acting as the guinea pigs on whom much of this material was tested. They contributed immeasurably to the clarity and style of the writing.

Finally, I wish to thank Rob for providing such a clear and definitive body of research with which to work, and for his patience in leading me through the labyrinthine halls of this subject. All these people and more share in the successful completion of this work.

P.A.A.

Introduction

Last spring, while stuck in morning rush hour traffic, a certain individual came up with an idea for a business. He had always liked water activities, so a marina business seemed attractive. He had dreamed for a long time of the day when it would be possible to begin a family business.

He sought advice from many friends before realizing that a professional facility planner would be needed in order to bring all the details into place. The planner suggested a written proposal of between 40 and 50 pages in length, detailing the new business venture. It was to include some marketing plans, financial spreadsheets, "rough-cut" facility sketches, implementation timetables, and other pertinent data.

This proposal would be geared to obtaining a lender's approval for a loan. A loan was necessary to keep the business going for the first three to five years of business activity, when profits were expected to be low. The down payment was to come from the individual's retirement account.

The planner preferred that the final proposal be in two forms, a hard copy and a soft copy. The hard copy would be the finished proposal typed out by the computer. The soft copy would be three computer diskettes. The first diskette would contain the written proposal, the second would contain the financial plans, and the third would contain the facility design.

According to the planner, the soft copy of the proposal (the computer diskettes) would allow for changes at a later date should the individual not want to terminate present employment in order to start the business. The hard copy of the proposal (the typewritten plan) would be immediately available should the individual choose to begin taking the proposal to lenders and government agencies for approval.

What follows is what any facility planner would consider in planning for any business. It is written to assist you in developing your own business proposal or, if you prefer to have that done by a facility planner, to help you understand the planning process that the consultant will employ. (The finished marina proposal is the case study given in Chapter 6.)

PART ONE
Site Selection and Analysis

1
Company Criteria

Corporate decisions are not very different from decisions that individuals make in their daily lives. Decisions like what to wear today, what to study first and for how long, and where to go to college are not fundamentally different from corporate decisions such as what to produce, where to produce it, and how much to charge. Such decisions go through similar processes of mental calculation to determine the values involved and whether or not a specific decision will result in the attainment of those values. In other words, does a given act or decision promote the accomplishment of some goal for the individual or the company? A rational approach would seem to indicate that any action that does support one's values is an action to be taken, whereas one that does not is probably a waste of time.

Although individuals often must make truly complex decisions, corporate decisions seem to be even more complex. More people and institutions are affected, and there is an increased number of interactive elements on which the decision is based. The more that is affected by a decision, the more complex making that decision becomes.

One very real fear that businesses must deal with, particularly in our modern world of mass communication and mass data availability, is one of paralysis by analysis. Because of the extraordinary amount of information available to us in the business world, there is sometimes a temptation to try to take everything into account, thus rendering the problem insoluble by virtue of its enormous number of variables.

It is not necessary to take everything into account in making a decision. Not every scrap of information is important to the decision, nor does every variable have a serious effect on the decision itself. The trick is to determine what the important elements, the criteria, for a successful project are going to be. For example, what does a company need to consider in making the decisions involving the creation of a new facility? What are the important issues? What is superfluous to the decision or secondary and of low-order priority? Deciding on these criteria is the first step in creating a new facility. They are, in fact, the first decisions that must be made.

I. DETERMINING EXPANSION CRITERIA

A. Short Term versus Long Term

The deciding criteria for expansion can be thought of as existing in two dimensions, one short term, or the more immediate needs of the company, and one long term, the long-range needs of the company. These criteria may be fundamentally different, yet both must be satisfied before making the decision to expand.

It may be, for instance, that there are immediate needs for expansion based on present facilities being utilized to or beyond their logical limit. At the same time, it may be that the future is expected to bring an overall reduction in demand for the company's product and that, in fact, what is required is a retooling of facilities to prepare for a shift to the manufacture of some alternative product or product line. In considering expansion, both needs should be taken into account. In many cases, present plant capacity is sufficient for present levels of demand although future trends indicate that more plant space and equipment will be needed. Expansion may become a matter of preparing for the future (long-term expansion criteria) rather than dealing with current needs.

Both short- and long-term needs must be dealt with, however. What criteria does the company have for the short run? How important are such considerations as efficient use of capital, utilization of plant capacity, and age of equipment? Will these considerations operate as constraints on future expansion? And what are the long-term criteria of the company? Is it essential to have increased productive capacity before the actual increase in demand occurs? What is the lag time between the decision to expand and the final creation of the additional facilities? How much lead time do you have? How quickly can new facilities become self-supporting or profitable? How do expansion plans fit into the overall strategy for company development?

Every company has a set of plans that it works with, from its grand competitive strategy to its day-to-day procedures. These plans exist whether they are officially enunciated or not, though it is to be hoped that they are. The key in deciding how to integrate the expansion needs of a firm into the general philosophy of that firm is to understand that both the short-term and long-term needs of the company must be satisfied, and that the firm itself is the deciding element in determining just what those short- and long-term needs really are. Any plan for expansion must reflect and support the long- and short-term goals of the firm. Otherwise it becomes an isolated project, separate from the rest of the organization (or, if you will, organism), that is disfunctional, useless, and in many cases, damaging to the health of the firm itself. The difference between cancerous cells and healthy ones is whether or not their growth adds to the growth and vitality of the entire organism. This is no less true of the organism known as the firm. Before such decisions can be made, it must first be determined whether or not certain criteria will be met.

1. Short-term criteria

a. Immediate Needs for Expansion. The criterion here is a question of how important expansion is to the immediate capacity of the company to supply goods and services to customers. If there is an immediate need for expansion to fill current orders, then a degree of urgency enters the decision-making process, automatically pushing the decision makers in certain directions. If demand has recently expanded at an unanticipated rapid rate, immediate relief may be needed to ease the strain on present facilities. This dictates that long, time-consuming approaches to expansion will not work, since immediate changes are in order. There will be no time, for instance, to engage in extensive economic analysis before proceeding. Things have to happen now.

b. Disruption of Current Activities. In an expansion situation, it must be remembered that the firm is in operation and needs to minimize the disruptive effects of any move or expansion into new facilities. The plans for the expansion must take into account that current operations need to be continued with as little disturbance as possible until the actual shift to new facilities occurs. In the case of an increase in the size of current facilities, this disruption is minimized by the fact that the firm can continue with its current operations until the expanded facilities are made ready, and as the increased space becomes available, workers and equipment can be moved little by little. If the choice is to move to a completely new façility, however, it becomes a bit more complicated in that the move may have to take place all at once. An inordinate amount of preparation may be required for a smooth transfer of equipment, offices, and personnel. Any choice on whether to and how to instigate an expansion will need to take preparation for transfer into consideration.

c. Efficient Use of Capital. Any expansion of facilities will entail the use of capital. In terms of immediate needs, how can the firm expect this expansion to affect the availability of capital for current operations? How will it affect working capital, liquidity, and other financial considerations? Can the company afford the immediate drain on cash, credit, and other financial assets in order to fund the new project? Current needs cannot be sacrificed for the sake of future expansion. A proper allocation of financial assets is imperative.

2. Long-term criteria

a. Company Goals. Is the planned expansion sufficient for long-term company goals? The expansion plans must reflect the long-term goals of the firm and contribute to those goals. Where the firm sees itself in the future will dictate what the expanded facilities will look like. Expansion is expensive and not something to be undertaken every few years if it can be avoided. It is important that the new and expanded facilities be sufficient to satisfy future needs, not just immediate ones. Among other things, this means a productive capacity sufficient to satisfy expected long-term demand for the product or service the firm offers.

b. Changes in Production Techniques. What changes in manufacturing techniques will be necessary or desirable in the new facilities? An expansion of present facilities or creation of new facilities offers an opportunity to modernize the production techniques employed by the company. Which of these possibilities should be pursued is a matter for long-term consideration. By creating a state-of-the-art facility the firm can forestall obsolescence, particularly with regard to the efficiency of its competition, and increase its own overall efficiency, but this may not always be necessary. Many factors shape the decision as to whether or not an extensive change in methodology is necessary to achieve the company's goals.

c. Specialization Concerns. To what degree should specialization and standardization determine the nature of the future facilities? The issue here is one of how much future change in the industry can reasonably be expected to occur. If change is expected to be extreme, a more generalized approach to construction of new facilities would be in order, offering increased flexibility of action to the company as the new changes take place. If the product is a standard one and few innovations are expected in the future, the firm would more logically be expected to opt for a facility with a high degree of specificity in equipment and a high degree of standardization (a narrow range of options as to specifications for the product) in the facility plan.

It should be relatively obvious that the short-term and long-term criteria noted above are far from exhaustive. The specific criteria that a firm will need to take into account are determined by the individual conditions under which the company operates, as we shall see in future chapters. Each firm has its own set of needs, problems, and conditions under which it operates. The crux of the issue is to be certain that the long-term and short-term goals of the firm, whatever they may be, are met in the facility expansion process.

B. Off Site versus On Site

A second decision that must be made in determining the firm's criteria involves whether to locate expansion facilities at the present site of operations or off site, at some new location. Again, there are potential advantages and disadvantages to each choice, the most effective method depending on the individual constraints under which the company is operating.

1. Off-site expansion. The advantages to off-site expansion stem from flexibility. By creating the new facility at a completely new location, the company has the option to choose exactly what it wants in terms of the various elements of the productive process. The site of the new facility can be chosen to maximize the efficiency of transportation, real estate, utilities, tax, community, personnel, and raw materials considerations as well as a host of other factors.

In addition, a completely new facility means a completely modern one, one in which there is a minimum of dated, highly depreciated equipment, an expected minimum of downtime, and a heightened level of efficiency.

On the negative side of locating in an off-site location is an expected higher cost. Whether the expansion represents a secondary facility or a completely new one into which the company intends to move, the expense of locating the site, purchasing land, and starting from the ground up (literally) can be expected to be considerably higher than what is involved with on-site expansion plans.

2. On-site expansion. The shortcomings of on-site expansion basically stem from a lack of flexibility. The capacity to expand on site depends on the suitability of the firm's present location for expansion purposes. Is there enough room? Are there local factors other than space that have initiated the expansion process to begin with? If the firm's present location is restricted as to space, a local expansion may be equally limited, or at least altered; the firm may find it necessary to build up rather than out. Additionally, expansion may result in disruption of present operations beyond what the firm is willing or able to tolerate. Machinery may need to be moved, thus increasing downtime and limiting production. Present buildings may not offer the most efficient arrangement in terms of the new expansion. The company is limited by what already exists in its efforts to create a larger, more effective facility.

On the positive side, on-site expansion minimizes costs of expansion in many cases. If there is sufficient space, the purchase of additional real estate may be unnecessary; if more space is required, the new investment represents a smaller outlay than in the case of a totally new facility. It is also easier to oversee and follow the development of new facilities that are on site.

Often, a firm will have an established relationship with local suppliers and other firms in the immediate area. Expansion on site means a minimal amount of disturbance to those relationships and eliminates the necessity of reestablishing such relationships with local suppliers and vendors at the new location. Beyond this, there is also the matter of personnel to be considered, including the need to train personnel for the new facility, the necessity of transferring personnel from one site to another (which can be quite expensive if the second site is not local to the firm's present community), and the necessity of laying off those who cannot or will not consider transfer to the new facility.

II. COMPANY APPROACHES

A. Ad Hoc Approach

Two approaches are normally available to a firm intending to expand its production capacity. One is the ad hoc approach and the other is tne formal approach. In the case of the ad hoc approach, a group of employees, usually though not necessarily at the middle and upper management level, is given the task of investigating the feasibility of and requirements for expansion. These individuals work together to gather data, develop criteria, and report results. The expansion project is not their primary work at the firm; they work on the project in addition to their normal duties. This is the case of an executive committee charging certain of its members to look into the feasibility of such a project, or of requesting investigation and supportive information for such a venture from the major department heads in anticipation of a plant expansion or move.

B. Formal Approach

Alternatively, the formal approach can be used if the issue is a major or an urgent one. In this method, a group of individuals may be hired, transferred, and/or requisitioned to act within the company as a formally functioning department, whose sole purpose is to develop plans and undertake studies connected with expansion of facilities. In this second approach, the project is given official status, a recognized position within the firm's organizational structure, as well as funding and support by the authority of the internal hierarchy. The primary difference between the two is the degree to which the individuals will be involved in the study and project development. The formal approach requires the project to be their primary duty. The ad hoc approach requires it to be merely a secondary set of duties that is subordinate to their primary raison d'etre within the company. In either case, in order to achieve a successful expansion of productive capacity, a functioning structure must be established.

III. RELOCATION CRITERIA

There are a number of different strategies available to a firm in making the decision to relocate. Actual relocation criteria will, of course, depend on individual circumstances. Yet the basic choices that the company has to work with will determine which of the strategies available to them is to be used. Depending on the nature of the product, the industry, and the market for the goods, a firm may opt for any of the following approaches.

A. Multiplant Strategies

These are strategies that entail expansion by increasing the number of facilities that a firm operates rather than expanding an existing one or moving to a larger facility. Multiplant strategies are useful when dividing the market up into geographical zones or regions or when encountering the effects of the law of diminishing returns. In the first instance, that of geographical multiplant considerations, the company may choose to locate plants in a number of different locations, each near or centered in a marketing region. The advantage of this strategy is that the production of the supplied goods is near their point of sale, thus reducing transportation costs, storage costs, and lag time between production and delivery. Other factors are also directly related to the distance between markets and point of product origination.

In the second case, where the law of diminishing returns is the primary issue, a company may be in a position in which any further increase in plant size would not result in further efficiency or increased output. There is, as the law implies, an upper limit to the productive capacity of a single productive unit. As one approaches the limit of a plant's capacity, it is often more efficient, both economically and in terms of production volume, to create a second plant of similar size rather than attempt to increase the capacity of the existing plant. A simple example will suffice to illustrate the point.

Imagine a cobbler who realizes a need for increased production capability due to a sudden rise in the demand for one particular type of shoe. Let's place our cobbler in a small kingdom in Europe some five hundred years ago, and let us assume that for some reason, the ruler of this particular kingdom takes a fancy to this special shoe.

Politics being what it is, there is an immediate rush among the local courtiers to purchase the same type of shoes as the king in order to gain some degree of favor, or at least to be noticed. (Is this apple polishing or just shoe polishing?) At any rate, because of the sudden interest of the king and then the courtiers in these particular shoes, one thing leads to another and suddenly everyone who is anyone (and some who aren't) are begging for a pair of these shoes! The fashion has suddenly changed and our cobbler's shoes are the "in" footwear for the kingdom.

Alas, our hapless cobbler can produce only eight pairs of shoes in a day, working hard all day long. There is an upper limit on the cobbler's capacity to create the product. So an assistant is hired in hopes of doubling the output. But what's this? Instead of doubling the output from eight pairs of shoes a day to 16, the cobbler finds that with the assistant, more than twice as many shoes are being produced. In fact, their production rises to 22 pairs of shoes! What's happening?

Actually, it's quite simple. The cobbler and the assistant are experiencing what is known as *economies of scale*, which, as you may know, is the result of cooperative efforts. By dividing the work, they have achieved a more than proportional increase in production. Downtime has been reduced through the process of specialization, say, into one doing uppers and the other concentrating on the production of heels and soles and the joining of the two. Economies of scale are the reason for increased size of production units in the first place. There is no fundamental difference between the rationale for what happened to our cobbler and what happens in business. Large production units tend to be more efficient due to economies of scale.

Let's get back to the cobbler. Since one assistant has worked so well, our cobbler hires a second assistant and then a third, for a total of four people—the cobbler and three assistants. In each case there is found to be a greater than proportional increase in output (in the form of shoes) as a result of an increase in input (in the form of cobblers). The second assistant raises production from 22 pairs of shoes per day to 36 pairs per day, the average production per worker going from eight to 11 and then to 12 units per day! Delighted at this result, the cobbler hires more and more cobblers, expecting economies of scale to carry their production ever upward. Is there no end to this process? Indeed there is. With the third cobbler assistant, output rises to 52 units, or an increase to 13 units per cobbler, and with the fourth assistant (a total of five cobblers cobbling away now) the production rises to 70 units, or 14 pairs of shoes per cobbler. (Production will no longer increase at an increasing rate with the increase in workers, but rather now at a constant rate.) With the fifth assistant, production rises to 84, again an increase of 14 units and with the sixth to 91, which is only an increase of 7 units! If we assume that each of the cobblers is equally competent, why is there a *decrease* in production with the introduction of the sixth assistant? For that matter, why is there not a continuing increase in production per worker? It seems each worker should be more productive than the next because of the economies of scale. Yet with the seventh assistant (eight cobblers), production averages only 10 per cobbler; with the eighth the average is 8, and with the ninth the average is 6. With the introduction of the sixth assistant, the cobbler found that net production actually decreased! But why?

The why is known as the Law of Diminishing Returns, which, simply stated, tells us that as units of productive capacity increase, eventually output will decline, first relatively and then absolutely (see Table 1.1).

By the way, did we mention that these cobblers were all working together in a room ten feet by ten feet square? By the time the tenth worker was added, the cobblers were probably cobbling each other more often than the shoes! Herein lies the problem. There is an upper limit on the efficiency that can be gained from a plant or productive process. As Figure 1.1 illustrates, at some point, it takes so much more input to create an additional unit of output that efficiency drops to zero and then takes on negative values. Just as the cobbler experiences this in one small shop, so does the firm in its plant, thus creating a need for expansion.

Diminishing returns is a physically observable phenomenon. There is an upper limit on the efficiency that can be gained from a single productive facility; when this limit is reached, the solution is to create a second productive facility. Hence multiplant production strategies are often worthwhile to the growing company.

B. Product Plant Strategies

This strategy is useful when an individual firm produces more than one product or more than one type of product. The strategy here is to create a facility for each separate product or product group, particularly if the products are fundamentally different in nature. The methodology used to create plastic toys and those used to create slingshots or yoyos may be quite different; attempting to integrate the production of both into a single facility could create problems in facility design, materials storage, inventory storage, control and testing, and so forth. By dedicating a single facility to the production of a single commodity, the firm is able to take advantage of the efficiencies of specialization, where dedicated production methods and dedicated machinery are used to maximize effectiveness. By reducing the problems inherent in the production process to those involved with a specific product,

TABLE 1.1 Economies of Scale and the Law of Diminishing Returns in Shoe Production

Total number of cobblers, including assistants	1	2	3	4	5	6	7	8	9	10
Total shoes produced	8	22	36	52	70	84	91	80	72	60
Average output/cobbler	8	11	12	13	14	14	13	10	8	6

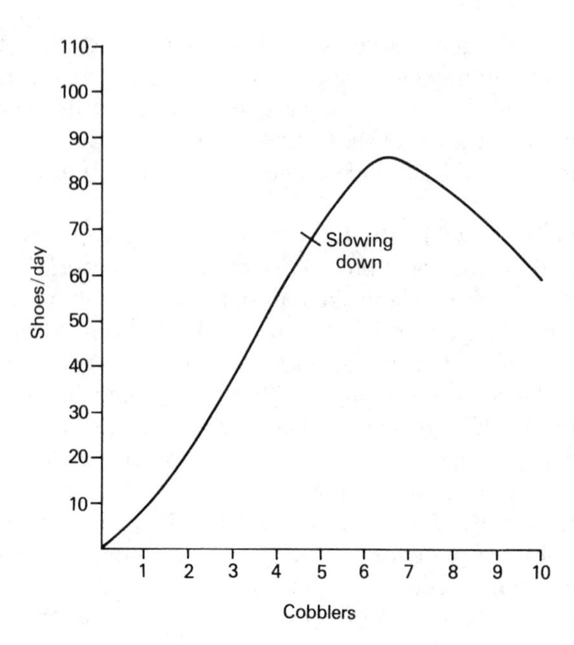

Figure 1.1 "Elf" Shoemaking Facility

there is more opportunity for control and a decrease in wasted time and effort. Further, if the products are fundamentally different as to inputs, such as the raw materials or power requirements for the productive process, the firm may locate its facilities close to the sources of specific needs, whether it be electrical power for the production of items such as aluminum, or raw labor for hand assembly of toys, electrical components, and the like.

C. Market Area Plant Strategy

This strategy is particularly useful in cases where being close to markets is advantageous to the firm. Perhaps the items produced are used in a particular industry that is located in a particular part of the country, such as textile machinery manufacturing plants locating close to textile mills in the south. Or perhaps transportation costs are prohibitive over long distances, and it is found to be advantageous to minimize these costs by locating close to the point of product use, as in the case of prefabricated houses, which are often located for regional markets. Or perhaps it is a matter of locating near a source of supply, which explains the predominance of steel industries and automotive plants along the Great Lakes, where the coal and iron ore is nearby and easily shipped by waterway. By locating near the market for the goods, transportation costs are minimized and lead time between production and delivery is reduced as is the need for inventory storage. If these are major problems, a market area plant strategy is often the answer.

D. Process Plant Strategy

This is a useful strategy for firms whose productive process involves a number of different individual processes. An example may be the case of a paint manufacturer, where individual chemicals are created prior to combining them into the final product. In addition, certain by-products created as the primary ingredients are produced may have commercial value in other contexts, either as a separate sales item or as an ingredient in other products, such as plastics or cosmetics or fertilizer. Since it is necessary that individual elements of the productive process take separate routes toward the final creation of the product and since many of the individual components may be products in themselves, it is often advantageous to isolate the production of the component parts in different facilities, finally bringing them together at a separate facility for final assembly. Thus an automotive manufacturer may produce only chassis at one plant and engines at another (or the individual components that make up these subassemblies), bringing them together for final assembly at an "assembly plant" designed for just that purpose. Through this approach, specialization is once more allowed to maximize efficiency by having the productive unit concentrate on a relatively narrow range of activities.

E. General-Purpose Plant Strategy

In this strategy, the plant is designed to be a general-purpose facility capable of producing a wide range of different products by selective use of machinery designed to perform a specific function. The secret to success here is to generalize rather than specialize. Machine shops are examples of the general-purpose plant. The equipment in machine shops is designed to carry out particular functions that may be of use in producing a wide range of products. The combination of machines and how they are used determines what is finally created. Milling machines are as useful for milling gears as they are for creating accurate edges on metal plates. Lathes can create spindles, cylinders, pistons, valves, axles, and a host of other items simply by changing their settings and specifications. It is the flexibility of the general-purpose plant that creates its usefulness, and in industries where the final product may change often, this is a much more logical approach to production. It simply does not make sense to create a dedicated facility to produce a product for which there is a limited market or one that promises to disappear soon. By producing by the job rather than by the product, general-purpose facilities are able to remain profitable.

Table 1.2 shows the four different types of facilities by industrial groups.

TABLE 1.2 Plant Charters by Industrial Groups

The table is a percentage listing of the four different types of (facilities) plants divided into industrial groups. For example, industrial group 1 has 30 percent product plants, 65 percent market area plants, 5 percent processing plants, 0 percent general-purpose plants.

	Industry Group Codes*								
Type of Plant	1	2	3	4	5	6	7	8	9
Product	30	25	43	83	65	54	58	80	73
Market area	65	70	50	11	21	38	33	8	3
Process	5	4	0	6	9	8	8	8	19
General purpose	0	1	7	0	5	0	1	4	5
Total (percent)	100	100	100	100	100	100	100	100	100

*Industry group codes explained:

1. *Agriculture-tied.* Some food processing as in meat, tobacco.

2. *Market sensitive.* Some food processing, as in dairy, printing, plastics, fabrication, can making, miscellaneous manufacturing.

3. *Forest-tied.* Lumber, pulp, and paper products.

4. *Labor-rate sensitive.* Textiles, apparel, furniture, leather, consumer electrical goods.

5. *Heavy chemicals/oil/rubber/glass.* High-volume chemicals, oil refining, rubber, stone, clay, glass, concrete.

6. *Special chemicals/metal.* Special chemicals and drugs, some metal fabrication.

7. *Heavy metals.* Primary metals, some metal fabrication.

8. *Industrial machinery/transportation equipment.* Most machinery, most transportation equipment.

9. *High technology.* Office equipment, computers, electronics, space vehicles, instruments.

Source: Roger W. Schmenner, *Making Business Location Decisions*, © 1982, p. 13. Adapted by permission of Prentice Hall, Inc., Englewood Cliffs, New Jersey.

IV. REAL ESTATE SITE CRITERIA

Among the other decisions facing the firm in its efforts to expand are considerations of the physical site for a new facility. A number of factors must be taken into account in choosing the site and in deciding on the most desirable combination of characteristics for that site. Whatever the site may be, it must represent the proper combination of capabilities for manufacturing the product and efficiency of operation. To that end, the company must consider the following.

A. Size of Site

Plants and equipment are expensive; the larger the facility, the more expense the firm will incur. For this reason, it is important that the site be of proper size, neither too small to do the job nor too large to be economically feasible. A careful analysis of space and equipment requirements is necessary to determine exactly what is needed in terms of physical plant size in order to ensure a proper use of funds. It must be remembered that company resources are limited and plans for expansion have to compete with other uses of funds within the firm's structure. Consideration must also be given to future expansion needs, being certain that sufficient space exists for any expected increases in space requirements as the firm grows. By including this element in plans now, the firm can forestall future needs to move facilities once more in the face of increased demand for its products or services.

B. Product Lines

The physical characteristics of the site of a new facility should reflect the products to be produced there. Different products have different requirements in terms of such factors as waste disposal, floor space, the ability or inability to create multistory structures, the accessibility of community services, and so forth. An appropriate site will supply these requirements with a minimum of inconvenience.

C. Raw Materials Available

Depending on the nature of the operation, raw materials either will or will not be a primary consideration in physical plant structure. In industries where a large amount of raw material is necessary to the productive process, or in cases where raw materials are bulky or unusual, it may be necessary to determine plant location and characteristics in accordance with this factor. For example, steel mills are located on or near the Great Lakes in this country. As mentioned earlier, that is no accident. Many mills locate there because of the raw materials they use (coal, coke, iron ore, and so on) which are bulky, heavy, and most easily transported by water and by rail. The Great Lakes' proximity to iron ore and coal deposits and the availability of water transportation and railroads make this the ideal location for such operations. Similarly, beer manufacturing may be located near sources of the proper type of water or in the midst of the grain belt, where barley and hops are readily available. Indeed, the regionality of whiskey production in the United States (Kentucky bourbon, Tennessee sour mash, or Southern corn liquor) is the result of a need to produce a local product that would use the abundance of locally available grains. Where transportation of raw materials is expensive or the availability of these materials is geographically limited, the production facilities are most often located close to the source of materials or close to the appropriate type of transportation.

D. Labor Force

Many industries are labor intensive. This is particularly true of certain agricultural industries and of industries requiring judgment in the manufacturing of the product. In such labor-intensive industries it is advisable to locate near a readily available supply of inexpensive labor. Labor does not necessarily refer to physical labor, or blue-collar types of jobs. Industries that require highly qualified experts often locate near a university town or close to other similar industries where the population has a high percentage of qualified workers. The trend since the 1960s for industry to move into southern states is partially due to the abundance of small towns with a ready labor force and no dominant industry in place. This is not a new phenomenon, however, as is evidenced by the abundance of textile mills in small southern towns such as those along the Georgia–Alabama border. There, the textile industry is and has long been dominant in an otherwise rural area.

E. Transportation Needs

Manufacturing operations require suitable transportation facilities not only for the hauling of raw materials but for the transport of finished goods to markets. Facilities producing goods for foreign consumption, particularly bulk goods, can often be found near seaports. Other industries locate near rail lines or, if their main form of transport is trucking, along major highways and interstates. Even small manufacturing companies whose marketing area is more than just local locate in industrial parks near interstates for easy inflow and outflow of raw materials and finished goods. Since the beginning of commerce some seven thousand years ago, whole cities have sprung up along transportation routes, whether they be river or oceanside, with the surrounding countryside dependent upon these commercial centers. The shift from agriculture to industry has in no way changed this distribution of population; modern transportation networks are an expansion of the same concept.

F. Utilities Availability

Along with raw materials and labor, there must be sufficient amounts of electricity, water, sewerage, and other utilities available at a facility's site. Part of the shift of industry into the sunbelt is due to abundant supply of utilities and inducements by local and state governments of reduced utility costs. An aluminum company may locate near a hydroelectric dam because of the inordinate amount of electricity required for the production of the product, or a nuclear power plant may locate near a river or the ocean because of readily available supplies of water. Before any serious consideration can be given to a site, the developers of a new facility must assure themselves that there are adequate utilities available for their needs.

G. Environmental Considerations

Impact on the environment is a factor that is gaining importance in the decision of where to locate production facilities, particularly in this country. Industrialization creates unavoidable changes in the ecosystem within which it is located. In recent years, the changes in the environment owing to high levels of industrialization have become dramatically evident. At one time, the United States was a virgin territory with such abundant resources and land that it was not necessary to concern ourselves to any appreciable degree with the ecological effects of our industrial activity. This is no longer the case.

Through legislation and industry awareness the impact of industrial development is closely monitored to ensure that we do not destroy our environment through our efforts to produce goods and services. The environment is fragile and easily altered. Any expansion project must take this fact into consideration in choosing a site. Water supplies, wildlife, and the delicate balance of nature can be irreversibly altered if care is not taken. Pollution of rivers and streams, the unintentional creation of erosion, the disruption of migratory patterns, and destruction of natural habits are only a few of the ecological issues to consider.

H. Interaction with Other Company Facilities

Manufacturing is often a complicated process that requires the interaction of a number of different elements. Each of these elements or subassemblies may be produced at a different location and brought together in a separate facility for final assembly into the firm's product. This has already been alluded to in the discussion of process plant strategies. The fact that the various plants of a firm using this scheme need to be interactive will reduce the number of effective site locations. The degree of importance of interaction and the nature of the interaction will decide such factors as the choice of transportation methods and the closeness of plant facilities to each other. For example, a chemical manufacturer may have one process for producing nitrates and another process for producing nutrients or insecticides and have these items produced at different sites prior to the final production of its fertilizers or plant foods. Since there is an interdependence among facilities, their interaction should maintain a high level of ease and efficiency that is reflected in the entire productive process.

I. Company Departmental Views

Departmentalization is a management technique by which a firm can organize itself into functional subunits, or departments. There are many different formats by which departmentalization can take place. Organizations may be departmentalized by function (production, accounting, finance, marketing, and the like), by geographic location (northeast, southeast, north central, and so on), or by customer (retail, wholesale, industrial, governmental, and so forth). Depending on how this departmentalization takes place, an individual production facility may find itself in a condition of complete self-sufficiency with all functions of business represented, or totally dependent on the "home office" for certain functions not related to production per se. Probably it will be somewhere in between. Wherever it is, the departmentalization structure under which the company is organized must be taken into consideration in choosing a site. With a reasonable degree of autonomy in decision making and action, a plant can be placed virtually anywhere. With tight controls in place by other departments of the company and low levels of autonomy, the facility may need to be where it can be easily reached, either physically or through communication links, and probably by both means.

V. EVALUATION OF SITE

The final criteria that must be developed before a site is chosen are financial. The site must be evaluated in terms of profitability to see if the profits received from the expansion are potentially high enough to warrant the project. Every project that a company undertakes, whether it be a marketing campaign, a research and development project, or a new facility, will have some return on investment. The final question becomes, in the light of all other facts gathered, will the new facility produce enough return on investment to warrant its creation? To make this determination, the firm needs to have the following facts.

A. Marketing Sales Projections

The first fact to be determined is whether or not there are sufficient sales increases expected in the future to warrant the new facility. The primary reason for expansion to begin with is the expected need for more product to sell, and this will be necessary only if there is a market for those goods. The firm needs to establish probable levels of sales (the potential market and the expected penetration of the company into that potential market) in order to determine production levels, future revenues, and cost factors connected with actually producing goods in the new facility. If there are not enough expected sales in the future (both long term and short term), there will not be sufficient justification for the increased productive capacity. It is on the basis of these projections that the new plant is seen as necessary and determination is made of how large it needs to be.

B. Estimates of Start-Up Costs

The initial outlay of funds for the plant must be determined. This represents the initial investment that the company must make in order to undertake the project—costs that are incurred before the first unit of product ever comes off the line. Start-up costs include those costs directly connected with the building of the facility and of staffing and equipping it. They include physical plant and equipment costs, licenses, utilities, administrative expenses, labor costs, initial inventories of raw materials and subassemblies (parts), and support function costs, among others. Without the willingness and ability of the firm to invest these initial funds, the project cannot be undertaken.

C. Construction Costs

A more specific cost to the company involves actual construction costs, either payments to contractors and subcontractors or direct costs if the company is in a position to undertake the physical construction of the facility within its own organization. Construction costs have to be planned for, to ensure that the proper flow of funds is available when needed. Normally, a firm will receive estimates of these costs through interaction with the architects and designers of the facility and through bids received from the various contractors who wish to undertake the actual construction of the facility. So variable are these costs, often dependent on factors such as weather, the availability of labor and equipment, and so on, that a definitive figure is difficult to find. In contractual agreements between contractors and the firm, great care is taken to define what provisions will be made for unforeseen delays, changes in prices of labor and raw material, and other unforeseen hidden costs. Waste removal, for example, may be added in at ballpark rates of from 15 to 20 percent, depending on the complexity and nature of the construction.

D. Five-Year Cash Flow Projections

Projections are made to ensure that the necessary funds will be forthcoming in the foreseeable future to provide operating capital for the facility and an acceptable level of return on investment as well. Cash flow projections are simply projected source and use reports that identify when moneys are expected to be generated by sales from the facility's output, and when and for what those fund will be used. Projections represent a detailed five-year budget to determine when (and if) shortages of funds will develop so that they may be planned for. As with any other plan, budgets become less accurate the farther they extend into the future. As the project is initiated and continues, the cash flow analysis must be continually updated to reflect historical financial information as it becomes available. Initially, the five-year projection has the same function as the marketing sales projection, that is, to determine if there will be sufficient funds to warrant the new facility's construction. Over time, however, it becomes useful for checking the accuracy of initial predictions and for amortizing costs in a more realistic manner over the life of the facility. (See the Kingdom Harbor Yacht Club case study in Chapter 6).

E. Present Value Analysis

Money has what is known as a *time value*. That is, at different points in time a given amount of money is valued differently. For example, if you were offered a choice of one dollar now or one dollar one year from now (and you were reasonably certain that the dollar would be forthcoming in one year), you would probably opt for the dollar now. That is because a dollar at this moment in time is more valuable to you than one dollar a year from now. Why wait if you can have it now? Alternatively, if you were offered one dollar now or five dollars one year hence, the future amount might seem a bit more desirable, though not necessarily desirable enough to convince you to forego the dollar now. Eventually, however, if the process of offers continues with ever-higher dollar amounts offered in the future as opposed to one dollar now, the time will come when you are finally willing to wait a year for your money. At that point, we have found what the future money will be worth in terms of today's money (that is, the money's present value).

Suppose you are indifferent about whether you have one dollar now or seven dollars one year from now. If that is the case, then the present value of seven dollars one year from now is one dollar in terms of today's dollars. Or, alternatively, one dollar one year from now is worth one-seventh of what it is worth now, an amount equal to approximately 14 cents.

In terms of our plant analysis, we must realize that whether we finance the construction and start-up of our new facility through expenditures of company funds or through loans from banks, we are spending money right now that will not return money until some time in the future. To see if the expenditure is justified by the return, we must compare dollars of the same value—that is, today's dollars with tomorrow's dollars. To do this, we convert the future income to present values (today's dollars) before comparison. Failure to do so artificially inflates the value of the future income stream that we receive from the plant's operations.

To illustrate, let us assume that the initial outlay for plant construction and other start-up costs is equal to $1,000,000 (one million dollars). Let us further assume, for the sake of simplicity, that there will be a single influx of revenues from the plant operation at the end of one year and that our criterion for the project is that we must make some profit in the first year or the project is to be rejected. If we determine that future cash flows indicate a return of $1,122,000 at the end of one year's operations, should we undertake the project? Without present value analysis, the apparent answer is yes. We appear to be making a profit of nearly $122,000. Are we? If the dollar that we receive in one year is worth less than it is now, as our present value analysis indicates, then the actual revenues received in today's dollars must in reality be less than the indicated profit of $122,000. But is this argument valid?

It is valid in light of the fact that there is a time value for money, and that time value is very real. We use it every day. It is time value, for instance, that determines the interest rate on loans. When we lend money, all that we are doing is agreeing to give up the use of our money in exchange for a larger amount later. As borrower, we agree to have a certain amount of money now and pay back a larger amount of money at some future date. How do we determine the highest acceptable amount of future money we are willing to pay for the right to have the funds borrowed now? How do we, as lenders, determine the smallest amount of money we must receive later in order to be persuaded to give up possession of our money now? We determine how much that future money is worth to us in present-day dollars and vice versa. When both parties agree on the present value of that future sum of money, the deal is struck and funds are exchanged. For this reason, it is simple to determine the present value of a future income stream merely by determining the percentage rate of loans. Whatever the going rate for borrowing funds is will give us a very good starting point for determining the present value of future money. Any decrease or increase from this figure is a matter of the personal beliefs of the individual firm, the going rate representing the present-value ratio of the economy in general.

Obviously, this is a greatly simplified explanation of the present-value process, since a large number of variables indicative of a specific situation may enter into the determination of the actual figures, yet it serves to illustrate the point.

Mathematically, it is a simple matter to determine the present value of a future amount of income if we know the interest rate that predominates in the economy. The formula for present value of a future amount of income is:

$$Pv = \frac{Pt_n}{(1 + i)^n}$$

where Pv is the present value, Pt_n represents the future value of the money after n years, n represents the number of years until the money will be paid, and i represents the going rate of interest. Given these data, we can determine the value, for instance, of $1,122,000 one year from now at an interest rate of 14 percent as:

$$Pv = \frac{\$1{,}122{,}000}{(1 + 0.14)^1} = \$984{,}210.53$$

for an actual loss of $15,789.47 (+$984,210.53 = $1,000,000.00).

To expand this model to take into account more than one year of returns, the cash flow projection can be used with the determined interest rate and present value calculated for each successive year, the resulting total present values equaling the present value of the entire future stream of revenues, as shown in Table 1.3.

By the application of the principle of present value to the numerical indicators connected with the facilities project, the firm can get a much truer estimate of expected returns on investment.

F. Formal Statement of the Rationale of This Site

The evaluation of the site must include a formal statement of all the factors that make this particular site suitable. By creating a formal statement of the findings of the study, the firm forces itself to put its ideas into a logical, cohesive form explaining the reasoning behind the site choice. Any missed factors or faulty logic is exposed through this process as the document is written, edited, and rewritten. In addition, at any future time, there will be a readily available detailed explanation of exactly how the firm has arrived at its decision. Through this process, the group responsible for choosing the site of the new facility can present its findings to all interested parties and be in a position to defend them, both by virtue of the rigorous construction of the document and by virtue of the organized manner in which the information is presented.

TABLE 1.3 Present Value of a Future Income Stream at 14 Percent Interest

Year	Pt_n	$(1 + i)^n$	$Pt_n/(1 + i)^n$
1	$ 338,420	1.1400	$ 296,860
2	$ 340,987	1.2996	$ 262,378
3	$ 401,000	1.4815	$ 270,672
4	$ 421,876	1.6889	$ 249,793
	$1,502,283		$1,079,703

G. Discussion within the Company of Possible Site Locations

Once the choices are available for comparison, the firm is then in a position to investigate the findings of the study and come to a conclusion as to which site among those possible is most satisfactory for the company's needs. In all probability, no one site will be ideal. It is rare to find a single location that meets all the criteria thought to be important for a facility location. However, by this in-depth comparison of possible candidates, the firm is able to choose the site that best suits its overall needs.

QUESTIONS

1. It would appear that in order to do business, a firm needs a site of some sort. Is this true?
2. Under what conditions would a firm not need a facility in order to function?
3. What are the main considerations in choosing a site?
4. In the decision to move from one location to another, what are some of the primary considerations that must be dealt with by the firm?
5. Is the environment an issue in site selection?
6. Is the environment a major issue in the decision to expand or move to a new location? What are some examples?
7. In recent years there has been an increasing move among manufacturing firms to locate in the Southeast rather than in the traditionally industrial Northeast. To what factors of site location can we attribute this change in strategy?
8. Why do you suppose so much textile industry has been located in the South? Why did so much heavy industry originally locate in the North and along the Great Lakes?
9. How are the conditions surrounding labor force issues contributory to site location decisions?
10. Given that the chief executive positions of a firm include Chief Financial Officer, Chief Executive Officer, Vice President of Manufacturing, Vice President of Marketing, and Vice President of Information Services, explain how each of these individuals could influence the location decision.
11. Name some possible ways in which a labor-intensive industry and a technologically intensive industry might differ in their site selection strategy.
12. What is the role of transportation in the facilities expansion decision?
13. How do multiplant strategies differ from single-plant strategies?
14. What is a product-plant strategy?
15. What is a market-plant strategy?
16. What is a process-plant strategy?
17. What forces cause the strategies noted in questions 13 through 16?
18. Explain the difference between an ad hoc approach and a formal approach to facility design.
19. How do start-up cost estimate techniques differ in design and purpose from long-term (five-year) cash flow projections?
20. Note the site of your educational institution. Why do you suppose it was chosen over other sites? How much of the decision appears to be purely economical (efficient) and how much from other causes, such as political or traditional?

2
Company Purchase Options

Among the decisions that a company must make when designing and creating a new facility is that of how to finance the acquisition of the necessary land for the project. There is a wide range of options available to the firm, each with its own unique set of advantages and disadvantages. This may on first inspection seem to exacerbate the problem of choosing a site for the new facility. It does, in fact, greatly increase the effectiveness of the firm's analysis by affording the firm the opportunity to find just the right combination of conditions to maximize the efficiency of the final decision. Even with the additional analysis that must be undertaken in the face of a wide range of choices, the aggravation is more than compensated for by the added efficiency of the final choice. Money is always in short supply in business decisions, and getting the most from the dollar spent becomes a prime consideration.

I. FINANCIAL OPTIONS

A number of financial options are available to the firm. The choice that will be appropriate for a particular firm is dependent upon the conditions under which the firm itself is operating. In some cases, there is cash on hand to fund an expansion project, although this tends to be a relatively rare occurrence in today's market. More often, funds are short and the need for expansion great, resulting in the necessity of borrowing funds for investment. This becomes a matter of cash flow, profitability, and interest rates as well as expected market conditions and availability of loanable funds in the financial market. The following are brief definitions of the primary choices a firm can expect to find. See Chapter 5 on types of financing for a more detailed discussion.

A. Lease Purchase

The lease purchase is an arrangement whereby lessees are allowed to purchase the property if they so desire, with all or part of the lease fees paid applied to the purchase of the property. The flexibility of lease purchase is advantageous to the firm looking to expand or create a new facility, because it delays the purchase decision without losing the capital already invested in lease payments. It also gives the firm the right to control the leased property for the duration of the lease period. At the end of the lease period the firm would have the option of either terminating the arrangement if the property or leased item is unsuitable for its needs, or purchasing it.

B. Purchase

The other alternative available to the firm is that of owning the property outright. This is probably the best course of action if funds are readily available. However, there are costs involved in the purchase approach, either in the form of interest or of opportunity costs to the firm, which are costs incurred by choosing to purchase some particular property rather than use the money for some other business asset or venture.

Generally, purchase is preferable to lease purchase as long as the choice of property is a certain one and funds are available. If there is any doubt to the wisdom of the purchase, delaying the purchase decision by the lease purchase agreement would be the best bet.

II. FINANCIAL TERMS

In order to understand the nature of financial agreements, it is useful to be aware of the terminology used in the industry. Whereas many of the concepts you will encounter will be familiar from common usage, some of the more recent conditions in the field of finance may not be. Some of the terms are further discussed in Chapter 5 on types of financing; therefore, these terms, *adjustable rate mortgage*, *carryback financing*, *wraparound mortgage*, and *purchase options* are only briefly mentioned in this section.

A. Adjustable Rate Mortgage

The adjustable rate mortgage (ARM) loan is one on which the interest rate rises and falls with changes in prevailing rates. The advantage of this kind of mortgage is that if interest rates fall, the monthly payment on the loan also falls. Unfortunately, it will also rise as interest rates rise. This type of mortgage offers a lower initial interest rate, and payments are governed by indexes that serve to insure fairness. There are several other features of the ARM, including a preset adjustment period, an interest rate cap, and a payment cap. This is a viable alternative for present-day financing.

B. Alienation Clause

The alienation clause in a mortgage or promissory note allows the lender to call the entire loan balance due if the property in question is sold or transferred to a third party. Basically, this clause protects the lender from loss of income or increase in experienced risk by way of a transfer of property from a mortgagee of high standing to one of low standing. It establishes the right of the lender to demand full payment rather than having to transfer the mortgage to some other party. Many lenders now insert a clause to allow themselves additional flexibility by creating the opportunity to increase interest rates through the renegotiation of the loan with the new customer. They are no longer tied to the terms of the original mortgage once the new title holder appears.

C. Carryback Financing

This is an alternative to 100 percent financing with a single lender. Essentially, it means that the lender consists of two sources, the financial institution and the owner of the property, who lends the remainder of the amount. The borrower pays two monthly payments, one to each party, and often ends up paying more than would be due from a single loan source because of the high risk the lenders carry. The advantage lies in the fact that the initial lump sum due is less than would be necessary on the open market.

For the firm that is short of cash and seeking to develop a new facility, carryback financing may be advantageous because of the smaller amount of money required up front. If the additional payment can be handled, this is a very viable alternative.

D. Equity Sharing

Equity sharing makes the individual providing the financing for the purchase co-owner of the property involved. Two approaches are most often used, shared appreciation mortgaging, where profits are split upon resale of the property, and "rich uncle" financing, which is a type of shared appreciation mortgaging where the monthly payments are lower but the "uncle's" share of profits is higher upon resale of the property. For a small company seeking financing, this may prove to be a valuable source of funds.

E. Mortgage Company

A mortgage company is a firm that makes mortgage loans to individuals and to other firms, and sells them to investors. They act as manufacturers, creating the mortgage and then selling it as they would any other product. Normally, mortgage bankers continue to service loans after they have been sold to investors, receiving a percentage of the payments for their services. By creating the loan and then selling it for cash, mortgage bankers are able to maintain their liquidity for a rapid turnover on their investment, and generate a large number of mortgage loans on a rather small capital base. Mortgage bankers are usually smaller firms, local in nature, and may have from one or two to dozens of principals. Savings and loans, commercial banks, and other financial institutions often perform this function as well as mortgage bankers.

F. Primary Mortgage Market

The *primary mortgage market* is the "place" where lenders make funds available to borrowers. This market includes savings and loans, mutual savings banks, commercial banks, and mortgage companies. Loanable funds come from a wide range of sources, including depositors in lending institutions, pension funds, mutual funds, insurance companies, and a host of others. In the primary market, the borrower sees the lender as the source of funds and the one with whom business is conducted, and the source of the lender's funds is from depositors.

G. Secondary Mortgage Market

In the case of insurance companies and many other sources, we are looking at what are called *secondary markets.* They represent a source of funds as large as the primary one. The primary market is not large enough to satisfy the demand for loanable funds in this country, and thus these primary lenders often sell their loans to other lenders, such as savings and loans, private investors, or pension plans and credit unions. Receiving cash for that transaction, the initiator of the loan is now in a position to make more loans, while the members of the secondary market are receiving the benefits of loans already made without the hassles and problems of initiating and servicing the loan itself.

H. Usury

Usury refers to the practice of charging a higher interest rate than is allowable under the law. Under the law, interest rates are held to a maximum through legislation of a ceiling on the interest that can be charged. The idea has questionable value in an ever-changing market in that the law soon ceases to reflect the reality of the business environment, thus creating unhealthy and unworkable constraints on the process of doing business. Therefore, the usury law is used basically as a reference point to help keep interest rates from getting out of hand. Most state laws governing usury have been altered to reflect the changing times, and many have clauses allowing for a reasonable degree of latitude, but it should be noted that they are often confusing and complex. If any question of usury exists, a lawyer should be consulted.

I. Wraparound Mortgage

The wraparound mortgage covers existing mortgages and is considered to be secondary to those mortgages. In this type of financing, the lender assumes a lesser position than the primary mortgage holder, which serves to provide the lender with a higher rate of interest. The borrower, however, pays a lower rate of interest because the mortgage blends interest rates from the two mortgages (the lower rate of the primary old mortgage and the higher rate of the new mortgage). The lender absorbs the old mortgage and wraps his new mortgage around the old one. The borrower makes one payment to the lender. The lender takes part of this payment and makes the payment for the borrower on the old primary loan. It is a viable alternative only in the case of a loan assumption and will not work when an alienation clause is present, in that the clause prevents the original owner from offering wraparound financing. Such an arrangement is advantageous to a firm when current market interest rates are high and the original owner's mortgage interest rates are low.

J. Purchase Options

Options represent the right to purchase or lease some piece of property, at a set price, at some time in the future. They provide control in that they prevent the seller from disposing of the property without first consulting the person who holds the option. Unfortunately, if the person holding the option is unable to pay at the time the seller decides to sell the property, the option money is kept by the seller. The advantage is that for what may be a very small sum of money, the firm holding the option is given first consideration for the purchase of the property. It is a good alternative for the firm that is not yet certain when or if expansion will be necessary, and for the small businessperson who is not yet ready to put down a large amount of cash.

K. Simple Interest Rate

This is also referred to as the *nominal interest rate*. It is the stated interest rate of a loan and represents the amount of interest paid on a loan at maturity. Interest is nothing more than rent for the use of someone else's money. As such, the simple interest rate is the rental charge attached to the mortgage or other debt agreement.

The calculation of simple interest is done by use of the formula

$$\$I = PRT$$

where I = interest, P = principal, R = interest rate, and T = time. The amount of interest dollars that will be due at the end of the loan period is the combination of the borrowed amount (principal), the interest rate (R), and the time period (T).

With mortgage loans, the payback of the debt is usually done on a monthly basis wherein interest and part of the principal are paid back so that the entire loan is paid off at the end of a number of months or years. The interest rate applied to the monthly payments is on the amount of loan outstanding and will therefore decline as the balance declines. To determine the payback, lending institutions and firms employ amortization tables that calculate how large monthly payments must be in order to fully execute a loan of a specific size for a specific period of time. The factors involved are the number of periodic payments, the interest rate, and the amount of the original loan. As an example, a five-year loan of $1000 at an interest rate of 10 percent would require monthly payments of $21.25 to be fully executed at the end of the loan. Should the rate increase to 15 percent, the required monthly payments would rise to $23.79.

Amortization tables often offer figures per $1000, allowing for an easily determined monthly repayment computation by merely multiplying the rate per thousand by the size of the loan divided by one thousand. That is, the monthly rate at which a given loan would need to be repaid would be:

$$\text{Monthly payment} = \frac{\text{face value of loan}}{1000} \times \begin{array}{l}\text{rate per thousand for}\\ \text{time period specified}\end{array}$$

Example:

$$\text{Payment} = \frac{\$80{,}000}{1000} \times 13.22 \text{ (10\% for 10 years)} = \$1057.60/\text{month}$$

III. FINANCIAL SOURCES

When people think of mortgages and loans, they usually conjure up images of banks, savings and loans, and similar institutions. In reality, there are a large number of different institutions available for the procurement of development funds, and a brief survey of the market reveals a surprising degree of diversity in the services that these institutions offer. The choices range from large institutions to small, private sources. Many of these sources are overlooked by the novice and are worthy of serious consideration.

A. Savings and Loan Associations

Originally designed for the creation of real estate loans to private citizens, savings and loans have grown to number more than four thousand in the United States and to offer a wide range of services. They are a major source of real estate loans, though they tend to concentrate in individual home mortgages. They are, as their name implies, an institution for the purpose of saving moneys and making loans. Unlike commercial banks, savings and loans have limited portfolios of loans and technically do not offer checking accounts, though their passbook savings approach is frequently designed to operate the same as any other payable-on-demand type of instrument. As a source of loans, the savings and loans offer long-term loans based on their deposits which are, for the most part, long-term savings deposits not subject to rapid and heavy variations with momentary changes in financial conditions. They also offer certificates of deposit to depositors at higher interest rates than can be obtained on passbook accounts, which make them particularly attractive. These certificates have specific deposit durations ranging from six months to several years; withdrawals may not be made without penalty before the end of the period. Due to the increased stability and reduced risk connected with these certificates, the availability of loanable funds for real estate is usually relatively high.

B. Commercial Banks

Commercial banks are traditional, full-service institutions where depositors deposit funds in both demand accounts (checking) and time deposits (savings). With these funds, the banking institution makes loans of all types, offering interest to the depositors for their deposits and charging interest to the borrower for the use of the monies. In addition, full-service commercial banks provide other services, such as trust management, investment management, factoring, financial counseling of various types, and so forth. There are more than fifteen thousand such banks in the United States, and the vast majority of them belong to the Federal Reserve System, which carries out the banking activities of the federal government and acts as a bank for commercial banks.

Since commercial banks receive the majority of their deposits from individual depositors through demand accounts and time deposits, they tend to prefer short-term loan creation to long-term lending. Real estate development loans and others that are paid back rather quickly are the preferred instruments for commercial banks. The firm seeking funds for facilities construction may find the availability of funds from commercial banks somewhat limited. If, however, a project is small in size or if the payback period (life of the loan) is of a reasonably short duration (five years or less), they can be a source of funds. Some banks do make long-term real estate loans of up to thirty years, but these are usually sold into the secondary market once made.

C. Mutual Savings Banks

Most mutual savings banks are located in the northeastern United States. Mutual savings banks are owned by the depositors, who share in the profits that the mutual makes from its loan placements. Laws tightly govern the operations of these institutions to protect the rights of the depositors, and as a result, only high-quality, low-risk loans are generally allowed. Real estate loans and those for facilities construction fall handily into this category.

D. Life Insurance Companies

A prime source of funds for long-term mortgages and loans, life insurance companies are in a unique position in that they have a very good idea of their sources of funds and of the schedule by which those funds become available. Through their actuarial studies, they are also fully aware of their outlay to policy holders in terms of collected policy funds and maturity paybacks. Consequently, they can budget for long-term investments with a high degree of accuracy, and can commit moneys to projects such as real estate purchases and other commercial ventures. Insurance companies finance office buildings, shopping centers, and industrial construction projects, as well as investing in high-grade municipal or corporate bonds (the latter of which is also a possible source of construction funds). Keep in mind that, as we have mentioned, payback of these loans is often on the basis of participation in profits as well as return of principal and interest, through the process described as equity sharing.

E. Mortgage Companies

Commercially oriented mortgage companies are lenders for high-risk investments. In cases where the borrower is without substantial credit history, is endeavoring to enter a high-risk project, or is unable to attain a loan at a more conventional institution, mortgage companies often supply funds. Because of the high risk involved in such transactions, mortgage companies are characterized by high interest rates and severe, tightly defined contract terms. The high level of default that these companies experience creates a need for this approach. In some cases, if funds are not otherwise forthcoming, entrepreneurs may find this type of lending institution to be the only one available to them.

F. Mortgage Brokers

Mortgage brokers are somewhat different from mortgage companies. Brokers do not actually make any loans at all. They act as true middlemen to bring borrowers and lenders together. Through their efforts, they receive a finder's fee in the form of points, which are paid at the time the mortgage or loan is contracted. Mortgage brokers usually do not service the loan either, and they tend to be small in size, local in market area, and oriented toward specific types of real estate loan placements. We can consider the mortgage broker to be a source of funds only inasmuch as it facilitates the creation of the loan by bringing the principal parties together, for a fee. For the firm seeking a loan in times of tight money or unusual circumstances, this can be a valuable service.

G. Bonds

Municipal bonds, on first inspection, appear to be a strange source of funds for private construction projects, yet they can be a possibility. Municipalities offer bonds to investors and can do so at a low interest rate relative to the going rate because of the bond's tax advantage. Most municipal bonds are tax free; the income derived by the investor from their purchase is not taxable income. As a result, investors are willing to buy them.

Cities are sometimes willing to lend construction funds to companies locating in their city, particularly if the construction adds to the revitalization of some part of that city or to the general economic welfare of the population. Under the right circumstances, a company may be able to secure a loan from the municipality in which it intends to locate, to get that loan at a reasonable rate, and to ensure the cooperation of the city involved in the deal. Access to such funds is decidedly limited.

Corporate and private bonds are also funding sources. Earning rates are higher than that of municipal bonds to offset the higher risk to the investor and the loss of a tax shelter that investors in municipal bonds enjoy. Corporate or private bonds may also be sought as a means of increasing the level of participation in a project. For example, the employees of a company or the membership of a church would be able to become involved with the financing of their facility without simply donating the money, since they would be repaid in full at the end of the bond time period. Local banks set up bond accounts into which the facility must make periodic payments to ensure that the bonds are paid off in time.

H. Pension Funds

Normally, pension funds are part of the secondary market for construction funds, operating as purchasers of mortgages and loans from primary banking sources. However, the 1980s revealed a trend in their activity extending beyond secondary real estate markets and high-grade bond markets to diversification of their portfolios into direct loans for such needs as commercial real estate and facilities construction. Although this is still a small market, it is a growing one.

I. Trust Funds

Like pension funds, trust funds are now branching out into areas other than stocks and high-grade government and corporate bonds. With the increase in fluctuations in both these markets, wide swings in price and yield have forced investors to seek alternative forms of investment that offer relatively high returns on investment and at the same time are relatively free from the risk of interest shifts, inflationary changes, and other market swings. As a primary source of construction funds, trust funds are still a limited possibility (home mortgages seem to be more popular), but they are still a possibility to be investigated.

J. Finance Companies

For the small firm, finance companies may be an ideal source of funds, in spite of the popular myths of fly-by-night outfits. Numbering in the thousands, finance companies primarily receive their income from business and construction loans. They generally charge an interest rate from 2 to 5 percent higher than the going rate, which is often viewed as acceptable by borrowing firms as it represents a business expense and reduces their level of taxable income. The tax savings will often offset the higher interest expense enough to make the loan a feasible one. Also, finance companies usually accept property mortgages as collateral, which reduces their "felt" risk in the lending of funds. This can mean a lower interest rate than they would otherwise charge, helping to bring the rate more closely into line with other lending institutions.

K. Credit Unions

These institutions normally participate in consumer loans for such items as automobiles and also to clear other debt. However, as is the case with trust funds and pension plans, they have recently expanded their portfolios to include other types of low-risk, stable projects. Availability of funds and applicability are similar to those discussed in the case of pension funds and trust funds. Primarily it should be remembered that this is a relatively new market for construction funds and that the participants tend to be quite conservative.

L. Private Individuals

If the size of the loan is not too large and if the duration of the loan is not very long, it may be possible to secure the necessary funds for facilities construction from private individuals. This is a very small market to deal with, and unless the firm seeking the loan has direct contacts with such an individual, it is quite difficult to secure this type of loan at all. In the case of a small or privately held firm, a relative or a group of relatives and friends might be enticed into pooling their investment resources in order to gather the necessary funds and then loan those funds either as a straight debt or on a participatory basis in which they are able to share in profits, equity, or both. Again, this market is limited and should be approached with caution, for the protection of both the lender and the borrower.

M. Stock Market

The selling of stock to secure funds for expansion is fundamentally different from the other methods mentioned in that rather than securing funds to be paid back, the firm engages in the selling of shares, that is, ownership, in the company itself. Stock issuance can produce the necessary funds, and this method of obtaining funds is attractive because they need never be paid back. This eliminates interest expense, the expense of servicing loans, and the restrictions that a loan agreement might impose. On the negative side, the company has diluted its ownership of the company by increasing the number of stockholders. In the case of a small company, this might prove unsatisfactory to major stockholders, who find their control of the company's operations diminished, while in a major, publicly held corporation the use of stock involves paper work, red tape, and approval from the various private and governmental organizations that control the issuance of stock. However, for companies on the grow that anticipate a major increase in size and operations it is often the best choice, and one that can garner much higher amounts of usable funds in the long run.

IV. EVALUATION OF FINANCIAL OPTIONS

Given the various sources of funds available to the firm, the question still remains as to whether a particular possibility is feasible or not. Each alternative arrangement that might be made with a financial institution is going to produce available funds and an accompanying obligation for some time into the future. In analyzing the wisdom of a project, or of a specific method of financing that project, it is necessary to first be aware of requirements the firm may have for their source of funds and terms of payback that the firm would find acceptable. This becomes a matter of financial analysis of the project itself, and it is to that analysis that we now turn our attention.

Three specific aspects of the project need to be analyzed to determine what the firm's requirements and capabilities are. These are the break-even analysis of production, the short-term cash flow situation, and the long-term cash flow situation.

Viewing each of these three in turn offers the firm a clear picture of what it can and cannot afford to do.

A. Break-Even Analysis

Break-even analysis is a method of determining the necessary level of output for the facility in order for the company to begin to make a profit. It involves analyzing income and outgo to determine at what level of production enough revenues are generated from the sale of the product to cover the costs of production. In reviewing the procedure, we will first look at a simplified view of the model and then shift to a more sophisticated and realistic model that better demonstrates the nature of the operation.

In simple terms, what is needed is to find the level of production at which total revenues (TR) is equal to total costs (TC). At this point, there is neither a profit being made on the sale of the product, nor is there a loss incurred due to not selling enough to cover costs. The firm has broken even. It is imperative that the firm produce beyond this level if it is to make a profit on the operation.

To determine this point, we must make some general assumptions about the functioning of the market. To begin, we define total revenues and total costs in order to be able to compare them. This is a simple matter.

Definition: Total revenue is equal to the price of the item in question multiplied by the number of units sold, or mathematically,

$$\text{TR} = (p)(q)$$

where TR = total revenue, p = price per unit, and q = quantity of units sold. Note that if this equation is graphed it yields a straight line beginning at the origin and sloping upward to the right at a slope equal to the value of p. It is a linear relationship as defined. This can be easily seen in Figure 2.1, represented by the sales line.

Definition: Total cost = fixed cost + variable cost

Here we have the total cost of production defined as the combination of those costs that do not vary with levels of production and those expenses that do vary with levels of production.

As we see in the definition, two types of costs are incurred in the production process. A brief definition of these terms is also in order. The latter, which we call variable costs, are costs that vary directly with the level of production and are directly dependent on the number of units produced

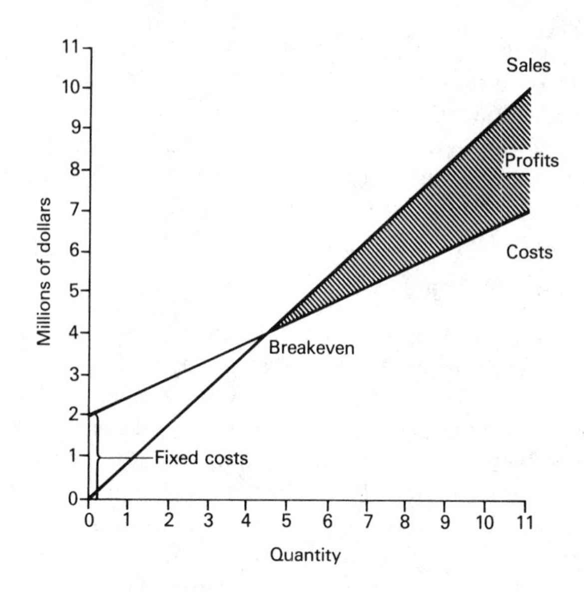

Figure 2.1 Break-even point graph.

for their existence. The raw materials that go into making a product will vary with production for obvious reasons. The more units one makes, the more raw materials one will use in the process. Power for the machinery and equipment necessary for the production process will also vary with production levels, being higher for more production and lower for less production.

Fixed costs, on the other hand, are costs that exist independently of the level of production, such as the mortgage on the property, the telephone charge, the office staff salaries, property taxes, and other expenses that remain constant in the short run as production varies. This much should be obvious. What may not be obvious is the way in which these fixed costs vary. A detailed analysis of this process is offered in Chapter 3 on venture analysis.

We can define variable costs as cost per unit times number of units produced (VC) (Q). Fixed costs, being nonvariable, would be a constant, or simply (FC). Combining these, we can create a deterministic formula for total costs, as related before, such that

$$\text{TC} = (\text{FC}) + (\text{VC})\ (Q)$$

which is also a linear equation represented by a line that intersects the y axis at (FC) and upward to the right at a slope equal to (VC) (Q) (see Figure 2.1, the costs line). Thus we have two linear equations representing total revenues and total costs. It should be a simple matter to determine where total revenues are equal to total costs (the break-even point) by simply finding the point at which these two lines cross on a graph or, alternatively, setting one equation equal to the other and solving for Q (since all other elements in the two equations are of known magnitude). This approach can be seen in Figure 2.1. Note that the intersection of the two lines indicates the minimally acceptable level of output for the product in question. Any production quantity below this intersection produces losses, as the figure for total costs is higher than that of total revenues. Beyond this point, the firm makes ever-increasing profits as the lines diverge, with total revenues always higher than total costs.

We wish it could be so easy. If so, companies could simply produce more and more goods in a single plant facility and as long as the market absorbed the output, profits would get higher and higher. In reality, this is not the case.

In light of the information offered on economies of scale and the law of diminishing returns in Chapter 1, it becomes plain that costs of production do not vary uniformly. The variable cost is not a constant of proportionality defining a constantly rising slope, but rather a variable slope curve in the shape of an inverted S. Considering what we know about input-output relationships from Chapter 1, we should not be too surprised.

To review, as additional units of input are added to the productive process (more labor, more raw materials, more machinery and equipment), the rate of output increases more than proportionately as a result of economies of scale. Each additional unit of input is more efficient than the last because of the synergistic effects of larger production units. This means the cost of the next unit of output is less than that of the unit before it, not the same. Variable costs actually decline over the initial levels of production.

At the other end of the curve, when the law of diminishing returns takes hold, as input increases, output increases less than proportionately, creating a cost of production for successive units that is rising, not identical to the unit before. Costs rise faster than production. The result, as indicated in Figure 2.2A, is an inverted S curve, where initially costs rise more and more slowly, then at a steady rate (constant returns to scale) and then at an increasing rate as the law of diminishing returns takes hold.

If we add to this variable cost curve the fixed cost, we simply shift the position of the variable cost curve upward until it intersects the y axis not at the origin, but at a point equal to the value of fixed costs (the position of the y intercept in the equation $Y = mX + B$, where B is the fixed cost figure or y intercept and m is the formula for the variable slope of the curve). The value of X is the quantity produced in this case. This is illustrated in Figure 2.2B.

By superimposing the still linear total revenue curve on this graph we can see, as illustrated, that the true break-even analysis will show not one but two break-even points where the total cost and total revenue lines intersect (see Figure 2.2C).

This has a rather profound effect on our analysis. It turns out that only in the narrow window between the two break-even points is it possible for the firm to make a profit. Any level of production below the first break-

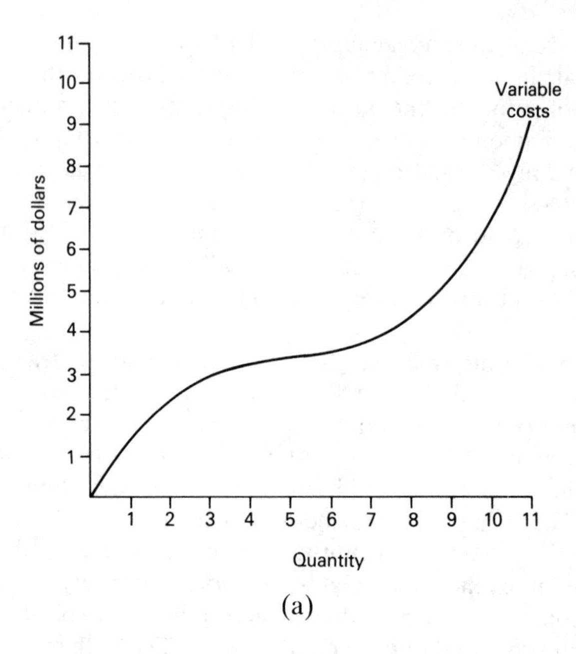

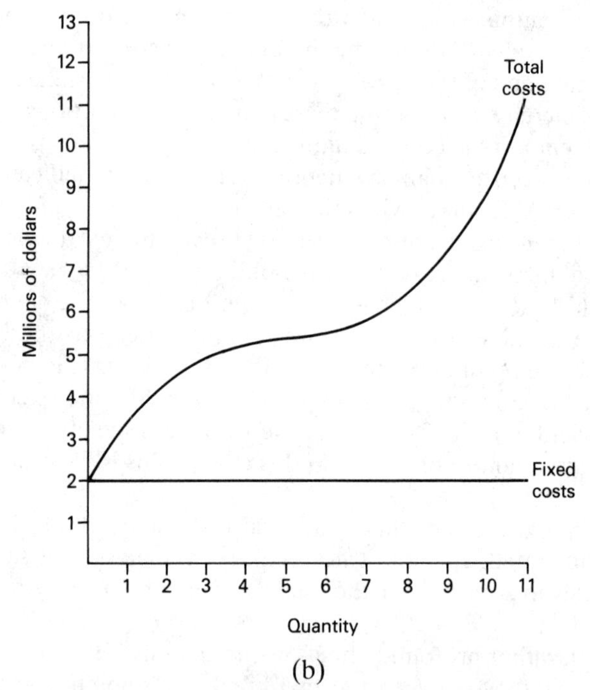

Figure 2.2 (a) Variable costs graph; (b) total costs graph; (c) multiple break-even point graph.

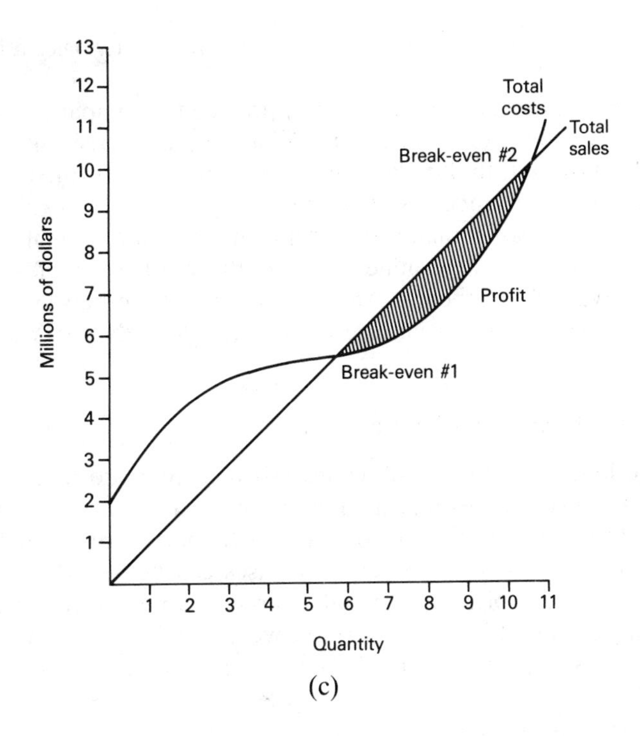

(c)

even point results in a loss rather than a profit, and the same is true of any level of production beyond the second break-even point. Maximum profit is reached at the point between the two intersections where the distance between total revenues and total costs is at a maximum.

The result is that in determining the size and production levels of the plant, management must take into account both minimum and maximum levels of production in the facility. Additional profits will not be maintained at any points past the second break-even point on the graph. See the problem section for examples.

B. Short-Term Cash Flows

Levels of production beyond the initial break-even point will probably not be immediately realizable. Also, start-up costs will be heavier in the beginning of the facility's operation than later. All of this must be taken into account to ensure that enough income is available to take care of the immediate costs of doing business and for day-to-day operations.

To do so, a short-term cash flow analysis is used. This method specifies when income is expected to be realized by the firm and when bills are to be paid, and compares those dates to determine if there will be adequate cash for operations to take place. The information is usually presented in a cash-flow diagram that details cash sources and needs on a daily basis for the short run.

Cash-flow charts are budgets. They provide the company with a schedule of sources and uses of funds to forestall any liquidity crisis that may arise. They also ensure that, in the planning of the new facility project, the firm is always in a position to meet its obligations.

With this forecast in hand, the firm is in a position to analyze possible financing schemes and determine which is the most workable among the alternatives available to them. An example of this can be seen in Chapter 6, where the five-year financial forecast for Kingdom Harbor Yacht Club is presented.

C. Long-Term Cash Flows

In the long run, the cash-flow analysis is identical to that of the short term except that the types of expenses incurred and revenues received are different. In the long run, such costs as yearly charges for insurance, maintenance, taxes, and new machinery purchases can be added to the spread sheet to give a clearer picture of overall operations for the foreseeable future. The company can use this to predetermine where funds are going to be needed and are not available from normal sources, and then plan for such contingencies.

In both forms of cash-flow analysis, the purpose is to be certain that cash is available when needed and, at the same time, that cash does not sit idle when it is not needed. Holding cash costs the company profits. As we said, there is an opportunity cost connected with large cash reserves due to the loss of income that could reasonably be realized by investing the cash in other projects, stocks, bonds, or reduction of debt. By holding the cash rather than reducing debt or holding some other asset, the company incurs costs.

For this reason, it is imperative that the firm be in a position to accurately forecast its cash needs. Companies must minimize their opportunity cost losses and at the same time insure against short-term or long-term losses from a shortage of needed funds. Cash reserves and cash availability are major issues in determining how best to maximize profits for the firm.

D. Backstop Positions

Forecasting is imprecise at best. What happens if the firm miscalculates? What can be done if there is a sudden major shift in conditions, either external or internal to the company? As with any enterprise, preparing for the negative contingencies as well as the positive ones is an essential part of the planning process.

Through the use of accepted techniques, competent entrepreneurs can closely predict needs, cash flows, and future market trends. Yet it is still an estimate with which we are dealing. Because of this, those anticipating a facility project will construct scenarios representing worst-case situations in which their best-laid plans fail to bear fruit, and they will plan for how to handle such eventualities. Such plans are known as backstop positions.

The process is simply a matter of determining the worst case that could come about as the plans are put into effect and then planning for how to handle them. Backstop positions are carefully constructed to reflect the most accurate information available. They do not reflect any pessimistic attitude among those initiating the project. Backstop positions represent a hedge against disaster. They are rational and vital to the success of the project. As contingencies arise, they can be dealt with if they have been planned for. To ignore such possibilities would be to court disaster for the project and possibly for the firm. For these reasons, backstop positions are created to round out the financial planning aspect of the project, supplying management with a clearer picture of what can and cannot be done as the environment in which they are operating changes.

QUESTIONS

1. Financing appears to be a major consideration in plant expansions and modernizations. Why?
2. What are the financial options available to firms in supporting their facility's efforts? How does each lend itself to specific firm conditions?
3. In this chapter, a number of financial options are presented. Which of these do you feel are most applicable to the facility expansion process and which are not? Why?
4. What types of financial institutions are available sources for the small and medium-sized firm seeking to secure facility expansion funds? What types of institutions are preferable for large firms?
5. What are the primary differences in choosing a stock offering form of financing versus an issuance of bonds or securing of some other form of business loan?
6. What is the fundamental difference in philosophy and approach to business expressed in choosing either the public stock option or the bond/debt option when financing a facility project?
7. There are different considerations to be dealt with in the creation of short-run and long-run cash-flow schedules. What are some of these differences?
8. What is a backstop position? How does the concept apply to facility development?
9. Should a firm wait until it can pay for facility expansion without risk before undertaking such a project, or should it accept reasonable risks based on expected future conditions? Under what conditions does one or the other of these approaches become the most acceptable?
10. In the airline industry, where an airplane can be defined as a facility for doing business, one firm asked its employees to accept a multiyear short-term salary cut to help finance operations, while another firm was given a multimillion dollar

airplane by its employee organization. What, if anything, do these two incidents suggest for future facility financing in this and other industries?

11. How does leasing rather than purchasing affect financing in the long run? In the short run?

12. Under what circumstances would it seem preferable to lease rather than purchase or build a facility from the point of view of financing the project?

3
Venture Analysis

As the title implies, creating new facilities is an ad*venture*, a journey, an undertaking involving risk. The entrepreneur should seek to know as much as possible about the risks and the roads that must be taken prior to beginning that journey so that problems can be anticipated and either eliminated or planned for. Contingency plans can be made for all foreseeable conditions, such as legal and environmental problems related to the project, thereby offering solutions before the problems appear (see Table 3.1 for a departmental listing of data).

Venture analysis involves six categories of investigation: facility alternatives, costs, expected returns, unforeseen problems, governmental concerns, and project utility. In this chapter, each of these areas is explored in some detail.

I. FACILITY ALTERNATIVES

A number of facility alternatives are available to the firm, each with its own set of problems and advantages. The choice that a particular firm makes depends on the applicability of that alternative and the degree to which problems can be minimized while advantages are maximized. These alternatives include facility expansion, relocation, remodeling, downsizing, merging, and closing.

A. Expansion

Expansion of an existing facility is perhaps the most obvious alternative available. Its advantages lie in the fact that the location of the facility remains the same and that, in many cases, production can continue while the expansion

TABLE 3.1 Major Factors That Shape Plant Location Searches

Item	Data Usually Come from	Department
Access to markets	Location of markets	Logistics
Access to supplies	Location of suppliers	Marketing
Community aspects	Site visits	Site team
Government aspects	Site visits	Site team
Competition	Industrial sources	Division
Environment	Environmental agencies	Staff
Company interaction	Own knowledge	Division
Labor	Department of Labor	Personnel
Site itself	Local development agencies	Site team
Taxes	Government offices	Tax
Financing	Local offices	Site team
Transportation	Railroads & trucking	Site team
Utilities	Utility companies	Site team

Source: Roger W. Schmenner, *Making Business Location Decisions*, © 1982, p. 33. Adapted by permission of by Prentice Hall, Inc., Englewood Cliffs, New Jersey.

is underway. If the type of operation does not change and if the expansion is simply an add-on to existing operations, this may be the most advantageous type of approach. The shortcomings involved with expansion stem from disruption of the normal flow of work, which may take place as construction of the expansion continues, and from the fact that the facility now has two ages, an older, perhaps less efficient component and a newer, more up-to-date component. In addition, this approach may result in higher costs from the necessary purchase of additional land, or in the overutilization of presently held property, resulting in less efficient design. At times, lack of sufficient land area for projects can be overcome by building up rather than out. This is an alternative dependent on the condition of the present building and its ability to take the load of additional floors.

In some cases, expansion of facilities may take place internally, within the confines of existing buildings. In this case, expansion becomes a problem of altered layout, introduction of heavier utilities requirements, or straining of existing utilities. There will also be a number of hidden costs, including lower productivity among workers and unforeseen delays.

B. Relocation

Relocation involves the complete movement of the operation from one location to another. The advantages in this approach are that a completely new facility can be constructed, that operations at the old facility can continue while the new one is under construction, that upon completion of the new facility the older facility can often be sold to offset some of the costs connected with the new facility, and that state-of-the-art methods can be employed to increase efficiency of operation and reduce operating costs. Disadvantages involve the necessity of searching for a new location and purchasing it, the possible necessity of dealing with zoning requirements and other local regulations at the new site that would not be incurred at the existing site, higher overall investment in real property than with the expansion approach, and disruption of personnel and support facilities at the time of occupancy.

Disruption is not always taken into serious consideration. With a relocation, new workers may have to be hired, and current employees may discover that they need to alter their lifestyles to accommodate the move. Travel time to work, choice of where to live, and shopping and banking habits may all change. The same is true for the company, which may find it necessary to give up long-standing business agreements with vendors and to seek out new vendors who are local to the new facility. All of these changes are costly and time consuming.

C. Remodeling

Through the remodeling process, the facility is modernized to increase efficiency and, in many cases, productive capacity. The advantages are that the changes take place at the existing facility, that the changes are internal for the most part with a minimum of additional construction taking place, that the work force can often be maintained with a minimum of retraining and restaffing, that the cost of the change is confined to machinery and equipment with minimal construction costs being incurred, and that the upgrading of production methods can often reduce variable costs significantly.

The disadvantages of remodeling or modernizing are due to the extreme disruption to operations that remodeling can produce. With just a step-by-step approach there is disruption and inconvenience. If the remodeling is a major one, the operation could even find itself shut down completely, either for an extended period of time or periodically. Another problem centers around the condition of the present facility in terms of age, the type of construction used in the original design, and the present availability of utilities, road access, and similar factors. Even with a remodeling the present facility may prove to be inadequate for needs, and the process, while initially appearing to be a viable solution, may be more trouble than it is worth. The law of diminishing returns sometimes simply overcomes a given structure. In addition, with a remodeling approach the firm may find itself locked into the confines of the existing facility, unable to take full advantage of other alternatives that might be available to them at a new location.

D. Downsizing

Downsizing can become an option to consider under several different circumstances. There may be a reduced demand for the product or service offered through the facility, or changes in technology may have made it easier to produce the item or service in question. A company may also consider downsizing as an attempt to increase efficiency by using smaller production units in a number of locations rather than in a single large facility. If the latter is the reason for downsizing, it is then a form of modernizing. In the case of reduced demand for production, downsizing is an attempt to create greater efficiency and cost savings by reducing the amount of physical plant and activities to match the new lower demand. Downsizing is a consolidation process. The advantages are reduced fixed costs as the size and type of equipment is reduced, reduced variable costs as idle machinery time and labor costs are reduced, a faster, more efficient mode of operation resulting in more production in a shorter span of time, and a more efficient use of space.

Disadvantages include a reduction in the firm's capacity to react to increases in demand should they occur, an increase in unutilized space if it is not put to some alternative use, and possible administrative problems resulting from a shift from single-facility to multiple-facility management.

There are certainly many opportunities to downsize in industry, times when bigness is more of a disadvantage than an advantage. It is an alternative often overlooked in the facilities-planning process but one that is increasingly finding acceptance as markets contract, expand, and shift.

E. Merging

Similar to downsizing, merging involves combining operations into a single location. The cause could be a reduction in demand for a product or possibly the increased efficiency afforded by new methods and new technology. This approach combines the advantages of large production units (economies of scale) with reduced room and time requirements through automation, robotics, miniaturization, and a host of other leading-edge production alternatives. In this way, the firm may find it possible to consolidate its operations to include a number of different or related operations under one roof.

The advantages are reduced physical plant and the reduction in plant-related costs that accompany that change, reduced need for coordination of efforts among various facilities, a better utilization of plant and equipment, reduced variable and fixed costs due to more efficient production methods, and a tighter, more easily supervised operation. Disadvantages include possibly reduced flexibility, greater transportation costs, disruption of supply sources, and the prospect of having to upsize at a later date if market changes warrant.

F. Closing

Facilities that are outmoded, expensive to operate, or furnishing unneeded excess capacity can simply be closed. Closing a facility can bring about a quickly realized saving of cost and, in some cases, can provide a source of funds for more profitable investments. Facilities are expensive to maintain, as are labor forces, inventories, and the general overhead connected with the location. By closing an unprofitable operation, a firm can cut operating costs by eliminating the fixed and production costs of that operation. If the facility is owned, the firm may also realize a recovery of capital investment through the proper disposal of the facility itself. If recovery of the capital investment is not seen as a desirable action, the owned facility can also become a source of income through a leasing arrangement with another firm or, in extreme circumstances, may represent a source of depreciation credit for tax purposes.

Closing a facility may have an undeserved stigma attached to it if it is perceived to be an indication of failure on the part of management to deal with the problems of that individual facility or with the overall operation of the company. This is unfortunate. A facility is a tool to be used by the firm. Like any other tool, when it is no longer necessary or capable of performing its intended function, it should be discarded, scrapped, or sold. In addition, those involved in the creation of the facility may be reluctant to see it abandoned or destroyed. As a result, this very real and valuable alternative may be overlooked or not given its due. Closing an unprofitable facility eliminates a drain on firm operating capital and adds to the ease with which the overall operation is managed. It releases funds for other uses and may represent the most efficient method for restructuring available to the firm.

It should be noted, however, that there can be some definite disadvantages connected with this approach as well. Closing a facility can result in a loss in investment rather than a recovery of capital. Although it may appear that disposing of the facility will garner capital due to appreciation of land values, the actual closing costs may prove prohibitive. Are there contracts with unions or individual managers that can lead to excessive relocation costs or severance fees? What costs are nonrecoverable? What penalties are incurred from premature cancellation of contracts? These are all questions that must first be answered.

In considering the impact of closing a location, the ramifications of the act must be seen in larger context as well. A firm, particularly one with an older facility that has been in operation for some time, has an investment in a community and in its employees as well as in land and equipment. Goodwill developed over years of operation can be destroyed very quickly when a facility is closed. There are ethical considerations to be dealt with here, involving the loyalty of the workers, the social and economic impact of the closing on the community as a whole, and the ethical obligation of the firm to the people it employs. Every firm has social and service goals for its employees and their community as well as to its customers, whether those goals are enunciated or not. To ignore them is to deny those social and service goals that define the firm itself. The consequences of closing a facility go beyond the apparent costs and savings. It is not an undertaking to be lightly entered upon anymore than it is one to be summarily dismissed on the basis of personal prejudice.

II. COSTS

A major element in the venture analysis of new facilities development is cost, including start-up costs and ongoing operating costs of the plant. Beyond these obvious cost categories, there are several others that must be considered in order to create a complete and realistic evaluation of cost expectation. These include the cost of moving, replacement costs of machinery and equipment, lost revenues, and the relationship between value and cost. In this section, both the basic and the more nebulous cost factors are discussed.

A. Start-Up Costs

These are initial, nonrecurring costs connected with the starting up of a new facility that are exclusive of normal production costs and overhead. Start-up costs represent fees and expenses that the company must pay to start the new facility in business. They include planning costs, architectural and initial construction costs, licenses, taxes and fees to local and state governmental agencies, legal fees, fees to consultants, and others discussed in more detail in this section, such as replacement costs and construction costs. Briefly, start-up costs are all those costs associated with the new facility that are nonrecurring and are incurred before the facility goes into operation.

In the planning stage, it is necessary to determine start-up costs for each of the alternatives being considered to ensure a fair evaluation and comparison of alternatives. Such costs will probably represent a considerable capital outlay that must be made up front, and can be recovered only over time as the facility begins to make a profit from operations. These costs are usually estimated either per square foot or in total from available industry information. Although much of this work is done for the firm by the architect and construction contractor, other elements must be estimated internally, such as the cost of moving from one facility to another.

Because start-up costs are often high, the choice of one alternative over another may come down to a matter of which approach allows for the initial outlay of funds, other factors being equal.

B. Fixed Costs

These costs differ from start-up costs in that they continue throughout the facility's operation and they do not vary in the short run. Economics defines a short-run time perspective as one in which at least some costs of operation do not change. These nonchanging costs are invariable and therefore fixed. Such costs include depreciation of building and equipment, costs of basic telephone service (if the number of lines does not change), monthly lease and rental charges on equipment, salaries of office and sales personnel (if their number remains relatively constant over time), and other factors not directly related to the manufacture of the goods and services in question. Fixed costs are those costs that remain constant from month to month or year to year. Mortgage payments, cleaning services, and maintenance contracts are other examples of this type of cost.

Some if not all of the previously mentioned examples are, in reality, subject to change under specific circumstances. The number of telephones installed at a facility does not always remain the same. Mortgages may have escalation clauses on interest rates, and cleaning contracts are periodically renegotiated at higher or lower cost levels. This does not negate the concept of fixed costs. It merely changes the length of time considered to be short term.

This explanation may suffice for the economist, but the practical businessperson needs a clear definition of fixed costs if effective budgets are to be outlined and overall costs determined. More practically, fixed costs are all those costs associated with the manufacture of goods or provision of services incurred by a company which do not vary with the level of production.

For instance, if a plant has a capacity of 100,000 units of production per day and that plant had cost a total of $1.78 million including equipment, the cost of operation would not change as far as plant cost is concerned if production went from 57,000 units to 68,000 units per day. There would be no change in the number of machines used or the number of buildings used or, in all likelihood, the number of clerks employed in the main office. These costs would remain unchanged and therefore constant in respect to the level of production.

Fixed costs are costs that exist whether there is any production of goods or services or not. They represent the base cost of operation and, at least in the short run, cannot be improved upon by efficiency improvements in manufacturing or in supplying the product or service of the firm.

C. Variable Costs

Some costs vary with the level of production. Variable costs represent the cost of direct labor used in manufacturing or supplying a service, the cost of raw materials, power consumption, variable depreciation reflective of levels of machinery use, warehousing, transportation, and other expenses directly incurred through the manufacturing process.

D. Opportunity Costs

Opportunity costs are incurred by alternative uses of funds. We define opportunity costs as the cost of not choosing the second best alternative. To be realistic, we need to determine the amount of this cost to know how much true profit is to be realized from the operation of this particular facility that would not be realized if we had chosen one of the other alternatives available. Economic profit is defined as profit that is unique to a particular choice. To determine that profit unique to the given choice, the opportunity costs, represented by the profit that would have been realized if the next best alternative had been chosen, must be subtracted from expected returns. For example, if the choice is between a new facility that is expected to yield an increase in profits of $1.5 million per year or an alternative facility that is expected to yield an increase in profits of $1.2 million per year, the cost of foregoing the opportunity to make $1.2 million must be taken into account. In other words, we would have made $1.2 million no matter which of the two choices we made, but because we chose the most profitable alternative, we make an extra $300,000, which is considered an economic profit. There are numerous opportunities for investment available to the firm. The goal is to choose those opportunities that will be most profitable.

E. Managerial Costs

It is possible to design a facility in order to enhance manageability. On the other hand, a facility that has been repeatedly expanded may surpass its usable capacity and become unmanageable. The cost of maintaining manageability can change depending on the complexity of the managerial system, even though the production operations remain the same. Systems that are expected to cover facilities that are too large or too diverse in their operations are headed for problems within their managerial structure. The costs of keeping management efficient are difficult to tally, in that they can include such intangibles as employee morale, process design costs, and costs of quality.

F. Moving Costs

In deciding among alternatives, it is necessary to consider the cost of moving under each, as it will have a major impact on the outcome of the operation. The costs connected with moving from one facility to another can vary considerably, depending on how much of the old facility is to be transferred. If all new equipment is to be installed in the new facility, for instance, there is a marked reduction in costs connected with moving from the old facility. If present equipment is to be transferred, costs can be quite high.

When the machinery and equipment are to be transferred from the old facility to the new, there is not only a cost involved with the actual removal and placement of machinery but also with the downtime associated with the move itself. For the period of time in which the machinery is down (being dismantled, in transit, set up and tested), no production can take place, meaning idle time for both equipment and employees. In addition there is an increased hazard of minor misjudgment that comes with any move, no matter how carefully planned; until the equipment is actually installed, some problems are simply not recognized. Even the most mundane of factors can become a major problem. A misjudgment of inches can result in delays, last-minute changes, and inconvenience.

Even if the movement of machinery and equipment is not a problem, a smooth transition is not guaranteed. Inventories must be considered, both for raw materials and for finished goods. Delivery schedules must be tightly planned and controlled to ensure continuity of operation. Records and files, office equipment, telephone service, and other normal services must be coordinated. The details can seem endless.

In determining expected moving costs, remember to include not only costs of the actual move but also expected costs from disruption of normal business operations and delays. Include as well a hedge against unexpected expenses not considered. The more simply and quickly a move can be accomplished, the lower these costs will be. At times it becomes more reasonable and cheaper to sell otherwise useful machinery and equipment and replace them with new equipment simply because of the costs of the move itself.

G. Lost Revenue

When a facility is moved from one location to another or responsibility for production and supply of goods and services is shifted, there will generally be a loss of revenue as a result of the move. A certain amount of downtime will be experienced in moving a location, whether the entire operation is to be uprooted and transferred to a new location, or just the employees and business files. During this downtime, there is low productivity if, indeed, there is any productivity at all. Lost productivity translates into loss of revenues from sales not made. For this reason, planning the movement of a facility must logically include consideration for the downtime involved, keeping it to a minimum, and finding alternative ways of handling the problems connected with downtime. Some firms stockpile goods in anticipation of a move, or preschedule deliveries of goods and services to allow themselves lag time during which to make the move. Another strategy is to temporarily operate at two sites simultaneously, making the move step by step so as to minimize the possibility of lost revenues. However the problem is handled, it must be dealt with in some fashion. Lost revenues resulting from an inability to supply customers can translate into not only short-term losses, but, more importantly, long-term losses from loss of goodwill, or the inadvertent loss of customers to competitors during the delay time.

H. Reproduction Costs

When a new facility is brought on line, it will bring with it the best of what has come before, that is, the ideas and methods that were effective in the past. This means reproducing conditions, specialized layouts within a facility, and other operational mechanisms in the new plant as they exist in the old. This reproduction means costs. Specialized equipment that may have evolved through trial and error must be reproduced at the new location. The clustering of related equipment may be unique and require special support structures. To maintain efficiency, a close adherence to established patterns of work may need to be maintained. In addition, new governmental regulations may require updating certain equipment or structures that would also result in added cost. The cost for this individualized work will be a factor in the creation of any new facility. These costs may or may not become a major factor in decision making, but because of the impact they can have on operations, they must be considered.

I. Replacement Costs

In considering the choice of replacing or moving machinery and equipment, a number of factors come into play. In addition to the cost of the item itself, costs that can be incurred include loss from the sale of the old equipment, administrative costs connected with its disposal, scrap value of old equipment versus depreciated value, financial costs, and downtime cost connected with normal equipment operations. In short, the decision to replace or transfer hinges on

1. whether a decent price can be obtained for used equipment when it is sold or if the loss in value for selling used equipment is too great.
2. whether the cost of managing and administering the sale of old equipment is prohibitive in terms of outlay for internal managerial time or fees paid to external disposal managers.
3. whether the scrap value for old equipment recoverable upon disposal compares favorably with the depreciated value of the equipment as presently accounted and the effect of new equipment purchases on depreciation schedules and short-term fixed-cost estimates for ongoing operations.
4. whether the financial costs of new equipment purchases are prohibitive compared to the efficiency obtained from the new equipment, whether financing is from internal sources or from debt capital (remember the opportunity cost problem of finding the best use of funds).
5. whether the cost of downtime connected with the removal and reinstallation of existing equipment can be justified by cost savings over the purchase of new equipment, which would result in a smaller loss of revenue and smoother transition.

J. Construction Costs

The cost of construction is a major contributor to the final decision in questions of facilities expansion. Should a firm choose to move from its present facility rather than expand? If it does, should it move to an existing facility that can be modified to satisfy its needs, or should the firm build its own? Obviously, a newly constructed facility will be more aligned to individual needs and therefore more efficient than a modified site, but what about the cost of such an endeavor? Construction costs vary from location to location. They vary with the economy, with interest rates, with the type of construction used, with the size and nature of the structure, and a host of other variables. In many cases, it is more cost effective to settle for a slightly less desirable physical plant than to have exactly what one needs. This is exactly what a firm must determine through an analysis of the construction costs and comparison of these costs from alternative to alternative.

K. Value and Cost

Finally, there is the issue of the relationship between value and cost. Value is a very subjective concept. This is as true in business as anywhere else. The concept of value deals with the amount of utility gained from the use or consumption of some product or service and, so the theory goes, price (cost) reflects the generally accepted value of an item or service to the society as a whole. Price is supposed to reflect utility, and it does. However, under individual circumstances, the utility of an item to one person may be quite different from what it is to another. This is also true of individual firms.

Businesses have an obligation to keep costs down when and wherever possible. This works for everyone: for customers who can realize lower prices, for the general population who realize a more efficient use of resources, for workers who can realize higher productivity, for investors who can realize higher profits, and for management who can realize a greater efficiency of operation. Yet value may not always coincide with low costs. What would be prohibitively expensive to one company may appear quite reasonable to another, depending on what it values and what its customers value. It is more expensive to produce a fine touring car than it is to produce a small compact automobile designed for basic transportation. Yet both properly exist in the market. Different people value different things. In the same way, the importance of minimizing costs may take second place to the importance of maintaining quality, or customer service, or some other facet of doing business. It is short sighted, then, to consider only costs in the analysis of ventures. Those doing the analysis of costs must also consider the purpose and the intent of the firm in exploring those possibilities available to them. Decisions are made on the basis of overall value, not simply cost savings and minimization. Under most conditions, cost is not the only key factor, merely one of many, and it is important to keep that perspective in decision making.

III. EXPECTED RETURN

The purpose of any new facility is ultimately to create profits. Since there are a number of alternatives from which to choose, the efficiency with which the capital is invested will dictate the return on that capital. Expected return on investment (ROI) is therefore of paramount importance in the investment decision. Analysis of proposed facility creation must yield information on this return and its suitability to the needs of the firm. When determining profitability, the firm will measure performance through four types of returns: short term, long term, minimum return, and bracketed return.

A. Short Term

Because of start-up costs and other expenses incurred at the beginning of a project, the short-term return on investment is usually smaller than in the long term. As we have seen from break-even analysis, until production and sales reach a certain minimum level, the firm can expect negative returns, or losses. In the short run this is acceptable, as long as there is sufficient working capital available to the firm to weather this particular storm.

Facilities, whether remodeled or new, represent a long-term investment. The firm expects negative or low profits in the beginning when start-up costs have to be absorbed by early profits and fixed costs have to be spread over relatively low levels of production. However, as the production increases and these costs are satisfied, true profits are expected to be realized. In the short run, the firm will willingly absorb these temporary losses as long as there exists the certainty that those losses can be recovered through later higher percentage profits.

Even so, there are still minimum acceptable levels of return in the short run. In considering the lowest acceptable level of return, the firm must take into account several items:

1. The amount of working capital available for the project. If the amount of working capital needed to remain in operation until profits begin to be realized is higher than the firm is logically willing to expend, it may kill the project.
2. Expected returns to investors. Even in the short run, investors expect some return on their investment. An inordinately low return may produce a reduction in confidence on the part of the stockholders of public corporations, with accompanying reductions in stock prices, availability of borrowing power for other projects, and so on.
3. The time value of money. If the recovery time is long enough to reduce seriously the efficiency of capital, the project may also die an early death. Overall profitability within the company may dictate an alternative use of funds.

B. Long Term

Long-term return on investment is nearly always higher than short-term return, because of the longer period of time for absorbing start-up costs and the ability to spread fixed costs over larger levels of production in the longer time frame. Yet this is not always the case, and not always desirable. It should be noted that in some industries, products have differential demand schedules that are higher in the beginning and then settle down to a lower, though more constant, state. Fad items and goods and services whose value is based on novelty are examples. Even goods and services that are more long-lived, such as restaurants, experience this phenomenon. In such cases, the facility's strategy may be to invest on the basis of high initial profits, and then to sell the unit when the demand begins to settle down and to move on to some other location and some other novel approach. In this way, a permanent facility can be viewed by a firm as having a short-term life and high "scrap value" not dependent on long-term prospects. As long as returns on investment are sufficient, there is nothing inherently unsound with this approach.

In the long term, the proposed operation must be profitable. Long-term return on investment must demonstrate an equal or greater profitability than other projects to win acceptance over them.

Firms have criteria for what is an acceptable level of return on an investment. These criteria are based on a number of factors, such as the expected cost of capital, the general return on investment for the industry and for the economy in general, the stated firm goals, and the need for retained earnings for other investments.

C. Minimum Return

This refers to the minimum return on an investment that the firm will accept. It is similar to the concept of break-even in that it gives a floor for the acceptability of an investment for the firm. Minimum returns are usually predetermined by policy, although they may vary from firm to firm.

In determining return on investment, the analysis of a facilities venture must determine the minimum expected return that the project will bring. The determination of this figure presents a worst case scenario with which the decision makers can make their determinations.

The advisability of an investment hinges not only on the possible returns that can be gained, but on how much risk is involved as well. The higher the risk, the higher the expected rate of return must be before an individual or a firm will be interested in the investment. By determining the minimum expected return on investment for the project, the firm is better able to judge the degree of risk involved and from that to determine how high expected rates of return would have to be in order for the project to be accepted. With a relatively high minimum expected return, a lower average (or most likely) return on investment might be more acceptable than in the case of a low minimum expected return on investment.

D. Bracketed Return

An alternative means of determining expected return on investment, and one that is the most reasonable considering the degree of uncertainty in estimating costs and profits, is to employ the concept of a bracketed return. The fundamental idea in the use of a bracketed return is that there is not a high degree of certainty about actual returns to be realized on an investment, but a range of returns that can be determined with accuracy.

The bracketed return combines the worst case minimum return concept with a "best case" maximum expected return to give a range of possible returns that is bracketed by the two extremes. See Chapter 6 for the Kingdom Yacht Sailing Club income forecast. Just as artillery homes in on a target by bracketing it to each side and then marching in on its position, so the bracketed return approach seeks its target by defining the outside possibilities with the expected return lying somewhere between them. The target return on investment in the bracketed return approach is neither the minimum nor the maximum, but some most likely figure in between. In determining expected return on investment, this most likely return figure is weighted more heavily than the minimum or maximum return figures, and an average is taken to arrive at a single figure.

As an example, if the minimum expected annual return on investment for some anticipated project is $995,000, the maximum expected return is $1,700,000, and the most likely return on investment is $1,110,000, the analyst might elect to weight the middle figure four times heavier than either of the other two, yielding a figure of [$995,000 + 4($1,110,000) + $1,700,000]/6 = $1,189,167. This figure would then be used to determine the expected return on investment. If it is assumed that the project has an expected capital cost of $20,000,000, the expected return on investment would be $1,189,167/ $20,000,000, or 5.95 percent per year.

IV. UNFORESEEN PROBLEMS

Along with the obvious determinations that must be made regarding the choice of available alternatives and the various cost factors, a host of considerations should be explored to identify unforeseen problems. Most of these have to do with environmental forces external to the firm over which the firm has no direct control. One of these external elements, government, is so important that we have chosen to present it as a separate concern. First, however, we deal with some of the other possible problems.

A. Neighborhood Concerns

A firm does not operate in a vacuum. The community in which it operates is a very real part of the firm's environment. It supplies services, a labor force, and community support. The relations the firm has with that community are important. The firm must therefore keep in mind many different neighborhood concerns about the impact the facility will have on those who live and work around it. How the facility affects road use, the crime rate, the noise level, the degree of strain on neighborhood services, and the physical environment are all elements in the decision to locate in a given area. Just as the neighborhood must be able to satisfy the needs of the firm, the firm and its facility must be able to satisfy the needs and desires of the neighborhood. There is an element of partnership between the facility itself and the environment surrounding it. Without this sense of partnership, a feeling of being part of the greater whole, there will be a lack of cooperation at least, and at the worst, the project could be a total disaster. The key is to choose a site that is appropriate to the needs of all concerned and thus eliminate the possibility of inefficiency.

B. Interest Rates

Cost of capital directly affects the profitability of any project regardless of whether the firm chooses to use external sources of funds or internal reserves. In either case, the interest rate is a factor. With external sources, it represents a cost for the use of someone else's resources. With internal sources of funds, the interest rate is an opportunity cost that is incurred by choosing to use one's capital rather than lend it to others at the going rate. In either case, interest rates are changeable.

Analysis of a venture should measure the probability that the present interest rate will hold, rise, or fall. A sudden change in the cost of capital could change a project from one that is profitable into one that is not.

C. Government Agencies

Encroachment of government agencies into the affairs of business is increasing. As much as it may be disliked by the business community, it is there for a reason: to protect the rights of the consumer, the public, and the firm itself. For whatever reason, self-regulation by the business community has been judged to be an inadequate means of protection. As a result, a number of agencies with the power to affect business activities have been established.

Agencies ranging from the Securities Exchange Commission (SEC) to the Organizational Safety and Health Administration (OSHA) can regulate what firms may or may not do. Although these regulations are well-known for the most part, particular circumstances may result in unexpected breaches of regulations leading to fines or facility changes that would be otherwise avoidable. Therefore, it is necessary to ensure that the proposed facility does not violate any such regulations.

D. Tax Laws

Tax laws may vary from one location to another. In locating a facility, favorable tax laws may be sought to improve profitability. Many states and municipalities are willing to offer concessions to attract new businesses. Others practice high taxation of businesses as a source of revenue and payment for the increased services that more industry demands. The nature and expected state of the tax laws of a location must be investigated before their effect on a project can be determined.

E. Zoning Laws

The purpose of zoning laws is to ensure the orderly development of a municipal area. In most instances, construction cannot begin on a facility that is to be located in or near a municipality until the approval of the zoning commission has been obtained. The commission is the governing board that determines the types and allowable locations of industry for their community. For any given location, it is necessary to know the general climate of this committee (how favorably disposed they are to this type or any type of facility) and the ease with which permits and zoning approvals would be issued.

F. Environmental Considerations

As mentioned before, a facility will impact a community in many different ways. One form of impact involves the changes in the physical environment that could result from the development of the facility. Major alterations in the ecology or changes in the nature of the environment resulting from the introduction of the facility into the locale must be taken into account. Manufacturing facilities often create potential problems in waste disposal, air and noise pollution, changes in the appearance of the surrounding area, disturbance of wildlife and of quality of life for others in the area, and so on. It is best if a facility can be located where its relative impact is low in these categories. Again, it must be remembered that wherever the facility is located, it will be part of a larger whole and must address its responsibilities to that greater whole if it intends to operate successfully. Damage to the environment can result in legal and social problems that seriously disrupt the successful operations of the firm.

G. Unions

Any facility that employs a sizable work force must address the issue of unions. Is there a local union already in place whose members are potential employees of the new facility? What is the history of labor relations in this area? Is it possible to operate without unions? Do you want to do so? What are the average wage rates in the area due to union influence, and how does this impact the expected cost of labor for the new facility? All these questions must be answered before the venture is begun.

The presence of a union in a given location is neither good nor bad in and of itself. What is important is how the local unions operate and the nature of labor relations at the local level. At times, the existence of a union in a location can ease the procurement of workers, particularly if the firm is already involved with unions. It is also possible that lack of a local union could have the same effect. Either way, the issue cannot be ignored. Once normal operations begin, labor relations will be a constant managerial concern.

H. Competition

In this category, we are referring not only to competition providing the same products and services as the proposed facility, but also to competition for available resources, such as locally supplied services, labor, construction materials, and transportation services. A highly competitive market where a great deal of growth is taking place can translate into shortages and higher costs of operation. The presence of competition producing the same goods and services as the proposed facility may, however, indicate a favorable climate for the new facility by virtue of the competition's choice of location. It is, for instance, not accidental that steel mills and automobile factories tend to be clustered in certain regions, or that high-tech electronics firms maintain facilities in tightly defined areas. They are there together because of the advantages the particular location offers. As long as those advantages outweigh the risks of competition, the facility will be more apt to profit from that choice of location.

I. Weather Patterns

The effect of weather on operations ranges from the obvious to the subtle. In the gross analysis of a physical environment, some weather patterns are obvious and easily determined, particularly for products and services that are sensitive to weather conditions. For example, if a firm intends to open nurseries, the design and nature of the facility are affected by the type of weather patterns in the area. Northern climes require more hothousing, while milder southern regions afford the nursery a greater use of natural light and warmth. In a similar manner, products that are sensitive to temperature and humidity variations in their manufacture would be most easily produced in areas where weather conditions will not seriously interfere with production. Products easily damaged by high humidity, for instance, might best be produced in arid desert regions, all other factors being equal.

In a more subtle manner, weather patterns could affect the efficiency of an operation through secondary cause-and-effect situations. If a manufacturing process is a continuous one (such as steel production, where the furnaces are virtually never shut down), short-lived weather phenomena may have serious consequences. Conditions that would not otherwise have negative effects on an operation could be ruinous if constancy is a prime factor. In a mild region, there may be only one or two days of snow per year, which would seem to offer little or no concern for the facility planner. But what if there were normally short-term power outages during these times and your process demands high load power on a continuous basis? Or perhaps the firm is operating a JIT (Just In Time) facility that maintains low inventory levels and ships to customers on a daily basis. In an area where there is little or no snowfall on an annual basis, a sudden unexpected ice storm or heavy snowfall could result in all industry in the area coming to a standstill. Municipalities tend not to bother with the investment in heavy equipment necessary to combat ice and snow conditions if that equipment will not be used more than a few days a year. It is cheaper for all concerned to simply weather the storm. But what does that do to production? How do workers make their way to work with roads closed? And without sufficient backup stocks (which is one of the prime cost-cutting elements in the use of JIT facilities), a two- or three-day delay due to poor weather conditions could be quite costly. In certain types of facilities, then, weather is an important factor.

J. Traffic Access

Traffic patterns, like the weather, have both obvious and subtle effects on production. For ease of receipt and delivery of materials it is important for a facility to have access to several transportation systems, preferably not only rail and surface highway but also air and in some cases pipelines as well. The ability to maintain a constant flow of raw materials and finished goods demands it. In service industries, the facility must be easily accessed by the public or at a minimum must provide proximity for service personnel making deliveries or performing services. If the operation is labor intensive, easy access for workers is also of importance.

Beyond this, the level of traffic in these systems is also of major importance. Clogged airports or overburdened rail systems can mean expensive delivery costs and delays. Highways that are not adequate for the flow of traffic at peak travel times can reduce the efficiency of deliveries and of the labor force. Heavily traversed roads also increase the possibility of traffic jams and accidents. At a more subtle level, the direction in which workers and delivery vehicles have to move at a given time of day to reach the facility can be important. By locating to the west of the main source of workers, for instance, a company can substantially reduce car accidents among employees and therefore avoid absenteeism simply because in both driving to work and returning home, a western location will mean the sun is always at the drivers' backs instead of in their eyes.

K. Economy

The economy in which we operate is a dynamic, living system. Like all systems, it is in a constant state of change. The study of this system is anything but determinate. Rather, the economist, a social scientist, studies the system statistically, looking for trends, tendencies, and probabilities. In economics, one deals with expectations, not definitives. Because of the inherent uncertainty of the system, the firm must remain alert to unexpected changes in economic conditions. There are simply too many variables to be certain of what will happen next. After all, the experts are only as good as their predictions are accurate. And, as the saying goes, whenever there are three economists in a room together, there are at least four opinions.

Knowing they will be faced with unexpected change, firms plan for contingencies. A sudden change in oil prices could either be to a firm's advantage or the cause of disaster. A war in the wrong country, a natural disaster, a new discovery in science, or a change in the mood of the people would all have some effect. Any such change in circumstance could either benefit or hinder a firm's efforts to sell its goods and services. Those in authority must rely on their best estimates of the situation, but they must also recognize the degree to which their products and services are vulnerable to unexpected economic change. What if interest rates were to take a sudden change? How would the capacity of the firm to fund the project be affected? What if demand for the product were to suddenly take a drop? If the drop in demand were permanent, could the facility be utilized for the manufacture of some other product? How expensive would such a changeover be? What is the possibility of the technology being used in the new facility becoming obsolete in the face of new technology or new discoveries? How recession-proof is the product? These are all economic questions that need to be considered.

Again, there are few definites in economics. The firm can only create contingency plans to deal with the most probable of the possible scenarios and explore the general direction they would want to take in some of the more extreme and less likely scenarios. This is not to say that disaster will strike, merely that one must be aware of the possibilities and plan for them.

V. GOVERNMENT CONCERNS

Because of government's involvement in business affairs at federal, state, and local levels, analysis of any facilities venture will include a study of the impact of government regulation and licensing on the project. Manufacturing facilities in particular find governmental impact to be a major area of concern. A number of governmental considerations are briefly discussed below.

A. Planning Boards

Most municipalities have planning boards whose function is to ensure the orderly and efficient development of their community. Planning boards determine the most desirable types of future growth for their community and the locations for various types of municipal elements, including schools and parks, shopping and residential districts, and industrial and business districts. Any proposal for a new facility will necessitate presentation to and approval by the planning board.

B. Zoning Boards

Zoning boards are similar to planning boards except that, rather than simply planning for further development of the community, the zoning board determines exactly what type of activities will be allowed in given subsections of the municipality (zones). A zone is defined in terms of the type of activities allowed there, ranging from single-family dwellings or multiple-family dwellings to light industry or heavy industry. Within a given zone, only the assigned range of activities will be allowed. Any variation from that must be sought through petition to the zoning board. If a variance is given, an individual facility will be allowed but the integrity of the original zone definition remains. Areas may also be rezoned, allowing for new types of businesses and facilities in an area where they were not previously allowed. Rezoning an area involves changing the definition of the zone and can be difficult to achieve. As with a variance, rezoning can be accomplished only by presentation of a petition and a vote of approval by the zoning board. Although details of the board's functioning varies from location to location, essentially they all operate in the same way.

C. County Commissioners

The county commissioners are the governing body of a county. They are the equivalent of a local congress or town council. If a facility is to be located in a county but not within the boundaries of a municipality, then the local government officials with whom one must work are county commissioners. This body has the power to legislate, collect taxes, and pass ordinances. Any regulations or waivers thereof will be handled by this body, which normally considers petitions by interested parties.

D. Federal Agencies

When developing a new facility, a firm must deal with a number of federal agencies. Depending on the nature of the facility and the firm, some of these agencies could include the Organizational Safety and Health Administration, the Environmental Protection Agency, and the Interstate Commerce Commission.

All of these agencies and a host of others are designed to protect the well-being of the public as well as that of the firm. Often, contact with federal agencies is so convoluted with details that it is best left to the legal department of the firm to ensure compliance. In the planning stage, the issue is whether or not regulation will place insurmountable obstacles in the development of a given site or program.

E. State Agencies

Often when a federal agency is involved, there is a state agency fulfilling the same need on the local level, and the state agency must be dealt with as well. The firm must be certain it has complied with regulations and must also seek the necessary licenses, permits, and payment of fees in order to set up operations. Virtually all states have extensive bureaucratic systems for the management of commerce within their jurisdiction. Again, the purpose is the protection of the public and assurance of continuity in development of industry at the state level.

F. County and City Agencies

As the firm begins to develop, local agencies tend to take a more direct and immediate role in the planning of the facility. Licenses, fees, and permits at the local level are an integral part of the construction process, as are registration with local commerce-controlling agencies and adherence to local ordinances.

VI. PROJECT UTILITY

The final step in evaluating the various alternatives for a facilities project involves the overall expected value of the plan in terms of utility. Utility refers simply to the benefit and satisfaction received by the firm from some activity. Once the different alternative facility approaches have been defined and costs for each alternative have been determined, and after contingencies have been dealt with for unforeseen possibilities and government interactions, what remains to be done is to determine exactly how much value will be received from the construction and utilization of the facility. Two factors are analyzed: the life of the building and the life of the product.

A. Facility Building Design Life

Part of the initial analysis of the project involves determining the type of facility desired and the best way to achieve that end. With the aid of architects, a facility design was developed that allowed for the most efficient use of resources. Yet was that design sufficient to meet the needs of a growing company? There is a need to determine the life of the building itself and how quickly, if ever, a still newer facility will be necessary. This analysis of building life has an interesting twist to it, as it may be that the design has too long, rather than too short, a life. A firm wishes to receive as much value for its investment as it can, and certainly that includes the requirement that the building have an expected life equal to the expected demand for its use. However, a building does not need to last longer than that demand, unless that extended life can be shown to be advantageous to the firm in terms of resale value, refit value, or value from capital gains. In general, the more permanent a facility, the more expensive it is to build. A firm would not want to erect a building with a life of 20 years if its expected use is no more than 10 years. Certainly, it should be fully functional for the 10 years of required use, but unless the additional 10 years of useful life can be achieved in a cost-effective manner, why sink the additional funds into the facility? Closely tied to this concept is that of the business product life cycle.

B. Business Product Life Cycle

There are numerous business cycles extant at any given time. There are seasonal cycles, monthly cycles, yearly cycles, and 20-year cycles, as well as some longer cycles tied to other factors. Just as there are general business cycles, demand creates life cycles for any given product, some being short and others quite long.

If a facility is to be created to supply some good or service, the cost and life of the facility need to match those of the product or service it is designed to supply. Even a cost-efficient facility is useless if the life cycle of the product is too short to allow for its full utilization. There are a number of ways around this dilemma, some of which were discussed before, such as creating the facility in such a way that it can be converted to the production of other products when the demand for the original product disappears. Another alternative is to create the facility in such a way as to make it salable at a profit at the end of the life cycle of the product. The question in such cases arises regarding the best use of capital; yet if it can be shown that the overall return on investment is sufficiently high and that there is a true need for the facility, the project may still be a viable one, even in the face of a short life cycle.

Similar arguments can be advanced for products with extremely long expected life cycles. It may be that no facility can be economically constructed that will last for the entire expected life of the product, particularly if that product is a standard item. However, in the face of changing technology and advances in production design, it is possible to judge a facility too permanent. Because the rate of change is high in today's business world, and promises to become even higher in the future, flexibility is generally more important than permanence.

QUESTIONS

1. The book refers to *ventures* in facility expansion and redesign. What does *venture* mean in this context?
2. Name at least six of the costs that must be considered when beginning a new facility venture.
3. Specify the primary considerations in analyzing a facility venture.
4. How do opportunity costs play a role in venture analysis?
5. Can construction costs be accurately estimated in a multiyear building project? Why or why not?
6. Differentiate between *minimum return* and *bracketed return*. How are the two related?
7. A full venture analysis looks at hidden costs as well as obvious ones. What are some of the sources of these hidden costs?
8. A facility is part of a larger system, in the context of both the firm and the community in which the facility is located. How does this affect the cost of doing business?
9. A firm anticipating the development of a light manufacturing facility encounters fundamentally different problems than a firm seeking to create a service center for consumer products. In each case, how would the venture analysis be affected by unions, weather patterns, and zoning laws?
10. What are some of the government agencies with which the firm can expect to interact when developing a venture analysis?
11. How does the expected life of a physical plant affect the venture analysis?
12. What is a business product life cycle?
13. What is the format under which your local zoning board operates? What is their publicly announced strategy for the development of your area?
14. Why are zoning boards necessary in community development?
15. Are zoning boards and other land-use regulatory agencies a hindrance to facility development, or are they supportive of such activities?
16. Name at least four ways in which facilities can be changed.
17. What do you believe is the most often chosen method of facility change? Why?
18. What do you believe is the easiest method for facility change? Why?
19. What is meant by the term *downsizing*? Under what conditions would such action be appropriate?
20. How does facility expansion differ from facility remodeling?

4
General Site Analysis

Once the decision has been made to create a facility, the next step is to determine its location. Particularly with national or international companies, the number of potential sites is extensive. Even regional or local companies can have a large number of possible sites for the location of a facility. Naturally, some sites are more suitable than others for a particular type of operation. It is the function of general site analysis to determine where those favorable sites are located, and which meets the criteria of the firm most completely.

In order to carry out a complete site analysis, the planners must consider how the facility itself may need to change over time (flexibility), the general demographic trends of the area and of the industry in which they are operating, specific geographical concerns, the business climate in the communities near the potential sites, and other influences ranging from how the site will affect the corporate image to the nature of tax structures at a given location. In this chapter, we explore these areas of concern.

I. FACILITY COMPONENTS OF CHANGE

There are four primary ways in which a firm can alter its facilities: expansion, contraction, openings, and closings. Each of these changes represents a response to the peculiar circumstances of the individual firm. Although much emphasis is placed on the processes of expansion and opening, the decisions to close or contract can be equally viable solutions to capacity problems.

A. Expansion

From the point of view of general site analysis, the possibility of expansion of present facilities is a desirable alternative because of the resulting reduction in actual search and analysis of potential sites. An expansion represents an alteration in the use of present company property, or the annexation of adjacent properties through the lease or purchase process. This narrows the scope of activities by limiting them to locations with which the firm is already intimately familiar.

A present location would be chosen because of advantages determined to exist at the point of original facility development. Whatever those advantages are, they probably still exist. With so many of the requirements of a location already in place, expansion of a present facility guarantees a continuation of the positive benefits that caused the original placement of the present facility in the first place.

The benefits of expanding a present facility go beyond just the apparent advantages already derived from the existing facility's location. In general, expansion tends to be cheaper than seeking a new location, and there are the additional advantages of not disrupting the flow of goods (expansions are generally less likely to interfere with ongoing operations than are moves to a new location), little disruption of the workforce, no need to create and establish new communications, supply, transportation, and support relationships in new communities, and in general, a shorter lead time between the decision to increase capacity and the time that capacity becomes available.

Some consideration must be given to the general condition of the present facility, however. Age is important in determining the value of expanding a present facility. If the age of the present plant is too advanced, it may not be worthwhile to expand it. The availability of additional space is also a restriction. If the present location does not contain sufficient unutilized space, the expansion may be stalled or an alternative developmental plan necessitated, such as building up rather than out. (This is not necessarily a disadvantage, however, in that in building up rather than out, the necessity of laying foundations is avoided and a reduction in building costs is realized.) Leasing property may be the only option if additional land is needed. Even if adjacent land is available for purchase, the lease option should be investigated.

In addition, expanding the present facility may put an undue strain on the capacity of the local labor force, communication networks, and transportation systems to meet the facility's support needs. As with any other site location, these aspects of the present site must be analyzed.

B. Contraction

At times, a company may find it advantageous to instigate a contraction of the present facilities. The reasons for wishing to do this can be many. A firm may find that it is more cost efficient to reduce the size of a facility because of reduction in the demand for the product or service being produced, changes in technology, or a shift in strategy to maintaining smaller regional production units rather than larger, centralized units. A contraction does not always indicate less business.

Whatever the case, the choice to contract productive capacity at a location offers some particular advantages and has some shortcomings. In terms of advantage, a reduction in plant size can result in numerous savings, particularly in terms of plant maintenance, depreciation costs, taxation, overhead, and administrative costs. Smaller units are generally easier to control, since there is not so much to control. That in itself can create a higher return on investment and higher efficiency of capital as well as efficiency of physical operation.

On the negative side, a contraction necessitates decisions about what to do with excess equipment, productive capacity, and physical plant. These excess items represent a considerable investment on the part of the company, and recouping that capital may not always be easy. Although firms may find themselves reaping profits from downsizing, in many cases the contraction may merely reduce long-term losses. In financial terms, contraction must deal with avoidance of losses as well as increases in long-term profits. Does one sell excess land, lease it to some other company, or let it stand? Does one tear down unused buildings or let them remain unoccupied? What still needs to be maintained? What needs to be sold off or destroyed? What does divestiture of these assets do to depreciation schedules, capital gains tax positions, and opportunity costs? By briefly considering the three scenarios for the contraction decision that were presented previously (slowing demand, changes in technology, and changes in company philosophy), it may be possible to see how the advantages and disadvantages of contraction occur.

In the first case, a reduction in demand for the product, contractions are a viable approach under two conditions: either the reduction in demand is long term or permanent, or, if temporary, an increase in profit or reduction in costs can be realized by the temporary contraction of productive capacity. If the reduction in demand is permanent, it becomes evident that the company has excess productive capacity that is not expected to be utilized in the future. This productive capacity represents a drain on resources because of overhead costs, maintenance costs, and opportunity costs. That is, the unused portion of the productive capacity of the plant must still be administered and maintained, yet it does not and is not expected to contribute to future earnings in any significant way. Opportunity costs arise because of the unavailable capital that this unused investment represents, capital that could be used for some other opportunity to make a profit if funds were not tied up in useless plant and equipment. Under such circumstances, a reduction in plant size through the disposal of the excess capacity means lower costs (no more administration and maintenance) and possibly higher profits (through the alternative use of funds).

If the reduction in demand for the product is temporary, there may still be possible improvements in profits through the temporary reduction in productive capacity, although this process is a bit trickier. Such a tactic implies that the productive capacity of the installation is flexible and easily increased or decreased over time. Start-up and shut-down costs should be minimal for this solution to work, and ideally, if contraction is temporary, the phased-out equipment should be either cheaply and easily replaced, or leased to other producers on a temporary basis so that revenues can be realized during the time of disuse, thus reducing opportunity costs connected with unutilized plant and equipment. A firm manufacturing machine parts, for instance, may be in a position to rent or lease out space to other contractors during periods when it does not need certain equipment for its own production process. When demand rises for its own products again, the firm absorbs the short-term leased productive capacity and goes back "on line." In either case, the purpose of the contraction will be a reduction in costs associated with unused productive capacity.

In the case of technological changes, a contraction in plant size may result by virtue of cheaper and easier ways to do the same job. Streamlining and refitting with more efficient, modern equipment can reduce the size and scope of operations in many cases. Improvements in output per machine or speed of operation can mean that equal or greater levels of output are possible with a reduction in equipment and physical plant. This form of contraction represents a technological upgrading of operations. As a quick example, consider the difference in operation between a firm producing mechanical watches and one producing digital watches. In the first case, a number of mechanical parts (gears, cases, dials, springs, and so forth) must be manufactured or purchased and put together. Taking inventory of the pieces, workstations, trained workmen, or sophisticated automated machinery, just to mention a few, are necessary for this process. The manufacturer of digital watches, on the other hand, works with what amounts to a case, a watch strap, and half a dozen electronic parts (if that many) which are assembled in a matter of moments. Admittedly, the two products are similar, not identical, but the same can be said of automobiles, houses, radios, and calculators. With all these items, as with the watches, advances in technology have resulted in a complementary reduction in the facility cost per unit needed to produce these items. Technology means efficiency, and economic efficiency means more output per unit of input with a reduction in cost.

How about a contraction in facility size due to a change in corporate strategy? There are many ways to attack a market, and different firms attack the same market in different ways. Some firms prefer large, centralized production facilities to reap the benefits of economies of scale. Others prefer regional or even local productive units to take advantage of the easier control the smaller size allows them. Still others opt for new productive formats, such as the JIT approach that minimizes the need for inventory handling and storage. A change from one of these strategies to another could result in contraction.

Many factors can serve to alter the strategy of the firm. Changes in costs of transportation, availability of raw materials, redistribution of markets, and changes in competition (both from other firms producing the same products and services and from firms producing new or different alternative goods and services) can all create a change in strategy on the part of management.

If transportation from a central location becomes too expensive, the need for opening regional facilities may create a contraction of the larger, original facility. Here, total demand for the product has not dropped, only demand for products from the single facility, which offers the opportunity, indeed the necessity of downsizing. If the raw materials needed for the production of goods and services are no longer available in a particular location, a firm may choose to limit the demands it puts on local supplies by reducing production at a given site and simultaneously opening other facilities elsewhere, or it may choose contraction as a first step to extricating itself from dependence on diminishing resources while searching for alternative methods of production.

In the case of a redistribution of markets, contraction of a facility may be advisable as the firm chases customers by shifting production to areas where demand for the products and services is greater. Finally, with changes in competition, the firm may find it necessary to alter its approach to combat the threat of losing competitive advantage by shifting capital from one arena of operation to another. The productive capacity of a marginally profitable division may be reduced to increase the efficiency of that area and free up funds to enter new, alternative markets. Consider the firm, for instance, that has fundamentally viewed itself as a manufacturing concern and shifts its emphasis, redefining its raison d'etre as a maintenance/service company. It may choose to begin deemphasizing its production role, changing the facility to one of support rather than primary firm activity. In this situation, contraction could very well be a logical step in the process.

C. Openings

Because openings entail the full range of activities undertaken in planning and creating a new facility, they are treated only briefly at this time. Opening a new facility represents either a change in location (when connected with an accompanying closing of some older facility) or an increase in productive capacity of a firm in a growing market. With an opening, the site analysis becomes rather extensive, though the experience of existing facilities somewhat eases the process. New locations require the greatest degree of scrutiny. The basic needs of the site are set for the most part by reflection on the advantages and disadvantages of present sites and the requirements of the process to be undertaken at the new facility. Once these guidelines are established, the process becomes one of locating sites with those characteristics and choosing from among the alternatives.

D. Closings

Analysis of the possibility of closing a facility undergoes much the same procedure as that of contraction, and basically for the same reasons. Facilities are closed because of a lack of demand for the product, because of a shift in either markets or emphasis, or because the firm can no longer rationalize the cost of keeping the facility open. In a closing, the facility is either liquidated without replacement or operations are shifted to some other location.

Ideally, facilities are closed before they become a drain on company resources. Long-range planning that takes into account all possible long-term and short-term solutions to expected problems as well as analysis of future business trends can provide the firm with a clearer picture of when and where to instigate closings. In a multifacility organization, the question often becomes not whether to close, but which to close and when.

Analysis under these circumstances deals with determining the relative efficiency of the firm's various installations and the cost involved in both keeping them open and closing them down. This is never simply a matter of individual plant efficiency. In a multiplant operation, each element in the system interacts to create the fabric, if you will, of the company. A firm must remember to consider the effects that closing an individual unit will cause, not only on production and local conditions, but also on other installations. Hidden factors, such as the ability to take up short-term slack or lag time in ordering, often play a role in this type of decision. Given the choice, a facility that is relatively less efficient than others may remain open while others are closed because of its importance to the overall operation.

Alternatively, closings may depend on the capacity of the firm to shift production to other locations. If a change in company strategy or an overall reduction in demand for the product is the reason for the decision to close, that closing may represent a centralizing move. In that case, the activities at the facility closing down will be taken over by some other facility that will perform the closed facility's function in addition to whatever its present function may be. This process of pulling in one's horns can result in considerable savings to the company in not having to maintain two separate production units.

In any closing move, the firm should be certain that it has explored all of the other alternatives and that closing is the best choice. Closings result in either idle capacity, a situation similar to contraction on a temporary basis, or divestiture of assets. Particularly in the case of divestiture, it must be remembered that those assets will probably be more expensive to replace in the future, and therefore should not be released unless funds are absolutely necessary.

How do the choices of close and move or close and not replace the facility apply to site analysis? In the case of closing, the problem becomes one of deciding the best use of the site that the company already has. To move to a new site implies that the chosen new site is an improvement over the old one. This requires a comparison of the two sites point for point. There is no reason to close and move if there is no advantage of the new location over the old. In the case of closing the site without replacement, the choice becomes a matter of whether or not the firm will experience improvements in operation and profits over both the short and long term.

II. TRENDS IN SITE SELECTION

A number of factors go into site selection. Depending on the particular concerns of the firm, these factors will serve to shape the scope and emphasis of the site selection process. A number of the factors that affect the search for new locations are listed in Table 4.1.

In recent years, there have been some changing patterns in the process of site selection. Some of these patterns have been a surprise to those watching the shift from traditional site selections to a more varied range of possibilities. Among these trends have been a tendency toward geographic heterogeneity and major shifts to suburbs, rural areas, and the Sunbelt.

A. Geographic Heterogeneity

One trend noted in American business is a tendency toward a more even distribution of activities over the entire country. Traditionally, certain industries have located in certain parts of the country. Steel and heavy manufacturing have centered in what is called the industrial North, near sources of supply and cheap water transportation. Textile firms have located in the rural South, in small towns along railroads and highways where cheap labor, cheap power, and generally lower operating costs have given them a comparative advantage in the market. But this is no longer true. The trend in recent years has been toward a shift in heavy industry into parts of the country where it is not normally found, and a decentralization of service industries, locating away from heavily populated urban centers. There are many reasons for these changing trends, particularly the trend toward movement to the South and to rural and suburban areas. Overall, these reasons add up to lower costs of operation and greater efficiency.

Perhaps not surprisingly, the trend is not one of shifting plant locations

TABLE 4.1 Influences on Site Selection Factor and Their Percentage of Importance

Percentage Rank	Site Selection Factor
74	Favorable labor climate
60	Low land cost
42	Near markets
35	Low taxes
35	On expressway
30	Rail service
29	Low construction costs
28	Low wage rates
26	College nearby
25	Low energy costs
25	Government help
13	Available land
03	Near other division facilities
01	Air transportation
01	Quality of life

Example: The greatest influence on site selection is a favorable labor climate. It has a percentage point weight of 74. The three factors of favorable labor climate (74), near markets (42), and low wage rates (28) for a total of 144 would outweigh the three factors of low land cost (60), low taxes (35), and low construction costs (29) for a total of 124 by 20 percentage points (144 − 124 = 20).

Source: Roger W. Schmenner, *Making Business Location Decisions*, © 1982, p. 51. Adapted by permission of Prentice Hall, Inc., Englewood Cliffs, New Jersey.

from one area of the country to another but rather one of expansion into new areas. Firms are expanding rather than relocating. The incidence of new facility expansion in the Sunbelt states from the Southwest to the southeastern states is up, with a number of northern manufacturing companies expanding by opening new facilities in small towns in this region. In addition, the incidence of foreign companies investing in the West and the South has helped to bring about the general repatterning of manufacturing in this country.

B. Movement to Suburbs

The ability of firms to locate in suburban areas depends on the ability of those suburban areas to supply the logistical support that the firm requires and on the relationship of the suburban area to major metropolitan areas. Once, industry tended toward centralization due to the benefits of economies of scale and of nearness to transportation and communications networks, labor supplies, and capital. However, because of ever-increasing changes in technology, it is no longer necessary to centrally locate in order to take advantage of these services. Telecommunications has allowed businesses to spread out from centralized locations. Computers and improvements in highways and the transportation network overall have reduced the necessity of living close to work. This increases the flexibility with which firms may operate and consequently increases the number of hospitable sites for plant location, many in nontraditional locations. A firm can move into suburban areas without suffering from a lack of support. These communities have the added advantages of generally lower taxes, cheaper power, and cheaper land costs.

C. Movement to Rural Areas

Rural areas offer a combination of benefits; chief among them is the matter of land cost. Because of the high price of land in urban areas, firms requiring large facilities are forced to look elsewhere. This has always been the case. However, during the 1980s, an increasing emphasis on rural land developed in no small part as a result of the increase in forfeiture among farmers and willingness to divest among rural landowners. The price of land is tied to a number of factors contributing to its value, including location, characteristics, and productivity. Per square foot, the productivity of land devoted to service and manufacturing is far higher than that of farmed land. This raises the price of land to a level that is attractive to the rural seller while still at a level below the cost of land in more urban regions. Coupled with the increase in accessibility of lands formerly too remote for practical consideration, the value of rural land to manufacturing and business has grown to be substantially higher than the value to agriculture, particularly in the Sunbelt. As with suburban areas, rural areas offer lower land prices, lower taxes, a ready-made labor force if located near small towns, and often cheaper power and support services. Yet they no longer have the disadvantages of remoteness, poor communications, and poor transportation.

D. Movement to the Sunbelt

The movement of industry toward the Sunbelt signals a general trend in the social megastructure of the United States. It is reflective of and concurrent with a general trend toward population movement out of the North into the South and West, most of which make up what we call the Sunbelt. This movement seems to stem from two sources. On one hand, the tendency toward decentralization in business has resulted in an increase in the expansion of productive facilities into new areas of the country. This expansion has been largely to the West and the South since industry, particularly heavy and technologically oriented industry, has traditionally been centered in the North and Northeast. In addition, due to the shift in the type of manufacturing done in the United States (a preponderance of high-technology and consumer industry rather than basic industry), there is not so high a degree of dependency on localized concentrations of raw materials and bulk transportation. Even financial services can be decentralized with a minimum of physical displacement due to the preponderance of telecommunications in that industry. And from the manufacturer's point of view, many Sunbelt states have the advantage of low unionization, removing an expensive element of doing business for large manufacturing firms. Not only do union shops have higher wage rates, but they also require the maintenance of internal mechanisms for constantly dealing and negotiating with this part of the labor force structure. No matter what the morality of the situation, firms would prefer not to have to deal with unions, given the choice.

As a result, as people have migrated to the Sunbelt with its wealth of jobs and more pleasant living conditions, so industry has moved to the Sunbelt to reap the benefits of the region's wealth of labor, lower costs, and available support services. As the heterogeneity process continues, a form of commercial dilution acts to reshuffle the spread of productive capacity more evenly in the country, which necessarily requires a movement away from areas of high concentration to areas of low concentration. The low concentration has been primarily in the Sunbelt area.

When various potential sites are analyzed, prime areas in which to look would be in the Sunbelt, yet northern locations should not be ruled out, due, in large part, to the southern migration. Cities in the North are now in a position to offer lucrative packages to companies willing to relocate there in an effort to bolster sagging economies. There may be bargains available to the firm in this area of the country, assuming equal availability of services; raw materials; and financial, communication, and transportation networks.

III. GENERAL GEOGRAPHICAL CONCERNS

There is a tendency in American business towards geographic heterogeneity, with a more uniform distribution of manufacturing sites throughout the country. Yet each area of the country tends to have its own specific problems and advantages. It is imperative to be aware of these differences in selecting the proper site for a facility. In addition, a specific area within a region can offer advantages due to specific characteristics. Some of the more obvious concerns are discussed in this section.

A. New England States

The chief considerations of location in the New England states center around weather, proximity to ports and other transportation centers, and labor force. Traditionally a poor farming area, the New England states have historically tended toward sea trade, fishing, and industry. This tendency, though somewhat modified, continues today. The New England states have a relatively high density of industry and consequently an industrial labor force, unions, and a highly developed transportation and communications network. Proximity to financial centers and a wealth of seaports are also primary assets.

On the negative side, the weather, which shifts widely from season to season, the cost of power, particularly other than electrical power, and the relatively high cost of land and construction costs could be hindrances.

B. Midwestern States

Midwestern states offer a combination of agriculture and industry, with industry traditionally centered along the Great Lakes and major river arteries, and vast reaches of farmland spreading out over the flat central plateau. Since land is plentiful in this area, particularly in light of the agricultural shifts of recent years, proximity to large markets, financial and communications centers, and transportation centers is more limited than in the densely packed Northeast. Industrial labor may prove to be a problem in this area because of lower population density, although shifts from small farms to large cooperatives, which have displaced sectors of the agricultural labor force, may offset the problem. The opportunity to develop as the chief industry in a given urban area is also attractive. The price of land acquisition varies in this area, yet the basic geography of this part of the country can offer advantages in construction. Power and water availability and costs also vary. Weather considerations include annual deep freezes, extended snowfalls, and the possibility of substantial variance in rainfall. In some areas, both flooding and high winds are a consideration, though danger from the latter, tornados in particular, is perhaps more dramatic than real.

C. Pacific Coast

Like an ongoing boomtown, the Pacific Coast continues to offer possibilities for expansion to industry. Along the coast itself there are a number of well-developed ports from Washington down to southern California. Financial, transportation, and communications networks are strong and the population is growing. Several megalopolises have developed in the region along the coast, particularly in California. The range of topology also varies widely from the low coastal shore line to the high mountains of the Rockies, a range that runs north–south through most of the region. This range can offer both advantages and disadvantages at particular locations, acting either as a source of resources or as a barrier to transportation or communication.

Specific areas tend to harbor specific types of industry, the obvious example being the Silicon Valley area in California, filled with high-tech computer and electronics firms. In Oregon and Washington the availability of land is greater, yet urban areas are less abundant and populations lower than in California. Climate runs the gamut from the wide variations in temperatures and conditions in the north, to the nearly constant warmth in the southern end of California. Specific problems include earthquakes, a very real threat though seldom experienced, pollution, which tends to be high on the coastal side of the mountains where the atmosphere has difficulty in flowing over the high Rockies inland, seasonal mud slides, and forest fires. Topography varies, with coastal plains, high arid plateaus, elevated forests, and high mountains. The highest mountain ranges in the country are located in this region. Population tends to center in the larger urban areas, while the coastal plains remain agricultural. In the south, much of the vegetable and fruit production of the country takes place in orchards and truck farming. To the north, lumber production, coastal fishing, and basic industry predominate. The chief advantages and disadvantages in the region center in the variable concentrations of population, with communications and financial nets in a relatively small number of locations. Such patterns of population can create either a dearth of available labor or congestion due to concentration, a situation that may yield unfavorable conditions for plant cite location. Chief advantages in the region are a favorable climate, closeness to ready markets, a growing population, and well-developed communications, transportation, and financial nets in the more highly populated areas. Disadvantages include congestion along the coast, pollution, and a very serious water problem, one that is growing and promises to continue into the foreseeable future.

D. Industrial North

The primary advantages in the industrial North center in the already present infrastructure for an industrial setting, the availability of land and water transportation, nearness to heavy industry and its products, a relatively high density of population for both labor force and market (though there has been a tendency toward decline of both industry and population in recent years that bears watching), and nearness to raw materials.

One problem in this region is that it is already highly developed, which could result in high land and construction costs. Others are a very high rate of unionization, and in many areas, pollution and high cost of living.

E. Southeastern States

Industrialization in the southeastern states is on the rise, a pattern that is expected to continue for some time to come. This is in no small part due to the attempts of the southern states to lure industry into their borders, and many domestic and foreign companies are accepting the invitation.

The advantages to a southern location have already been discussed in reference to the trend toward movement to the Sunbelt, of which the southeastern states are a part. More specifically, the southeastern states offer relatively cheap land in rural areas, low taxes and power costs, a readily available labor supply in small towns and municipalities where no single industry dominates, very low unionization, and excellent weather. For the most part, the Southeast is an untapped resource for industrial development, though this is changing.

On the negative side, some locations offer limited transportation networks and poor local financial and technological support. In addition, the Southeast lacks the high concentration of educational facilities found in the industrial North, which can create a problem for high-tech industries. There are many competent institutions of higher education in the Southeast, but they tend to be more spread out than in traditionally industrial regions. For a company whose research and development needs are high, this could be a consideration.

F. Transportation Hubs

Transportation is a consideration no matter where a firm locates. It is difficult to sell goods that cannot be easily taken to market, or to make goods for which raw materials are not readily available. The more efficient the transportation network, the easier it is to accomplish efficient production and distribution.

The fact remains that lower-cost land and more abundant site areas may not be close to large metropolitan areas. Nearness to a transportation hub is a serious consideration and one of the primary factors used by industry in determining site location (as shown in Table 4.1).

Two primary approaches to this problem are either to locate near the transportation hub, an action that either will or will not increase costs, depending on the transportation hub and type of product in which you are dealing, or to ensure easy movement of goods from the plant's location to the transportation hub nearest to the facility. In the case of rail lines, this second approach would entail a spur line, and that may or may not be possible. In other locations, trucking, pipeline feeds, and air service may suffice to connect a given location with the nearest transportation hub, depending on the nature and value of the products being moved.

In general, the nearer one locates to a transportation hub the higher the cost of creating the facility, both in land and construction costs, and the advantages must be weighed against the inconvenience and cost of transportation encountered by a more remote site. It should be remembered that access to transportation centers and proximity to transportation centers are not always the same thing.

G. Environmental Areas

Two factors are of importance with regard to the environmental structure of a given site: the environmental conditions of the area and the impact that locating a facility will have on that environment.

The facilities planner must consider the environment of a site to ensure that it meets the needs of the facility to be placed there. Such considerations as pollution, water supply, days of inclement weather per year, soil conditions, topography, local flora and fauna, and even the temperature of ambient water supplies can be a concern. It must be determined what environmental conditions are required for the operation of the facility and whether or not a location matches those requirements. This is particularly true of requirements that are imperative, such as available water or raw materials.

Second, there is the effect that the facility will have on the environment of the site. These considerations include airborne and water-borne pollution, disturbance of local fauna and flora, effect of plant operations on soil erosion, noise levels, and so on. If the environment is going to be seriously affected by the facility, it may become more costly to protect against negative effects than it is to look for another site. The cost of a water treatment plant, high smokestacks to dissipate airborne gases, or noise barriers to protect nearby residential areas could be considerable.

H. Quality of Life

Again, we have a two-sided issue, mainly how the quality of life in a location compares with those offered at other locations, and how the presence of the facility will contribute to or detract from the quality of life already present at the given location.

In the first case, the site planner must consider conditions in the local area of a potential site to determine if it is going to offer a sufficiently high quality of life to entice workers to locate there. Usually, if a region is sparsely populated or if there is a low quality of life for the local residents, there is a reason. High desert land, for instance, can be quite inexpensive, and for good reason. It may be a miserable place to live, with hot, dry days and cold, windy nights, scarce water supplies, and isolation from centers of culture and population. This is not exactly the kind of spot that people rush to for jobs. Large metropolitan areas can be equally unappealing, with overcrowding, high crime rates, congestion, and high prices. No single environment can boast an inherently higher quality of life than another. Yet obviously, some environments are better than others. The quality of life in an area must be considered, and the highest quality of life that can be obtained should, on balance, be sought.

Quality of life, which is what economists are talking about when they refer to a high level of general economic welfare, is simply a matter of how well people live. It includes considerations such as availability of work, quality of education, availability of cultural events and institutions, cost of living, availability and cost of housing, crime rates, community solidarity, social structure, socialization patterns, and per capita income.

On the other hand, simply because a given site has a relatively low quality of life to offer does not mean that it is a poor choice. This brings us to the second side of the issue, specifically, how the location of a facility will affect the quality of life as it already is.

The number of jobs offered in a community rises as a result of a facility's locating there and can add to the quality of life in the area, particularly if the area is presently depressed. In addition, an increase in educational possibilities, overall economic improvement in the region, and the possible attraction of other business into the region can have very positive effects on the community as a whole. The increased revenues from taxes means an opportunity for local officials to improve conditions, and the support industries necessary for handling the needs of an increased labor force means growth as well.

It should be noted, however, that the arrival of another industrial facility in an area that is already highly developed can cause strain and hardship on the capacity of that area to offer needed services. This may actually decrease the overall quality of life in the area. Further crowding, more congestion, and higher levels of pollution, even in the case of the most conscientious planning, can cause serious erosion of economic well-being for those working and living around the site. There is a duty on the part of firms as well as a decided economic advantage to do what they can to ensure that the quality of life is enhanced by their presence rather than strained.

I. Labor Market

The quality and size of a labor force, as well as its cost, are decisive in facilities planning. On the one hand, the planner must ensure that the site has a sufficient supply of labor to take care of present and forecasted needs, and on the other hand, that it is not too expensive, or that there is not too much competition for those labor units. It should be emphasized that quality of labor is as important as quantity of labor, and that the market must be able to supply sufficient numbers of the right type of workers with the right type of skills for the firm, rather than just raw abundance. If a company opens a highly technical facility, it needs highly competent workers. If they can be secured at the local level, the necessity of importing competence at considerable cost can be avoided. Again, this element must be balanced against other favorable or unfavorable conditions to come up with the most efficient, economical mix of characteristics for the chosen site.

J. Distance from Present Facility

One of the reasons for expansion of facilities is to locate closer to markets and thus farther from present facilities. Yet a balance must often be struck between the choice of a distant location to capture or enter new markets, and the choice of a location close to present facilities for the advantages of support that that approach would afford. The primary issue in this part of the decision-making process lies in determining how important it is to have that support and to what degree a large distance separating the two sites will be a problem. Facilities located widely apart will probably be required to maintain a high degree of self-sufficiency, whereas those located within one region will not. Is the facility to be a stand-alone operation, or will it depend on other facilities for parts, personnel, or technical support? These are the issues with which to deal in this particular context.

IV. COMMUNITY BUSINESS CLIMATE

Within the community surrounding any site there are certain common issues that must be addressed. Rather than being specific to any one location, these are issues that will be encountered anywhere and must be investigated before deciding on the suitability of the site. They are an integral part of the general site analysis process. Although they have been discussed in examples elsewhere in this chapter, they bear reemphasis at this point.

A. Taxes

The level and type of taxation encountered by a relocating firm vary from one municipality to another. Highly industrialized areas tend to have a wide range of industrial and business-related taxes in their tax digests, allowing them to draw funds from the primary employers of the area. Sparsely industrialized regions, on the other hand, tend to be kinder with regard to taxes to large enterprises, both as an inducement to development and by virtue of not formerly having the need for such tax structures. The planner should determine local tax conditions and the probability of future changes in those conditions. It is sometimes possible to negotiate tax advantages in communities wanting new business or experiencing depressed labor markets.

B. Investment Credits

An investment credit is essentially a tax break offered by the federal government to companies who are involved in the construction of facilities. These tax savings have been known to be considerable and are designed to encourage expansion in the private sector. Since the 1988 tax law changes, however, the benefits of investment credit savings have decreased, and it behooves the facilities planner to determine the extent to which such changes affect any investment credits received.

C. Educational Support

In the case of highly technological businesses, the need for educational and university support may be considerable. Even a relatively standard or low-tech facility needs adequately trained and educated personnel. Training centers, universities, colleges, and community colleges are often a primary source of such training. In addition, management consultant firms and the suppliers of equipment may offer training services to the firm. The remoteness of a site from these types of services must be considered if training and retraining will be a factor in facilities operations.

D. Investment Support

The degree to which the local community is willing to invest in the establishment of a facility can be pivotal in the location decision. Communities hungry for the influx of industry will often make sizable contributions to the process by willingly upgrading highways, sanitary facilities, and the like to encourage industrial location. Furthermore, the availability and cost of investment funds may be more favorable locally than elsewhere, which has the added advantage of solidifying the feeling of partnership between the incoming firm and the surrounding community. How willing the local community is to involve itself in this part of the site creation process can make the difference between a favorable and an unfavorable site.

E. Relocation Services

A further indication of the enthusiasm with which a community pursues economic expansion is in its willingness to offer what is known as relocation support. Short-term waivers of certain taxes, fees, and other expenses make a community more attractive to industry, as communities are well aware. Beyond this local governmental support, the question arises as to how willing the local business community is to support the move and how capable they are of offering short-term housing, long-term housing, rapid development of power and utility, and other services needed to create an orderly transition for the firm and its employees to the new location.

F. Governmental Support

General governmental support involves the willingness of local government to encourage and aid in the process of developing a local facility. In some communities the local governmental attitude is hostile or at best lukewarm toward new industry, whereas other communities welcome and support such a move. It is essential to have a good rapport with local government and to receive its support of any move to locate in the community. Without it, the process of locating in the community can be seriously hindered, if not effectively blocked. Planners need to determine the political climate of the area and, if possible, to learn the experiences of other firms who have moved into the community.

G. Labor Pool

Labor involves two dimensions, availability of supply and quality of supply. A facility must be staffed adequately if it is to function successfully; the adequacy of the staffing involves not only the size of the labor pool but whether the labor pool contains the proper mix of expertise and specialties to get the job done. Obviously, not all people are equally qualified to do all jobs. Hence, the question arises as to whether or not the local labor force can supply people capable of doing the specific jobs available in the new facility. The planner needs to determine and match needs and labor supply in terms of educational level, kinds of specializations, degree of transience, and other factors that are skill dependent. If there are shortfalls in the labor supply, the planner must determine if they can best be corrected by importation of workers from outside the area or by education of the local workforce. The costs of these solutions must also be ascertained. A study of the basic demographics of a potential site, usually available from local and federal governmental sources, can go a long way toward developing this type of information. Other potential sources of information are the local chamber of commerce, the secretary of state of the individual state, state commerce administrations, and federal labor and business statistics sources, such as the Department of Commerce.

V. OTHER INFLUENCES

A number of other factors that do not fit into the above categories affect the general site analysis as well. Briefly, these include:

A. Corporate Image

The image of the firm, both locally and nationally, will affect the ability of the firm to locate in a given community. The more positive the image, the more local cooperation can be secured and the more favorable the mix of factors that can be obtained. If there are negative image factors present, such cooperation may not be forthcoming. A firm that has experienced an industrial accident at another location, or which is viewed as dealing in some dangerous product, such as nuclear fuel or military explosives, may find local communities less than enthusiastic about their presence in the community's midst. Such conditions may require considerable expenditures in public relations costs to turn the political and social climate around. On the other hand, a company viewed in a positive light because of demonstrated ethical and economically responsible behavior can find its way well oiled and easily traversed.

B. Climate Changes

Shifts in climate do occur, particularly in reference to large areas, and these changes in climate can affect the value of a given site. Weather can be a surprisingly important consideration. Changes in weather patterns may be temporary (on the scale of decades), yet still render sites obsolete that were once ideal. This can be due to changes in annual rainfall, unusually long periods of heat or cold, or destruction of topsoil, any of which may have an effect on the manufacturing process. For example, the destruction of topsoil may result in a problem with excessive dust, which could be quite troublesome for a facility that needs to operate in a dust-free environment. Or a facility with temperature-sensitive goods may find itself unable to help reduce summer heat through foliation of the location site because of eroded topsoil. Whatever the circumstances, it is important to remember that a production facility is a long-term project. What is expected to happen in ten to twenty years should be of concern today.

C. Neighborhood Changes

Demographics teaches us that the structures of urban and rural areas are not stagnant. Populations ebb and flow over the landscape, concentrating first here and then there, depending on a number of predictable and not so predictable factors. It is important to test not only the present conditions in an area but the trends as well. A neighborhood that is traditionally industrial but is losing its industry may offer savings in building or renovation costs now but may be costly later as the labor force moves to outlying, less congested areas. Traffic patterns, movement of support businesses, and willingness of local government to deal with inner-city problems can all undermine otherwise attractive locations.

D. Foreign Investment

This factor can act either as a source of funds and support or as a competitor for labor and resources in a given area. As foreign investment increases in the United States, the competition for good sites will rise, forcing higher costs and a more aggressive approach for American firms. Alternatively, the opportunity for joint ventures and for moving into a relatively low-cost area where the development of foreign firms offers increased economic strength may be exactly the element that sways a firm toward a given site rather than away from it. The impact of foreign investment can be positive or negative, depending on how it occurs, and must be considered in the planning of a facility.

E. Union Pressures

It would be naive to believe that unions and unionism do not have an effect on location decisions. Unions are, from the point of view of the firm, a mixed bag of headaches and blessings. Whereas a union allows for a stable, predictable set of labor force conditions, the need to bargain and negotiate and the employment restrictions that unions have the power to enforce can be costly, time consuming and generally disruptive in a business climate that is constantly changing at an ever-increasing rate.

One of the options open to a company is to locate in areas where the unions are not strong and where the open-shop approach to industrialization is used. This could be a blessing in terms of flexibility. If a firm operates in an industry that is totally unionized, it requires close consultation with labor in order to carry out facility relocation.

F. Restrictive Tariffs

Tariffs are fees or taxes designed to protect local markets from foreign competition. Indeed, the entire motivation for initiating an expansion or relocation may be to avoid high tariffs that keep a firm out of a particular market. As such, savings in tariff costs of doing business in a region may be so great for firms locating in those regions that they are willing to absorb large initial and operating costs to achieve those tariff savings. If the potential market is large enough and the profits substantial enough, a firm will undergo the large costs of facilities development in order to achieve those profits.

QUESTIONS

1. What are some of the effects of geographic placement on a facility?
2. When developing a facility within a specific community, a firm has a number of concerns to consider. Name four.
3. Where is the Sunbelt? Why is it an important consideration in today's facility location decisions?
4. In the present environment, there are three trends in plant relocation. What are these trends?
5. What is meant by *geographic heterogeneity*? What is the impact of this concept on plant location decisions?
6. What is *site analysis*? What major subtopics does it include?
7. Educational support is seen as an important element of successful site selection. Why? When would it be most important and when least important?
8. How does the size and composition of the labor pool affect business decisions for a computer manufacturer? For a banking firm? For a tire manufacturer?
9. How does a specific location, say, in the midst of Silicon Valley, affect corporate image for the computer manufacturer? For a weapons manufacturer?
10. What are the primary weaknesses and benefits of location in the New England States? In the Southeast? Along the Pacific Coast?
11. Local investment support is important not only as a source of corporate funds but also to community relations. Why is this?
12. What is meant by "quality of life"? How does it affect site decisions?
13. A new facility can result in ill will within a community. Why do you suppose this happens? How can it be avoided?

14. For what reasons would a firm lean toward a rural location for a new or relocated facility? What are some of the drawbacks? Does this combination of advantages and drawbacks seriously limit the applicability of such a strategy?
15. What is the present state of the investment credit concept in the United States? Why do you suppose this is?
16. Under what conditions would closing a facility appear to be the most appropriate option? Assume that the firm in question is not experiencing financial difficulties.
17. What types of major facilities exist in your local area? What types are conspicuous in their absence? Why?
18. Assume that you are a manufacturer of work clothing that is sold in large quantities to other industries. What criteria would you believe to be most important in choosing a site for a new facility?
19. Among other positive benefits, new facilities bring with them additional jobs, increased tax money, and in many cases, higher land values. This being the case, why would some communities not want to encourage new businesses moving into the area?
20. Any new project may encounter corporate inertia from within the firm. What elements of a site analysis would you emphasize in order to overcome this possibility?

5
Types of Financing

For any investment project, whether it be a new facility, research and development, or the purchase of new equipment, an efficient means of financing must be found. In its simplest form, this decision may be simply a matter of allocating funds within a budget, or of deciding whether to use existing funds or seek borrowed funds. Beyond that, the wide range of alternative methods of financing should be explored in order to determine the proper combination of debt and equity that best fits the requirements of the purchaser.

In facilities planning, the amount to be financed is generally a relatively high amount, and since it represents a major purchase, careful consideration must be given to the means by which financing happens. Whether one chooses to buy outright, a rather unlikely choice for large projects, or to borrow the funds, there are costs involved. These costs take the form of interest, either from directly incurring interest expense through the loan agreement, or opportunity costs, derived from not using the invested funds for some other project.

In this chapter, we explore the various options open to the firm and attempt to shed some light on the advantages and disadvantages of different means of financing. It should be remembered that there are no panaceas in financing a facilities project. Different circumstances yield different needs, hence the large number of possibilities.

We investigate first the decision to lease or to buy or build, then the decision to purchase for cash or finance, and also the various ways by which the financing can take place. Finally, we briefly explore conventional wisdom as to how best to handle specific classes of facilities.

I. ACQUISITION ALTERNATIVES

A. Nonpurchase Alternatives

1. Rental. Rental agreements are usually for short periods of time, or for periods of time that are unspecified. The fact that there is not a definite time period of obligation (beyond a weekly or monthly period) adds flexibility for both the lessee and the lessor. A firm that anticipates the need for an item or a facility for a short period of time or which, perhaps, is not certain of the length of time that the item will be required may find it best to rent. Obviously there is high risk for the lessor in this case, since there is no assurance of income from the rental for any specified period of time; this added risk is usually reflected in a higher rental charge. If the rental item is an extremely costly one, this risk premium can be substantial. The rental agreement usually includes protective provisions for the parties involved, specifying what can and cannot be done with the property and under what circumstances continuance or interruption of the rental agreement can occur. Items such as cranes, forklifts, warehouses, and temporary manufacturing facilities needed for specific, finite projects are items that might be best rented. The prime considerations here are time period and the degree to which future needs can be adequately predicted.

Rent based on sales less is an arrangement by which the rental fee paid is dependent on the size of the sales volume connected with the facility. In other words, the higher the sales volume, the greater the rental fee. It is a usage-based agreement. This kind of arrangement has the advantage of limiting the lessee's liability in times of slow sales, when money is apt to be tighter and therefore a lower rental figure advantageous. It shifts at least a part of the rental fees from fixed costs to variable costs, which protects the company in cases where there is great uncertainty or variation in production. This is particularly useful in seasonal industries where production varies widely depending on the time of year or availability of materials.

In the case of equipment, rental fees are often tied to how much the equipment is used. High or low levels of sales imply high or low levels of production (and use of machinery and equipment). The principle is similar to that used in the rental of copying machines, where the firm pays a base rate plus so much per copy.

This arrangement also expands the rental agreement in times of high sales, when the facility or equipment is more profitable or more extensively used. This reflects the negative side. Since the rate of rental slides with production, the more successful the firm is at selling its products, the greater the expense of equipment and facilities. Rather than receiving the advantages of economies of scale from higher levels of production on a given piece of equipment, the company finds costs rising as production rises as an effect of the shift in rental fees.

2. Leasing. Leasing is little more than a long-term rental agreement. It is essentially the same for the firm as it is for the individual who leases an automobile, an apartment, a condominium, or a home. In the case of the firm, the lease involves long-term rental of plant and/or equipment according to terms mutually agreed upon by both parties. Often, the lease includes provisions for disposal of the equipment at the end of the leasing period, though this is not always the case. A firm that leases delivery trucks, for instance, may have an agreement included that it will purchase the trucks at the end of the lease period from the leasing company. Also, a firm leasing a facility may find included in the lease agreement a provision for returning the property to its original state at the end of the leasing period, including removal of partitions, walls, and so forth. In addition, leases usually include provisions for maintenance of the property, restrictions on use, the rights of the lessee to alter the property, and options for renewal. Other normal provisions include those for increases in lease fees, adjustments for inflation, and terms under which the lease may be terminated prior to its full term.

The *lease–purchase* option differs from that of a straight lease. With a normal lease arrangement, the building and equipment revert back to the owner at the end of the lease period and become subject to renewal or lease for other purposes. In that approach, any end-of-lease fees and purchase options are centered in the depreciation of the equipment or facility in question and their salvage value. It is assumed that the leased real property will be at the end of its useful life at the end of the lease, or that the lessor has no wish to sell the item and therefore includes no purchase clause in the contract to begin with.

The lease–purchase approach is fundamentally different, however, in that it allows the lessee the option to purchase the property, with all or part of the lease fees paid applied to the purchase of the property. Again, the emphasis is on flexibility. The lessee has the option of treating the agreement either as a long-term rental situation in which the goods revert back to the lessor, or as a time-purchase arrangement in which the goods are bought on time, the monthly fees being interest plus principal towards the final purchase of the item. Should the lessees decide during the lease period that it is no longer advisable to continue the arrangement, they may cancel the agreement in accordance with the terms of the lease agreement, and the equipment, plant, and so on revert to the lessor. Or, at the end of the lease–purchase period, the lessee pays a final amount equal to the remaining investment of the lessor in the real property and title is completely transferred to the lessee. The advantage of the lease–purchase agreement lies in its ability to allow the lessee to decide at a future date whether to take permanent ownership of the property or not. Depending on the terms of the agreement, that decision can be made at any time, at specified times (as, for instance, at the end of each lease year), or at the end of the lease–purchase agreement.

B. Purchase Options

All of the foregoing options are based on the rental of property by the firm. They are variations on the same theme—that is, the choice by the firm to use someone else's property to produce its goods or services and the payment of rent for the right to use that property. The other alternative is outright ownership of the property. See Chapter 2 for more on this subject.

In the case of purchasing, the firm chooses to take ownership of the plant, equipment, and other property necessary to the productive process. This is particularly advantageous if funds for instigation of the purchase are readily available to the firm, either from its assets or in the form of credit and borrowing.

In this approach, costs are somewhat different from those incurred through the rental process. With a purchase, the cost is in the form of either interest or opportunity costs to the firm. If the purchase of the property is financed through a loan function, either by a direct loan for the property itself or through the issuance of company debentures (bonds), there is interest to be paid on the borrowed funds. This interest is actually rental for the use of someone else's money. In a sense, then, the additional productive capacity is still financed through the rental process, although this time it is the rental of capital rather than the rental of the goods themselves.

If the firm has funds available internally and purchases the plant and equipment outright, it experiences an opportunity cost connected with funds usage.

Opportunity costs (which not too surprisingly closely coincide with the going rate of interest at the time of purchase) stem from the fact that had the company not chosen to use funds for the purchase of plant and equipment, it could have used those funds for some other purpose. Since the primary goal of the firm is to maximize profits (theoretically), the funds would have been used to generate those profits. By using the funds available for the purchase of plant and equipment, the firm has foregone the opportunity to make a profit doing something else. This is a very real cost. At a minimum, the firm had the option of lending the funds to someone else, and this would have produced interest income at the going rate. This interest income is not realized if the funds are used by the firm, and thus the firm incurs the cost of that foregone opportunity.

Certainly the choice to use the funds for the purchase of plant and equipment is a wise one as long as the return on investment from the production and sale of company products exceeds the going rate of interest, but that does not eliminate the cost of capital in any way.

Additional conditions exist as a result of outright purchase of plant and equipment. The items have a limited useful life and must therefore be depreciated, creating a reduction of assets over time. With the depreciating process comes an accompanying tax break as a result of writing off the used-up portion of the asset's worth. Purchase also shifts the risk of obsolescence and inefficient use of the assets in times of low sales from the lessor to the firm. This risk is generally offset by a corresponding lower total cost over the life of the asset in question, whether it be plant facility or machinery and equipment.

In deciding the financial option best suited to the firm, all possibilities should be explored and the relative cost of each evaluated. Only through an analysis of all of these possibilities can a company be certain of maximizing the efficiency of their investment dollar.

II. FINANCIAL ALTERNATIVES

If rental, lease, or cash purchase are not viable options for the firm in acquiring a facility, then financing the purchase becomes the probable solution. Fortunately, many different methods of financing exist, most of which we discuss in this section.

A. Adjustable Rate Mortgage

This is an outgrowth of the rapidly fluctuating interest rates experienced in this country during the late 1970s and on into the 1980s. It is a mortgage loan on which the interest rate rises and falls with changes in prevailing rates. Because of the rapid changes in interest rates, it was discovered that lenders could not accurately predict the probable changes in long-term cost of money and that, since depending on shorter-term loans would not suffice for the real estate market, a preferable alternative was to vary the rates of the mortgage itself rather than risk the loss of profits on long-term debt commitments.

The advantage of the adjustable rate mortgage (ARM) is that if interest rates fall, the monthly payment on the loan also falls. During periods of falling interest rates, this is a great benefit to the mortgagee. Unfortunately, if interest rates rise, so do the payments, and since interest rates tend to rise in periods of tight money (limited loanable funds) and inflation, the additional amount of payments come at a particularly inopportune time to businesses whose own adjustments to prices and revenues tend to lag slightly behind the change in dollar values.

It should be noted, however, that this type of mortgage generally offers a lower initial interest rate (reflecting the lower level of risk experienced by the mortgager) and that the changes in interest rate and thus payments are tied to publicly constructed indexes that reflect true changes in the economy. This assures an element of "fairness" in the way in which changes in interest rates are determined. It should also be noted that to the interest rate is added a margin for the lender's cost of doing business. The margin acts to compensate for the risk of loan loss as payments rise. These are normal expenses incurred by the lender due to the amount of paperwork and the risk inherent with this type of loan and should be considered reasonable. There are several other features of the ARM that one should be aware of. First, there is a preset adjustment period for the changing of interest rates and payments, usually in increments of one, three, or five years. Also, there is an interest rate cap, or highest interest rate beyond which the mortgage cannot go, which normally includes provisions for both how high an individual adjustment can shift the interest rate and a lifetime cap in highest total interest rate allowable. Finally, there is a payment cap that limits the size that monthly payments can become, and an accompanying negative amortization clause that adds to the total of the mortgage outstanding if the payment cap results in the lender's not having to pay all of new monthly charges.

B. Graduated Payment Mortgage

This type of mortgage represents a fixed-interest rate loan with a monthly payment that starts out low and then increases through time. It is particularly useful with facilities that are expected to generate low revenues when first opened and higher levels of sales and revenues later. The idea is to put off high payments until the facility has gone through the initial start-up phase with its accompanying higher costs, and to raise payments later, when the cash flow has increased and is more adequate to handle the mortgage payments. Figure 5.1 shows a comparison of the graduated payment mortgage and a typical ARM, or adjustable rate mortgage.

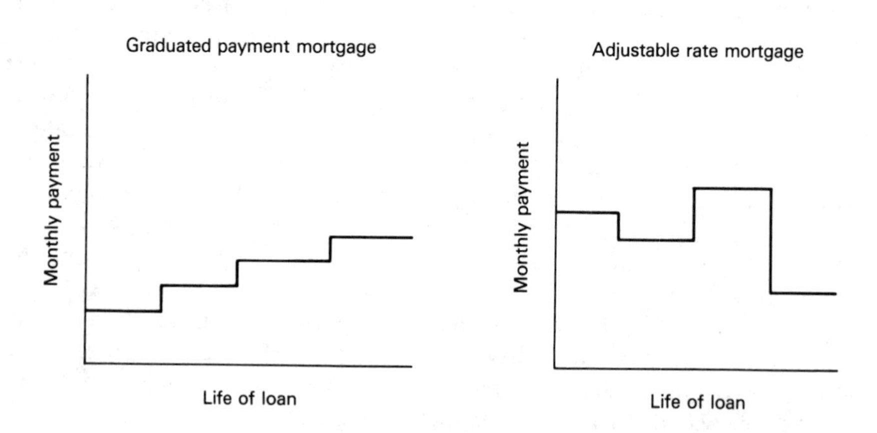

Source: Reprinted with permission from *Real Estate: An Introduction to the Profession* by Bruce Harwood, copyright © 1986 by Prentice-Hall, Englewood Cliffs, N.J.

Figure 5.1 Advantage Comparison Graph

C. Shared Appreciation Mortgage

With this type of mortgage, the borrower receives a lower interest rate on the loan in return for giving the lender a portion of the property's appreciation through time. The advantages to the borrower are centered in the smaller outlay of funds on a month-to-month basis in the form of lower mortgage payments, something that may be indeed attractive if there is a shortage of working capital. The attraction to the lender lies in sharing the benefits of the increase in property value as it appreciates over time. If such appreciation is expected to be high, a lender may well be induced into accepting this type of mortgage, thus relieving the borrower of high payment commitments for the life of the loan.

D. Package Mortgage

Here we have a mortgage secured not only by real estate, but by personal property as well. In the case of a package mortgage, the borrower may have need for more funds than the property itself can logically cover. Another possibility is that the lender may feel the necessity to further secure the loan with additional property for reasons of risk, uncertainty in the market, and so forth. In such cases, other nonreal-estate assets may be pledged to cover the discrepancy.

E. Blanket Mortgage

A blanket mortgage is a mortgage that is secured by more than one piece of property. That is, if the lender is insecure with the mortgage just covering the property in question, or if additional funds are required for construction that can not be secured in some other manner, the borrower may pledge other property in addition to the parcel to be purchased as collateral for the mortgage of the new facility. Blanket mortgages for expansion are not the norm, but neither are they rare; given the right set of circumstances, they can be quite useful.

F. Reverse Mortgage

With a reverse mortgage, the lender makes monthly payments to the borrower, who later repays in a lump sum. Also referred to as a reverse annuity mortgage, this is not as bizarre a proposition as it first appears. With reverse mortgages, the borrower receives an annuity payout of borrowed funds rather than a lump sum distribution, and the loan itself is secured by property already owned by the borrower. In these terms, we have nothing more than a second mortgage (often referred to as an equity loan these days) with an annuity pay-out. Upon the death of the borrower or the sale of the real property used to secure the loan, the total of annuities paid out are tendered to the lender in a lump sum.

In terms of facilities planning, this type of financing certainly is not predominant in the market. However, as a means of securing a steady flow of working capital, or as a means of offsetting mortgage payments for one property against another during times of low cash flow, it can be useful. Envision the firm that has funds tied up in property that is presently unproductive but expected to appreciate, and which also has a need to build a new facility. It is feasible to use the reverse annuity mortgage to generate capital through loans against the idle property, to be used for funding the operations or mortgage payments of the new property, without disposing of the present property in the process. If appreciation of the owned investment property takes place as expected, repayment of the annuity loan upon sale of the investment property can be easily handled.

G. Construction Loans

Also known as interim loans, construction loans are similar to annuity loans in that the borrowed funds are distributed periodically rather than in a lump sum. The difference is in the intent and length of the loan. In the case of a construction loan, the funds are not all payable at once because the property securing the funds is not sufficiently valuable to cover the size of the loan. As the facility is constructed, the value of the property rises and the payment of funds goes up. Interim loans are designed to cover construction only, and the lender is usually not willing to tie up funds for an extended period of time. Normally, at the end of the construction process, the construction loan is paid off by securing a permanent loan called a take-out loan, designed for that purpose. The name take-out comes from the fact that the loan takes indebtedness out of the construction loan and replaces it with indebtedness in a longer-term permanent loan structure. Construction loans are common in the development of facilities because they reduce the amount of capital that needs to be invested from company funds and allow a firm to pay for property as it is created.

H. Blended-Rate Loan

This is actually a means of refinancing by which the present interest rate on an existing loan is combined or blended with current rates. Blended-rate loans are more prevalent in the home market than in business mortgages but are sometimes available from lenders holding fixed-rate mortgages. They are particularly attractive to firms that wish to respond to reductions in the going rate without totally restructuring their debt position.

I. Buy-Down Mortgage

Although this type of financing is also unusual in an industrial setting, at times it may be encountered. Buy-downs exist where market rates are so high that builders and property holders are unable to attract buyers. They may choose to pay a lump sum to the lender in exchange for a reduced interest rate to potential buyers, either for a short period of time (three to five years) or for the life of the loan. It does cost the builder/owner money to do this, but in distressed markets or in times of high interest rates, the builder may see it as a viable way to move property and free funds for more lucrative investment. The firm seeking a new facility site or structure may find such a buy-down opportunity in a market and may be able to take advantage of it, reducing its interest rates and cost of capital considerably in the long run.

J. Equity Mortgages

This arrangement is possible when the firm seeking financing has equity in some property or business asset. In this case, the firm offers its equity as collateral for the loan. This loan carries with it a very low interest rate and is easy to obtain. The lending institution lends money only up to 80 percent of the equity value in order to decrease its exposure in the event of the loan's falling through. This has become a very popular form of second mortgage financing for houses, with the loan money often used for beginning small businesses. Should the business fail, the borrower could stand to lose both business and home. However, if the venture is successful, equity mortgaging provides a very inexpensive means of financing the business.

K. Carryback Financing

Technically, this is a note accepted by a seller in lieu of cash. Particularly popular in the case of land sales, this type of financing has advantages in that the seller receives a promissory note rather than cash, which guarantees the purchase price plus interest. When the purchaser is assuming a mortgage, it creates an additional payment to the original owner. Payment of principal is then spread out over a number of months or years, producing income for the seller. This is often preferable to lump sums in that it affords the seller the tax advantages of not receiving all of the payment in a single year. With carryback financing, the amount involved is often higher than it would be had the deal been consummated for cash, due to the time value of money and the risk incurred by the seller that the money may not be paid at all, should some catastrophic event overtake the borrower. The shortcoming lies in the fact that if payment is required in toto at some point in the future, the lender may find it difficult to convert the note to cash quickly and easily, thus reducing the liquidity of the paper and its value in the process.

From the point of view of the firm seeking to develop a new facility, this kind of purchase of land may be advantageous when cash is short, or when the cash that is available is more appropriately used for the purchase of other real property, such as plant and equipment. The primary benefit is that the firm can obtain much better terms through the seller than from the market. If the cash flow indicates that the additional debt expense can be handled, carryback financing can be a very viable alternative.

L. Wraparound Mortgage

This mortgage includes existing mortgages and is considered junior to those mortgages. The concept is that the lender will assume an existing loan and take on an additional loan in an amount reflecting the difference between the full loan and the original loan. By doing so, the lender receives a higher rate of interest than on the original loan and the borrower pays a lower interest rate than would be found on a totally new loan. This is similar to "leveraging up" in the stock market. Unfortunately, it is a viable alternative only in the case of a loan assumption and will not work when an alienation clause is present, which allows the lender to call the entire balance due upon transferral of the property.

M. Subordination

This is a process by which the seller agrees to subordinate all rights under the mortgage, that is, to take a lesser priority of payment among debtors. As an example, if a firm is seeking to buy a piece of property and construct a facility on that property, purchasing the property is the first necessity and creates a primary mortgage with the person selling the property. However, if the funds required for the construction of the facility itself can be secured only with a first mortgage (if the construction loan lender will not agree to a second mortgage position), the sellers of the property may be enticed into subordinating their rights and agreeing to take their mortgage as a second mortgage. It is usually necessary to agree to pay a premium of some sort to the sellers of the land, either in the form of interest or of a higher selling price, which serves to offset their higher risk. Be that as it may, subordination arrangements can save an otherwise unworkable arrangement.

N. Contract for Deed

Also known as a land contract or installment contract, this is a method of selling property in which possession of the property passes to the purchaser, yet the seller retains actual title to the property until the loan agreement is fully executed. This is not different from any other form of installment purchase, except that land is used as collateral. The ownership of the property remains in the hands of the seller until it is fully paid for, yet the purchaser has the right to use the property.

In terms of financing a facility, this kind of contract may be quite attractive to the seller, particularly if the seller is selling land on which the purchaser will build. As the value of the property increases, the security of the loan increases as well and risk decreases. The drawback to the seller is that capital is still tied up in the property with only a down payment, installment payments being received through time.

O. Options

Options represent the right to purchase or lease some piece of property, at some time in the future, at a set price. Options are valuable as a hedge against future events and as a means of ensuring that anticipated future actions can be taken, should the need arise. A firm deciding on a new facility may decide that it needs 50 acres now for its facility and that at some time in the future, it may require an additional 50 acres for expansion. When the expansion will be necessary and whether it will be necessary is not known for certain, but the company would like to have the option of doing it if it becomes necessary. It may then choose to negotiate for the new site with the individuals from whom it is purchasing the land. This will allow them to purchase, at some future time, an additional 50 acres contiguous to the present acreage, and to do so at a previously agreed-upon price. In this way, it does not have to purchase the additional acreage now, and for a price (the cost of the option), it has been able to hedge against higher future purchase costs by locking the seller into a specific selling price for a specific period of time.

The advantage to the seller in this case is that funds are received for the option without having to sell the land. The option is compensation for the fact that the land may not be sold to some other party, in that the seller has agreed instead to sell it to the person buying the option, at the predetermined price, and at the given future date.

III. FACILITY TYPES

The kind of facility being created has an effect on the choice of financing, because different combinations of characteristics yield different advantages in using one type of financing over another. A manufacturing facility, for instance, normally has a longer life than, say, an office facility because of its higher capital outlay and higher expected degree of functional expectancy. A manufacturing plant is expected to use equipment and maintain design far longer than the rather transient arrangements in an office environment, though the latter has a higher degree of flexibility in location and arrangement. In analyzing the relationship between type of facility and type of financing, certain characteristics can be noted as key factors and serve to form the basis of such an analysis. Among these factors are initial capital outlay, duration of expected usefulness of the facility, transience in the location and use of the facility, expected rate of variance in equipment, function, and output, flexibility of the facility, type of customer served, and importance of transportation and communication networks. In light of these factors, six types of facilities are listed below with a brief analysis of how their characteristics might influence the choice of their financing.

A. Manufacturing

A manufacturing facility, as noted above, generally requires a high capital outlay, either in the form of corporate investment of extant funds or through borrowing. Manufacturing facilities can be leased or even rented, if extremely short-run production is anticipated, such as on-site support manufacturing. Yet the predominant approach is to own a manufacturing facility and expect it to be around for a considerable amount of time. This is generally the most efficient way of obtaining an acceptable return on investment. In general, heavy manufacturing facilities have a long life, high capital outlay, a stable location with expected long-term use, relatively stable technology not expected to totally change in a short period of time, low design flexibility devoted to single product or product group production, stable, well-defined customer groups, and high dependence on transportation and communication nets that are established and maintained over long periods of time. This is a very stable type of facility and can be considered the most permanent of the types we discuss.

With this combination of characteristics, purchase and therefore mortgage financing is the most appropriate. The individual type of mortgage that is most favorable depends on the exact nature of the individual industry.

B. Service

Service facilities can be large or small, permanent or transient. They may require a great deal of storage space or virtually none, facilities for repair and production or little more than a service counter, a great deal of office space or minimal office space, and so forth. They are, in fact, as individual as the service to be performed. The general pattern in service facilities is that they are less permanent than heavy manufacturing and require less capital outlay. Service implies convenience, which generally means closeness to the customer. Since customer bases may change through time, there may be required a higher degree of flexibility on the part of the service organization in order to move with the customers it serves. A more transitory type of facility results. Lease and lease–purchase options under such circumstances can be quite favorable. This is particularly true in light of the general need for office and counter space inherent in a service facility, which may attract the firm to a retail area or office park environment specifically designed for a service operation. The characteristics of a service facility include shorter life than a heavy manufacturing facility, relatively low capital requirements, a fairly stable location life, variable though usually easily changed technology, fairly high flexibility of facility use, and fairly high customer stability. In service industries, timing is all-important. Consequently communication and transportation nets are very important.

For such a set of characteristics, either long-term lease or purchase is suitable, keeping in mind the shorter expected life of the facility itself and its resale capability. If the need to move every few years as the market shifts is a factor, yearly leases or medium-term leases may be a preferable option, providing the move itself can be made easily and economically.

C. Retail

More than any other type of facility, the retail facility needs to be in the right place at the right time. Customers must be able to conveniently and easily reach a retail facility and are quite capricious about their loyalties. For that reason, retail facilities locate in areas close to the population. Financing for small retail facilities such as mall locations or shopping district locations is usually on a rental or lease basis, often with the rent contingent on the sales volume of the individual store. Larger retailers may construct their own facilities and finance with conventional mortgage instruments. Still another alternative is to build–sell–lease back. In this approach, the retailer constructs the facility to suit certain specifications and then frees capital for other projects by finding an investor willing to purchase the property with a contract that guarantees a lease to the retailer who originally built the facility.

Characteristics of retail facilities include low to medium capital overlay, a life of five to 20 years, relatively long-term location, stable technology, high flexibility of physical facility, stable customer group, and high importance of transportation and communication nets, particularly between customer and facility. Of these characteristics, most can be found in any standard storefront or shopping area, and the higher necessity for flexibility (in terms of life, location, and use of space) is reflected in higher flexibility in financing the project. Hence the tendency toward short-term types of financing.

D. Wholesale

Wholesale facilities require less flexibility than retail facilities and can usually maintain a single location for longer periods of time. Wholesale purchases are generally bulk purchases, and the customers tend to exhibit higher degrees of loyalty. The wholesaler requires more storage space, less (if any) showroom space, and high access to transportation and communication networks. Wholesalers generally also require more square footage than the retailer, and more heavy equipment. The result is a fair amount of capital outlay, a high degree of stability, and low need for design flexibility. Purchase and long-term financing are therefore feasible. Wholesalers tend toward the standard mortgage debt instruments or long-term leases, if their needs are standard and require no special construction.

E. Warehouse

Warehouse facilities are general in nature and, with the exception of specific products that require specialized handling (chemicals, grain and farm products, radioactive materials), can easily utilize a standard storage structure. Since these structures exist in abundance in all metropolitan areas, with high concentration around transportation facilities, the firm will normally have little difficulty in finding a suitable site. Purchase or lease of such property is simple and straightforward. If the facility is expected to have a long life, purchase with either cash outlay or mortgage financing is the most efficient means of acquiring the site. If there is a shortage of available properties in an area, however, or if the market is expected to shift in the foreseeable future, a long-term lease is often the best answer. Because of their abundance in areas of high industrial/transportation concentration, a ready supply of leasable facilities can usually be found. Since the primary characteristics in warehousing are transportation and space, any number of existing facilities can normally be found that are suitable to the needs of the firm.

F. Office

General characteristics of office facilities include relatively low capital outlay, short to medium life, stable though not necessarily long-term location, changing technology, high design flexibility, variable degree of customer stability, high dependence on communication nets, and low dependence on transportation nets.

Offices tend to be generic in nature. That is, they all tend to have the same general characteristics, a standardized array of functions to be performed, with an infinite number of variations on how those functions are arranged within the individual facility. Offices can be located anywhere as long as they are close to one or another of the elements of a company's operations. Often the importance of the proximity of their location to customers is no greater than that of their proximity to manufacturing plants, seats of government, suppliers, or whatever group with which the office interacts.

This high degree of flexibility will often make available a wide range of choices for stable office locations. Of all the facility types, offices can most easily be relocated with a minimum of disruption, although, as anyone who has ever experienced an office relocation can tell you, the process is far from painless. It is merely that office facilities are best able to change and shift as circumstances dictate.

For this reason, unless the office is a centralized facility commanding the operations of a large multidimensional firm, the space is most efficiently leased or rented. The technology of office environments also adds to the flexibility of the facility, and this broadens the range of acceptable building configurations available to the firm. Standardization is high in this environment, and departmentalization and arrangement for primary and support structures can be easily accomplished. Lease plans for office space vary in their nature and can be based on anything from expected length of the lease to square footage.

IV. PRODUCT LIFE CYCLES

In both manufacturing and service industries the life cycle of the products being offered will vary and the life of facilities with them. This is such an important integral part of the financing decision that it bears expansion here.

Physical products vary widely in their appeal and in their importance to the economy, and as such they have variable lives. Not only do the expected lives of the goods change, but the desire of the public to purchase and use these goods changes as well. In manufacturing, this means finding a way to create the goods at a reasonable cost, taking into account their expected life in the marketplace along with other cost items. It does a firm little good to produce and sell goods with an expected popularity of only a few months if it requires years of operation to receive an acceptable return on investment. These so-called fad items are viable products only if they can be produced cheaply and easily and then forgotten. On the other hand, the manufacturing of a stable and staple good, such as steel, ensures that the investment will pay off before the market tires of the item being produced. In deciding on the type of financing to be used, it is imperative that the planner be certain of the life of the product, or alternatively, that the facility have a high enough degree of flexibility to be utilized efficiently in the production of other goods once the present item loses its market.

The service industry has a similar though slightly different problem. In service, there is a naturally higher degree of transience built into the product. Services are less physical, less tangible, and they tend to be immediate. Lawn services, laundry services, and personal services such as lawyers, doctors, and accountants are temporal in nature. There is the additional problem of changing demographics, changing customer needs, and often capricious consumption patterns. The supplier needs a high degree of flexibility.

In terms of facility planning, the planner must be as certain as possible that the present set of characteristics of a service are sufficiently constant to logically warrant expenditure for a facility to accommodate those characteristics. Long-term leases do not make sense if the market is about to shift location. Buying or constructing a specialized facility is not a viable project if the service it offers will no longer be needed five years down the road. There are a world of examples of how true this is. In Cocoa Beach, Florida, an incredible boom in rental housing and construction arose around the founding of Cape Kennedy Space Center, along with numerous support firms moving into the area. The boom collapsed in the budget cuts following the Apollo program, leaving landowners and service firms with no customers for their products. Alternatively, a successful reading of circumstances is illustrated by the move to satisfy legal requirements recently experienced in the state of Georgia, which in some counties requires exhaust emission inspections prior to licensing of automobiles. The inspection services are offered by private firms that are certified and licensed to do so, yet demand for the inspections centers in the period from January to mid-April when the licensing of vehicles takes place. This temporary rise in demand for the service has resulted in temporary buildings springing up in parking lots and otherwise closed service stations to take care of the temporary increase in demand. They are short-term facilities by design, low in cost (some are little more than pipe frames with plastic covering and temporary electrical service for the testing equipment), easily constructed and easily torn down. In this case, the facility meets the life of the product. In the Cocoa Beach example, it did not. Reading the life cycle of the product is crucial, particularly in a service industry.

QUESTIONS

1. Name five different types of mortgages.
2. Name three different types of rental agreements.
3. What are the primary differences between rental agreements and leases? How can each of the two options be to the firm's advantage?
4. Name five different types of facilities.
5. What are the advantages of cash purchase? What are the disadvantages?
6. In choosing to finance rather than purchase for cash, are the advantages and disadvantages simply the reverse of those noted in Question 5, or are there additional considerations?
7. What is the difference between lease and lease–purchase arrangements? Under what circumstances would the former be preferable? The latter?
8. Firms need financial alternatives available to them. Why?
9. Define and explain the concept of subordination.
10. What are options and how do they work?
11. What are the primary advantages of options?
12. What is a graduated payment mortgage? How does it differ from an adjustable rate mortgage?
13. Name eight financial alternatives available to the firm.
14. Explain negative amortization. Why do you suppose it exists?
15. What is a wraparound mortgage?
16. What is a payment cap? How would its removal affect corporate risk? Lender risk?
17. Of the various types of facilities noted in the chapter, which do you suppose generally requires the heaviest investment? Which the least?

18. How does opportunity cost help account for the tendency to finance rather than cash-purchase facilities?
19. Would you expect the interest rate on a long-term facility loan to be greater or smaller than on a consumer loan? Why?
20. For the small firm or individual proprietorship, what are the most viable financing alternatives? Why?

CASE STUDY

6
A Service Facility: Kingdom Harbor Yacht Club

INTRODUCTION

This facility planning proposal is a condensed combination of proposals for several different facilities that are to be developed. Some financial details have been omitted that would normally be included in a proposal being sent to a financial institution. For example, the lending institutions would require personal net worth statements from all the principals of the development corporation. They must be able to show their ability to absorb some of the debt load should the project be hindered or interrupted for any reason.

Another example would be the governmental proposal that began this planning process. Because this proposal involved government land and the necessary passage through government land to reach the water, the government was active in the proposal process. In fact, the government was the primary reason for the rejection of one of the proposed facilities.

Even though this is a case digest of several proposals, all the elements necessary for financial approval of a loan are present in the case study, except for the examples noted.

The proposal took many months to create and probably could have taken more. I aimed to finish the proposal at a point that would allow me to begin construction at a logical time. I believed it was important to begin construction early enough to ensure completion by Memorial Day, the beginning of boating season. This meant that construction should begin by Thanksgiving in order not to miss the best income months, which for the location considered are from late spring through early fall. Although the proposal may have been slightly shortened because of this time constraint, I felt it was more important to proceed with the plans than to wait another year.

In preparing these proposals, I found myself involved with many different activities almost simultaneously. I was looking for possible sites with real estate salespeople in three different counties; checking zoning codes in these counties; meeting with subcontractors to discuss storage buildings; talking to officers at the bank; meeting with land owners; talking with lawyers, accountants, and interested investors; and, finally, setting up meetings with the U.S. Army Corps of Engineers to determine the extent to which they would allow their property to be altered. Proposals involve more than just paperwork; they demand action.

In all, I have developed plans for five marina facilities, each one somewhat different from the others but all similar in many ways. My plans for a marina facility are at a temporary standstill while I search for a suitable combination of location, amenities, and financial opportunities. Yet I have kept moving toward the goal by using a certification facility, where I continue to build the club membership roster with sailing students whom I instruct. I now have a membership ready and willing to build the club, rather than building a club first and trying to fill it with members later.

As an unexpected advantage, I am also gaining information from club members about their preferences in a marina facility. I have discovered which aspects of the club are attractive to members and which are not. This allows me to adjust my plans before actually building the facility. The membership currently numbers in the hundreds and is continuing to grow. Eventually those members will have a marina facility built that is tailored to meet their needs and desires. The point is that the planning process is very important. Without proper planning, even the best idea can become a disaster.

It is also important that the planning process remain dynamic and flexible. Because computers give me flexibility and the power to change plans quickly, I work with five computer software applications. A well-equipped facility planner will have at least the following software packages available.

1. A computer word-processing program
2. A computer financial spreadsheet program
3. A computer aided design program
4. A computer construction planning program
5. A computer construction cost data program

My co-author and I believe that in the future we will see more and more well-motivated and highly productive small businesses. They will be started by people who are now working for larger firms. These people need to realize that proper planning is imperative in order to succeed in business. They should start planning *now*, before they leave their present occupations.

This case study will show you how to touch all the bases before you start looking for funds and become deeply involved in the planning process. The amount of money required in this proposal may seem relatively small, but the essential elements covered in the table of contents are applicable to even the largest of facility proposals.

Small entrepreneurs are most apt to spend years compiling a plan because they will typically be using more of their own money and do not want to lose it on a bad idea. These are the people we wish to encourage and help to put their plans into action.

KINGDOM HARBOR YACHT CLUB

FACILITY PLANNERS
Rob James and Associates
P.O. Box 2132
Cumming, Georgia 30130

ARCHITECTS
Millard, Inc.
270 Carpenter Drive
Atlanta, Georgia 30328

FINANCING
Forsyth County Bank
Cumming, Georgia 30130

GENERAL CONTRACTORS
M. J. Vallez and Associates
Cumming, Georgia 30130

ATTORNEYS
McIntyre and McMahon
Sandy Springs, Georgia 30328

ACCOUNTANTS
Jimmy Myers, CPA
Cumming, Georgia 30130

CONTENTS

MARKETING STATEMENT

Over the years, the families of America have come under unusual strain. Parents are separated by their jobs from their children, who are being raised by schoolteachers and day care centers. To offset this growing distance between family members, we believe a family-oriented club would give families a chance to spend time together.

Kingdom Harbor Yacht Club will provide the facilities necessary for family recreation near the water. The club will be located only 45 minutes away from a major metropolitan area, Atlanta, Georgia, which will make even weekday use possible. It will offer boat storage, a swimming pool, tennis courts, fishing boats, sailing boats, and rustic cabins. The club will have all of these available to the membership at family rental rates.

The clubhouse will be used to instruct the membership in sailing, fishing, swimming, and water skiing. The clubhouse was designed small (see rough cut scale sketches) to allow for small class sizes, easy maintenance, and intimate family gatherings for the members. Although just one clubhouse is planned in the beginning, others would be built as the need arises.

We invite you to read this proposal and then arrange a meeting with us to talk further about financing our project. We would like to call your attention to the marketing research and financial analysis located herein. Much effort was spent to make these as accurate and complete as possible. As you know, the federal government will guarantee up to 100 percent of this loan should enough jobs be created at this facility. That is why the fifteen-year job analysis sheet has been included with this proposal.

We invite you to be part of our dream and come get involved with Kingdom Harbor Yacht Club, a safe port for the family tossed by life's storms.

MARKET ANALYSIS

MARKET ANALYSIS OF STORAGE SLIPS

Storage Area	*Enclosed*	*Covered*	*Uncovered*	*Drop Charge*	*Occupancy*	*Comments*
Off water:						
ADMA Storage	$600	$300	$180		100%	
GA 400	$900		$240		100%	
Lan Mar Storage		$320		$20	90%	
Boat care		$270			100%	
Gulver Boat Store		$280	$200		90%	
Shady Grove	$660				90%	
Habersham		$920			94%	
Lanier Harbor		$500		$15	98%	$300 initiation fee
Kingdom Harbor		$588		$10		$100 annual dues

MARKET ANALYSIS OF COTTAGES

	4/1–11/1	*11/1–4/1*	*Weekend (3 days)*	*Comments*
Lake Lanier Islands with tub	$775	$650	$375	56 cabins stay full in season
Blood Mountain	$525	$525	$225	Recently opened, cabins are full
Vogel State Park	$336	$308	$ 95	Since 1930, 11-month wait in season
Kingdom Harbor with tub	$595	$495	$295	Pool, tennis, clubhouse, fishing, and boating
Charter boats	$499	$399	$260	American Sailing Association certified sailors

MARKET BROCHURE

KINGDOM HARBOR YACHT CLUB

Kingdom Harbor is meant to be a family retreat area, whether it be for a couple of hours, days, or weeks. It offers the membership pool, tennis, and clubhouse privileges. For an additional nominal fee, a member can use a storage slip, the vacation cottages, or the party room.

Membership is open to anyone, anywhere. Only when the facilities reach capacity will permanent membership be restricted to a replacement basis. When space is available, temporary members are allowed to use the cabins and facilities on a trial basis. Permanent members are always given first preference in case of scheduling conflict.

Membership fees:	
Initiation	$300
Yearly dues	$100
Boat storage	$585
Cabins	
Monday–Friday	$295
Friday–Monday	$300
Party Room	
8-hour day	$50
4-hour night	$25
Lessons/hour	$15
Charters/8 hours	$49
Pool	Free
Tennis	Free
Fishing	Free

Office hours:
October 1–May 1, 8:00–6:00 Friday–Monday
May 1–October 1, 8:00–8:00 Sunday–Saturday

Permanent members have key to gates.

ABOUT KINGDOM HARBOR COTTAGES

All cottages are fully equipped for housekeeping, including stove and refrigerator, with all necessary cooking and serving facilities, equipment, bed linen, and blankets. Guests need to bring only food. All cottages are air-conditioned.

RESERVATIONS Reservations for all cottages at Kingdom Harbor will be accepted on an 11-month advance notice basis. For example, at any time during the month of September 1990, a person may make a reservation for a cottage to begin at any time within the next 11 months, or up to and including the entire month of August 1991. A reservation will not be firm until a deposit is received and accepted by Kingdom Harbor. The deposit is due within five (5) days from the date the visitor receives the reservation notice. If the deposit is not received within the allotted time, it constitutes automatic cancellation.

Deposit $125.00

Check-in time 4:00 P.M., Monday or Friday. Please present your reservation deposit slip when checking in. **Balance of payment on cottage is due upon arrival.**

Check-out time 10:00 A.M. An additional fee will be charged after this time.

Time held Reservations will not be held after 10:00 A.M. of the second day, and both reservations and deposit are forfeited unless Kingdom Harbor has been notified of late arrival. Kingdom Harbor will not accept more than one night's deposit.

Occupancy limitation Cottage may not be used overnight to accommodate more than normal bed capacity. Maximum occupancy of cottages is 6 people.

Occupancy limit 2 weeks.

Late arrivals Kingdom Harbor must be notified of late arrival. After 5:00 P.M., late arrivals will be allowed to register only under emergency conditions.

Registration Register at clubhouse upon arrival.

Refund Deposit will be refunded upon a minimum of 72 hours notice given to the club prior to arrival; however, a $25 cancellation fee will be deducted from deposit. Deposit receipt must be returned to Kingdom Harbor for refund. Any unused portion of the reservation period will be refunded if minimum occupancy has been satisfied (minimum occupancy **one weekend**).

Firewood Firewood is *not* furnished but may be purchased at a nominal fee.

Eligibility Club members, permanent and temporary only. An adult must accompany all unmarried cottage guests under 22 years of age.

Visitors Visitors to the cottage area are welcome; however, the number of visitors and registered guests will not exceed the cottage capacity at any one time. Visitors are required to leave by 10:00 P.M.

Pets Pets are not allowed in cottages or cottage area. Kennels are available in Cumming.

Cottages should be left in good condition. Dishes and cookware should be washed and put away. Beds should be left unmade.

Equipment, furnishings, or bed linen may not be carried outside the cottages.

Excessive damage will be charged to the member renting the cabin.

KINGDOM HARBOR YACHT CLUB
P.O. Box 2132
Cumming, GA 30130

FAMILY MEMBERSHIP APPLICATION

Name ______________________ Home Phone ____________
Address ______________________ Office Phone ____________
City ________________ State ____________ Zip ________
Occupation ______________ Company Name ____________
Spouse Name ______________ Office Phone ____________
Occupation ______________ Company Name ____________
Children's Names and Ages ____________________________

Membership fees include pool, tennis, and clubhouse privileges.

Do you wish to store your boat at the member rate? YES/NO
What kind of boat is it? ____________________________

Would you be interested in spending a weekend or week in the cottages at the member rate? YES/NO

What dates would interest you? ______________________

Would you be interested in the following member activities?

______ Learning to sail? ______ Charter a sailboat?
______ Learning to fish? ______ Charter a houseboat?
______ Learning to waterski? ______ Charter a pontoonboat?
______ Learning to swim? ______ Charter a skiboat?
______ Learning to play tennis? ______ Charter a jetski?

My family and I agree to abide by the rules and bylaws of the Kingdom Harbor Yacht Club as posted and amended. I realize that my initiation fee is nonrefundable and my yearly dues are to be paid by April 1 of every year or my membership is forfeited.

Signed ____________________________ Date ____________

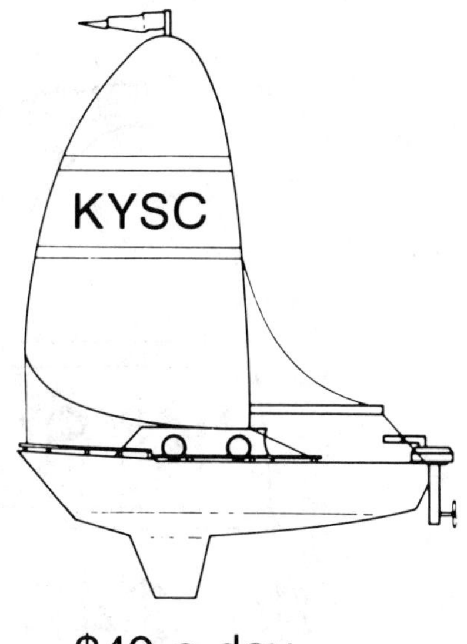

$49 a day

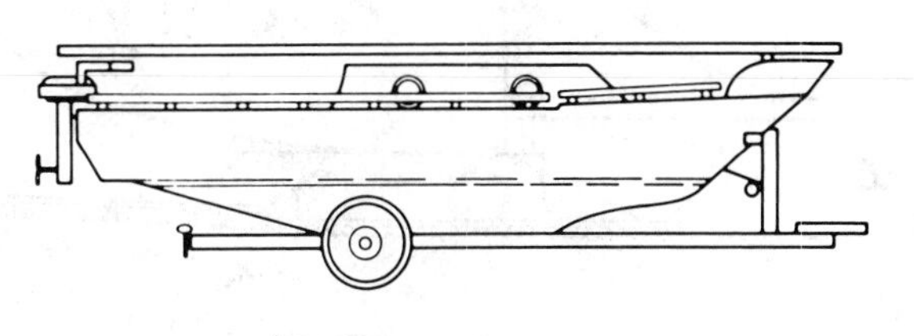

$75 a day
(on trailer)

"AD SHEETS"

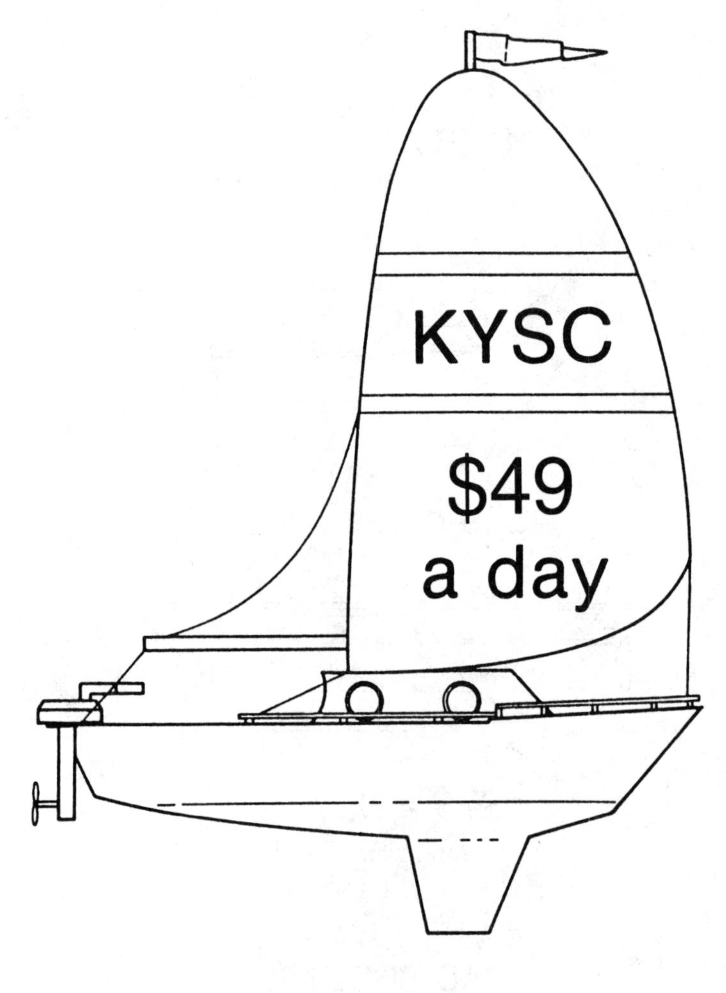

"AD SHEETS"

MARKETING ADVERTISEMENTS

KINGDOM HARBOR YACHT CLUB MEMBERSHIP RATES

Sample Rates:

$300 Initiation fee, one-time, nonrefundable
$100 Yearly dues, pool, tennis, and clubhouse
$400 First year

After first year, membership rates can change to maintain clubhouse, pool, tennis courts, and roads.

Special Introductory Offers

Boat storage: Rental

$300 Initiation fee
$588 Storage for year paid by April 1
$100 Membership dues, waived first year
$888 Year 1

$588 Storage
$100 Membership dues
$688 Year 2

Storage prorated to April 1
@$49/month + initiation fee
$10 per splash in or out

Vacation cottage: Rental

$300 Initiation fee
$295 Rental week
$100 Waived dues
$595 Year 1

From 4:00 P.M. Friday
to 10:00 A.M. Monday
$300 per weekend

"GREAT ESCAPE WEEKEND"

$295 Rental for week
$100 Membership dues
$395 Year 2

From 4:00 P.M. Monday
to 10:00 A.M. Friday
$295 per week

"FREE DAY SPECIAL"

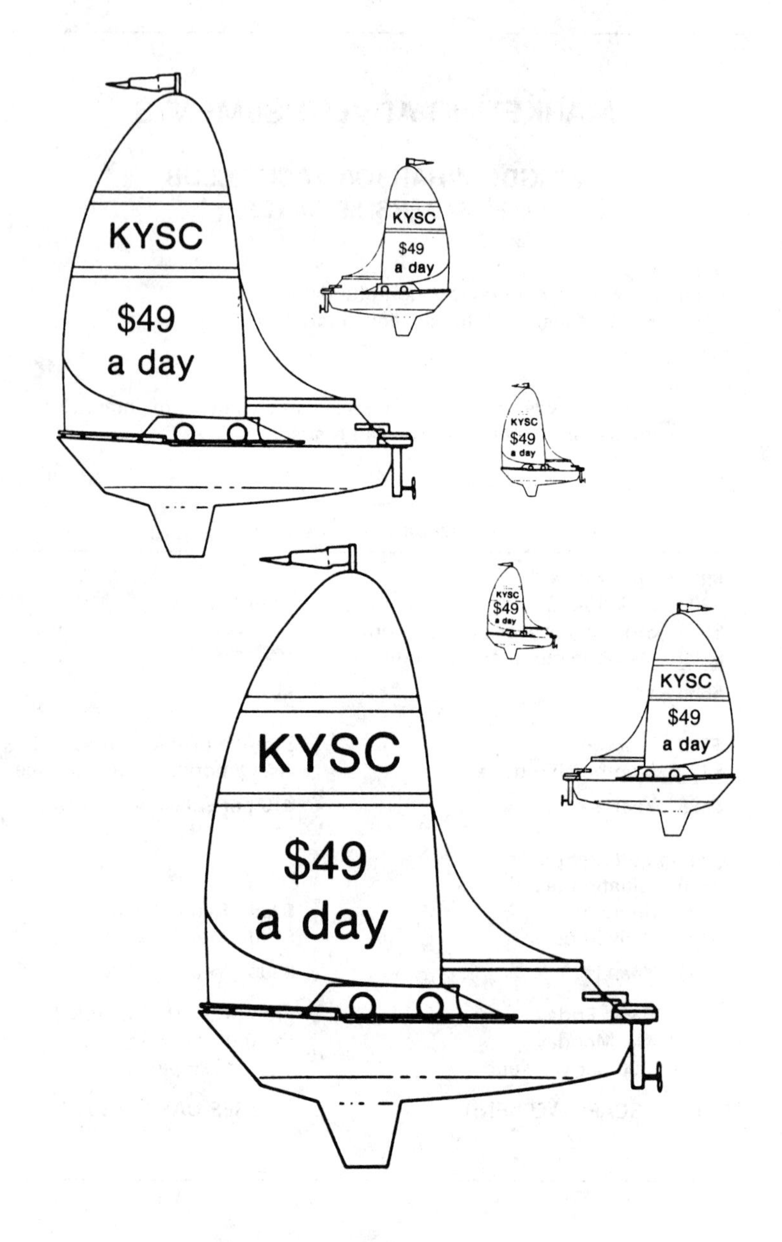
KYSC
$49
a day
KYSC
$49
a day
KYSC
$49
a day
KYSC
$49
a day
KYSC
$49
a day
KYSC
$49
a day

"AD SHEETS"

ROUGH-CUT SKETCHES OF CLUBHOUSE

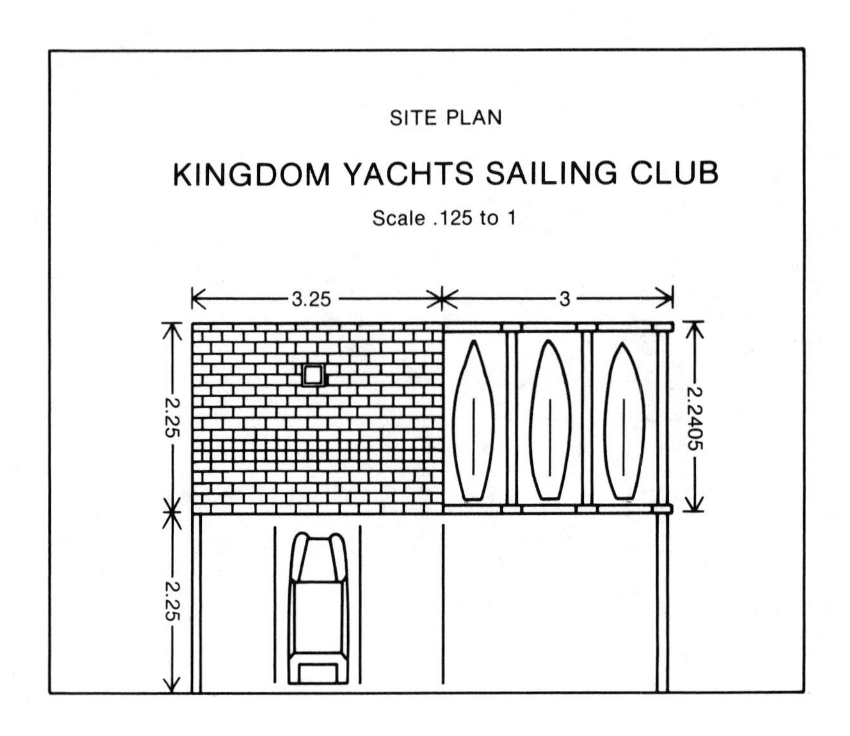
SITE PLAN
KINGDOM YACHTS SAILING CLUB
Scale .125 to 1
3.25
3
2.25
2.25
2.2405

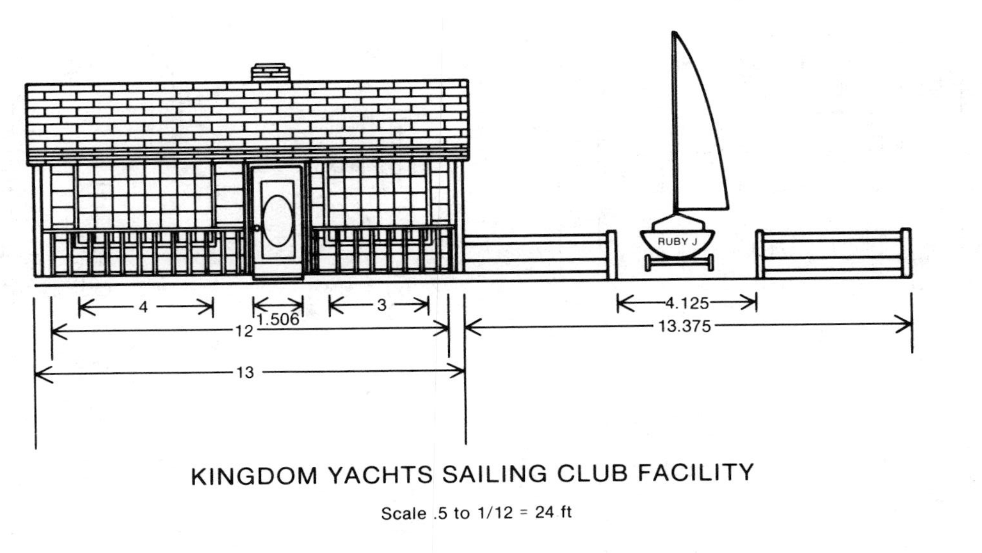

KINGDOM YACHTS SAILING CLUB FACILITY

Scale .5 to 1/12 = 24 ft

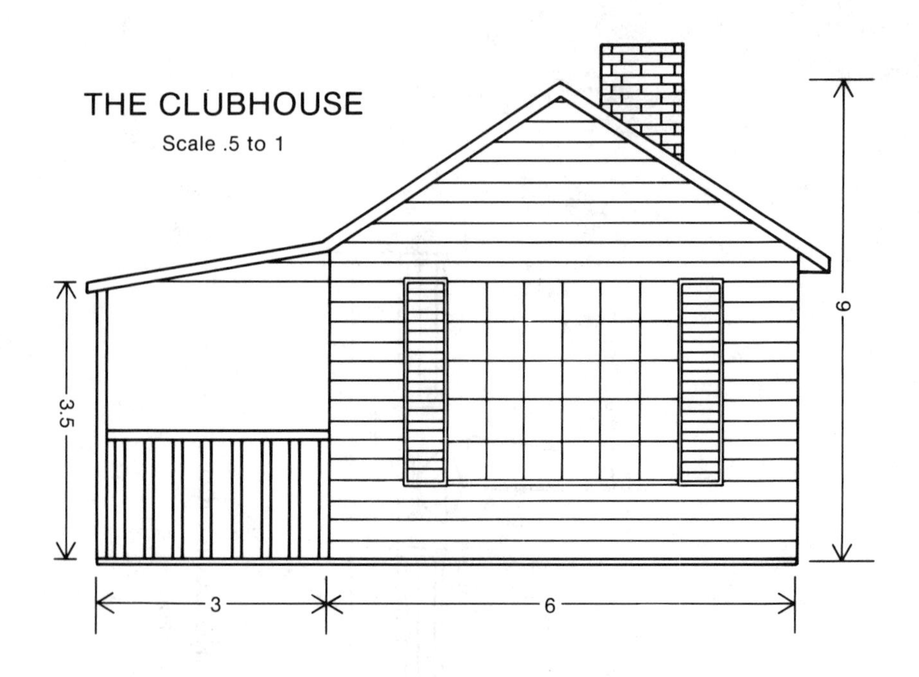
THE CLUBHOUSE
Scale .5 to 1
3.5
9
3
6

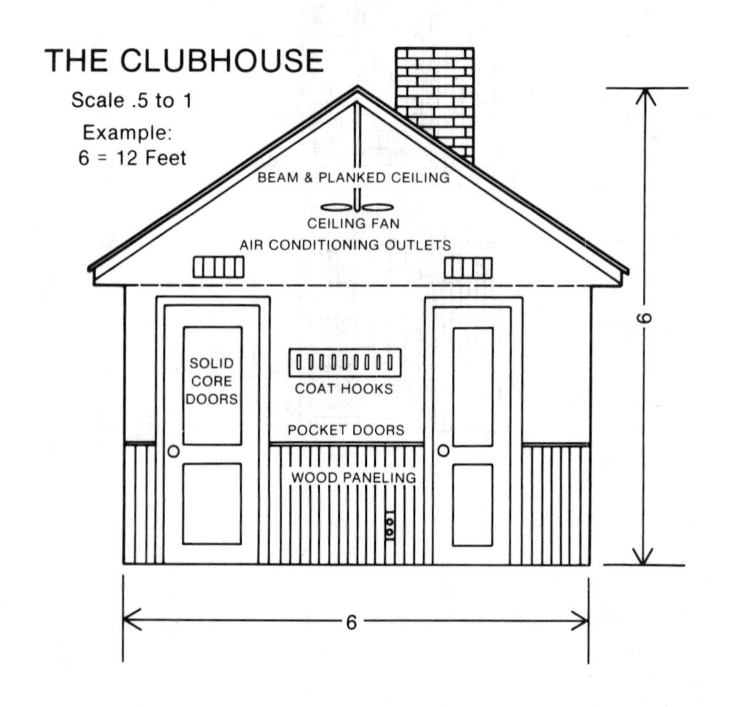
THE CLUBHOUSE
Scale .5 to 1
Example:
6 = 12 Feet
BEAM & PLANKED CEILING
CEILING FAN
AIR CONDITIONING OUTLETS
SOLID
CORE
DOORS
COAT HOOKS
POCKET DOORS
WOOD PANELING
9
6

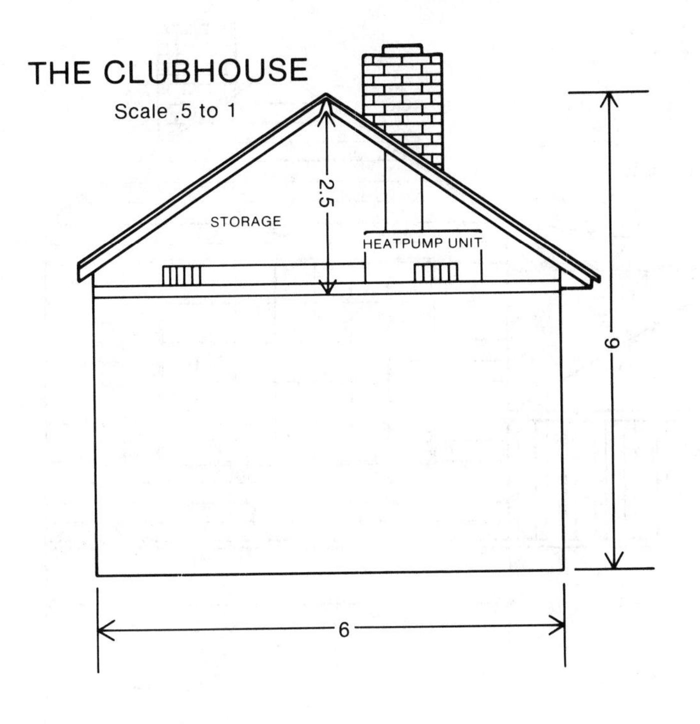
THE CLUBHOUSE
Scale .5 to 1
STORAGE
2.5
HEATPUMP UNIT
9
6

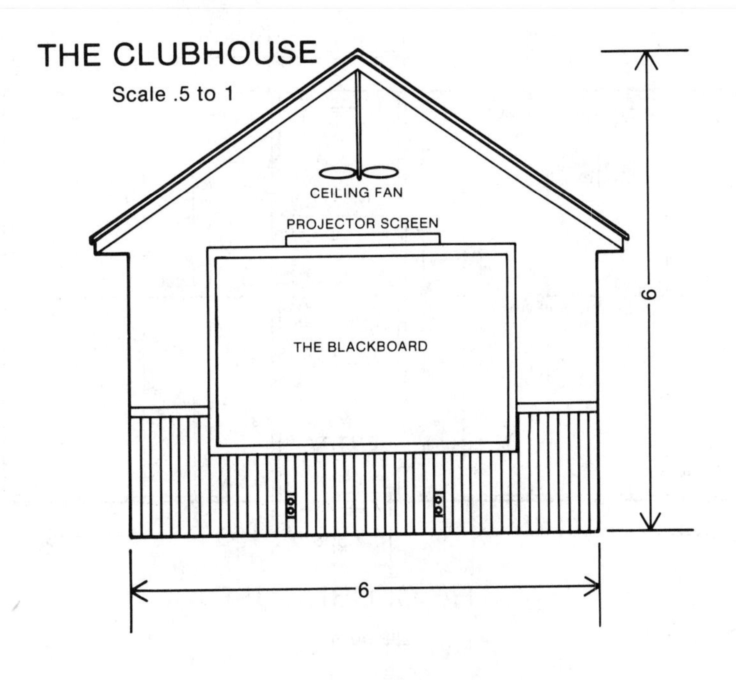
THE CLUBHOUSE
Scale .5 to 1
CEILING FAN
PROJECTOR SCREEN
THE BLACKBOARD
9
6

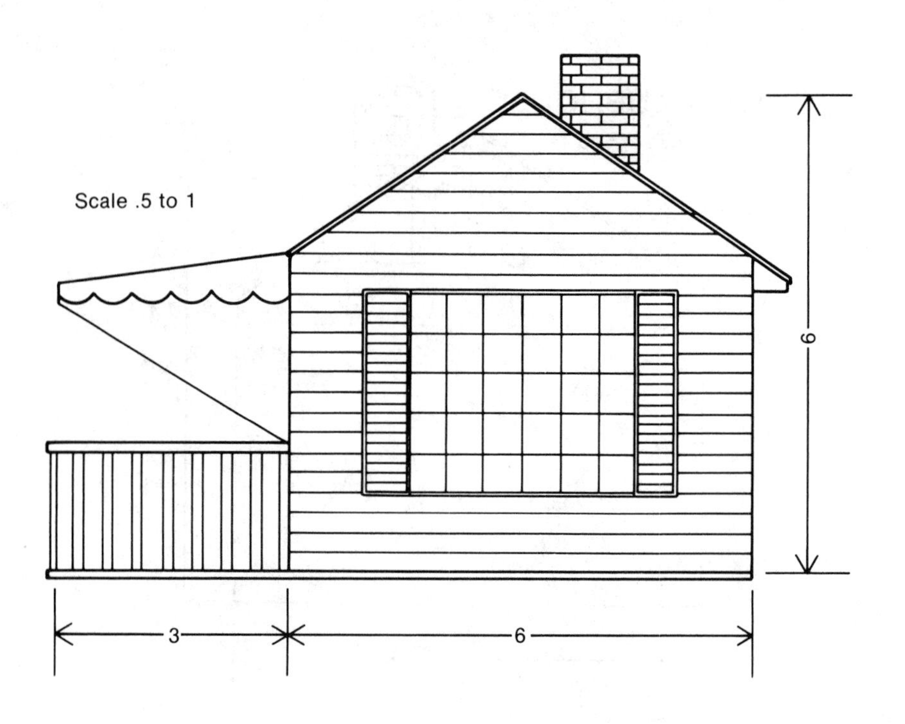

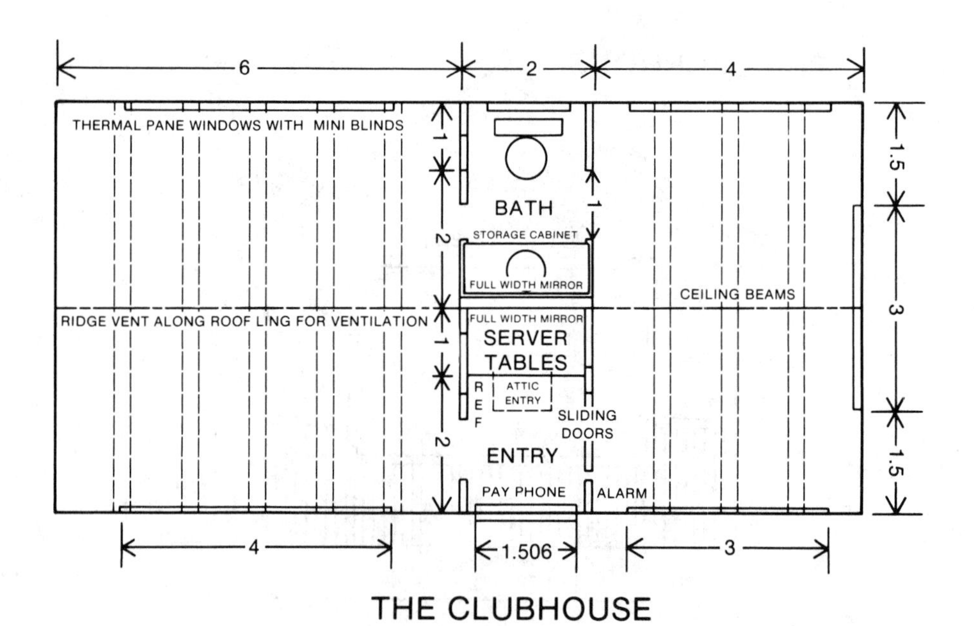

THE CLUBHOUSE

288 sq ft

Scale .5 to 1

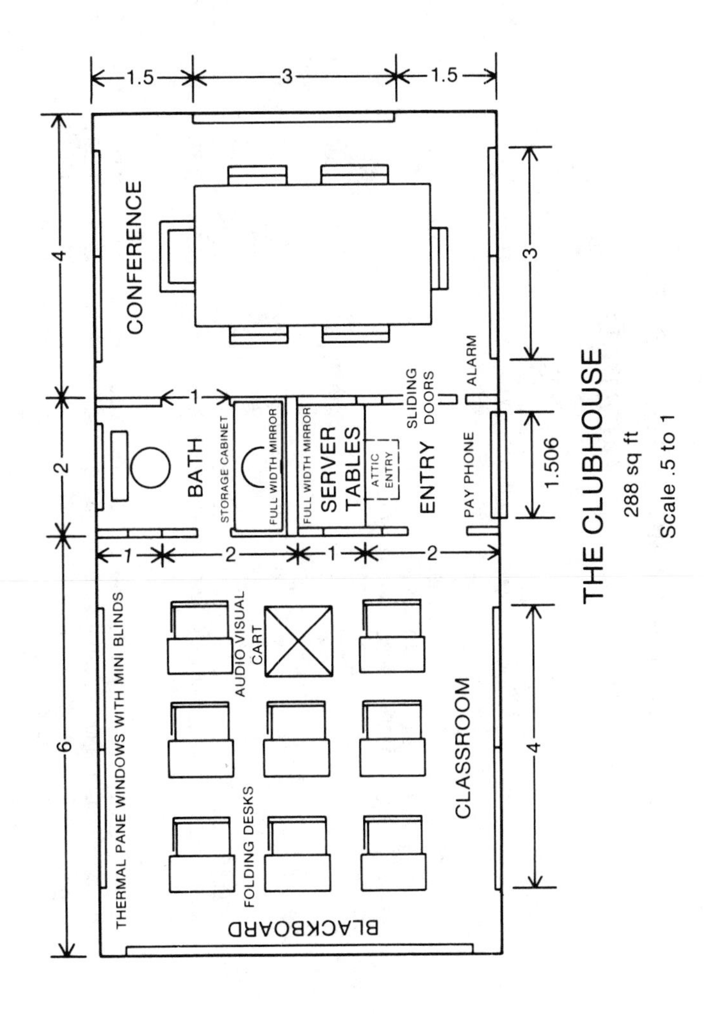
1.5
3
1.5
4
2
6
3
1.506
4
1
2
1
2
1
CONFERENCE
BATH
STORAGE CABINET
FULL WIDTH MIRROR
FULL WIDTH MIRROR
SERVER
TABLES
ATTIC
ENTRY
SLIDING
DOORS
ALARM
ENTRY
PAY PHONE
THERMAL PANE WINDOWS WITH MINI BLINDS
AUDIO VISUAL
CART
FOLDING DESKS
CLASSROOM
BLACKBOARD
THE CLUBHOUSE
288 sq ft
Scale .5 to 1

FIVE-YEAR FINANCIAL FORECASTS
KINGDOM HARBOR YACHT CLUB
First-Year Financial Forecast

Account	*Jan*	*Feb*	*Mar*	*Apr*	*May*	*Jun*	*Jul*	*Aug*	*Sep*	*Oct*	*Nov*	*Dec*	*Totals*
Income:													
Cottages					4320	17850	14280	17850	4320	2880	1440	1440	64380
Storage	5880	11760	17640	17640	29726	30379	24499	6533					144057
Member storage	9000	9000	9000	9000	30000	12000							78000
Member nonstorage				3000	3000	6000	6000	3000					21000
Sandwich shop					500	800	1000	800	500	200	100	100	4000
Boat charters					500	800	1000	800	500	200	100	100	4000
Schools					200	200	100	100	200	200	300	400	1700
Gas and ice					600	600	600	600	600	300	100	100	3500
Other income					200	300	200	300	200	100	100	100	1500
	14880	20760	26640	29640	69046	68929	47679	29983	6320	3880	2140	2240	322137
Expenses:													
Loan payment		10000	10000	10000	10000	10000	10000	10000	10000	10000	10000	10000	110000
Payroll	3300	3300	3300	6100	10615	10615	10615	10615	10615	5680	5680	5680	86115
Taxes	660	660	660	1276	2270	2270	2270	2270	2270	1184	1184	1184	18158
Insurance		1500	1500	1500	1500	1500	1500	1500	1500	1500	1500	1500	16500
Electricity		600	600	600	800	1000	1000	1000	800	600	600	600	8200
Phone		100	100	100	200	300	300	300	200	100	100	100	1900
Equipment				500	1000	1000	1000	1000	1000	1000	1000	1000	8500
Supplies					800	700	600	500	800	400	300	200	4300
Miscellaneous	550				200	300	500	500	800	200	500	200	3750
	4510	16160	16160	20076	27385	27685	27785	27685	27985	20664	20864	20464	257423

Second-Year Financial Forecast

Account	*Jan*	*Feb*	*Mar*	*Apr*	*May*	*Jun*	*Jul*	*Aug*	*Sep*	*Oct*	*Nov*	*Dec*	*Totals*
Income:													
New members					4000	4000	4000	4000					16000
Cottages	1440	1440	1440	2880	4320	17850	14280	17850	4320	2880	1440	1440	71580
Storage	50960	50960	50960										152880
Member storage	8600	8700	8700										26000
Member nonstorage			7000										7000
Sandwich shop	100	100	100	200	500	800	1000	800	500	200	100	100	4500
Boat charters				200	500	800	1000	800	500	200			4000
Schools	200	200	200	100	200	200	100	100	200	200	100	100	1900
Other income	150	100	100	200	300	400	500	500	400	200	100	100	3050
Gas and ice	100	100	200	300	600	600	600	600	600	300	200	100	4300
	61550	61600	68700	3880	10420	24650	21480	24650	6520	3980	1940	1840	291210
Expenses:													
Loan payout	10000	10000	10000	10000	10000	10000	10000	10000	10000	10000	10000	10000	120000
Payroll	5680	5680	5680	8010	10615	10615	10615	10615	10615	5680	5680	5680	95165
Taxes	1184	1184	1184	1697	2270	2270	2270	2270	2270	1184	1184	1184	20151
Insurance	1500	1500	1500	1500	1500	1500	1500	1500	1500	1500	1500	1500	18000
Electricity	600	600	600	600	800	1000	1000	1000	800	600	600	600	8800
Phone	100	100	100	100	200	300	300	300	200	100	100	100	2000
Equipment	1000	1000	1000	1000	1000	1000	1000	1000	1000	1000	1000	1000	12000
Supplies	800	800	700	700	800	700	500	600	400	800	700	400	7900
Miscellaneous	100	200	400	300	200	800	200	200	600	200	700	200	4100
	20964	21064	21164	23907	27385	28185	27385	27485	27385	21064	21464	20664	288116

Third-Year Financial Forecast

Account	*Jan*	*Feb*	*Mar*	*Apr*	*May*	*Jun*	*Jul*	*Aug*	*Sep*	*Oct*	*Nov*	*Dec*	*Totals*
Income:													
New members					4250	4250	4250	4250					17000
Cottages	1440	1440	1440	2880	4320	17850	14280	17850	4320	2880	1440	1440	71580
Storage	50960	50960	50960										152880
Member storage	10850	10850	10850										32550
Member nonstorage			8750										8750
Sandwich shop	100	100	100	200	500	800	1000	800	500	200	100	100	4500
Boat charters				200	500	800	1000	800	500	200			4000
Schools	200	200	200	100	200	200	100	100	200	200	100	100	1900
Other income	150	100	100	200	300	400	500	500	400	200	100	100	3050
Gas and ice	100	100	200	300	600	600	600	600	600	300	200	100	4300
	63800	63750	72600	3880	10670	24900	21730	24900	6520	3980	1940	1840	300510
Expenses:													
Loan payout	10000	10000	10000	10000	10000	10000	10000	10000	10000	10000	10000	10000	120000
Payroll	6075	6075	6075	8460	11085	11085	11085	11085	11085	6075	6075	6075	100335
Taxes	1336	1336	1336	1860	2438	2438	2438	2438	2438	1336	1336	1336	22066
Insurance	1500	1500	1500	1500	1500	1500	1500	1500	1500	1500	1500	1500	18000
Electricity	600	600	600	600	800	1000	1000	1000	800	600	600	600	8800
Phone	100	100	100	100	200	300	300	300	200	100	100	100	2000
Equipment	1000	1000	1000	1000	1000	1000	1000	1000	1000	1000	1000	1000	12000
Supplies	800	800	700	700	800	700	500	600	400	800	700	400	7900
Miscellaneous	100	200	400	300	200	800	200	200	600	200	700	200	4100
	21511	21611	21711	24520	28023	28823	28023	28123	28023	21611	22011	21211	295201

Fourth-Year Financial Forecast

Account	Jan	Feb	Mar	Apr	May	Jun	Jul	Aug	Sep	Oct	Nov	Dec	Totals
Income:													
New members					4500	4500	4500	4500					18000
Cottages	1975	1975	1975	3415	4855	18385	14815	18385	4855	3415	1975	1975	78000
Storage	54165	54165	54165										162495
Member storage	13000	13000	13000										39000
Member nonstorage			10500										10500
Sandwich shop	100	100	100	200	500	800	1000	800	500	200	100	100	4500
Boat charters				200	500	800	1000	800	500	200			4000
Schools	200	200	200	100	200	200	100	100	200	200	100	100	1900
Other income	150	100	100	200	300	400	500	500	400	200	100	100	3050
Gas and ice	100	100	200	300	600	600	600	600	600	300	200	100	4300
	69690	69640	80240	4415	11455	25685	22515	25685	7055	4515	2475	2375	325745
Expenses:													
Loan payout	10000	10000	10000	10000	10000	10000	10000	10000	10000	10000	10000	10000	120000
Payroll	6375	6375	6375	8760	11385	11385	11385	11385	11385	6375	6375	6375	103935
Taxes	1396	1396	1396	1920	2498	2498	2498	2498	2498	1396	1396	1396	22786
Insurance	1500	1500	1500	1500	1500	1500	1500	1500	1500	1500	1500	1500	18000
Electricity	600	600	600	600	800	1000	1000	1000	800	600	600	600	8800
Phone	100	100	100	100	200	300	300	300	200	100	100	100	2000
Equipment	1000	1000	1000	1000	1000	1000	1000	1000	1000	1000	1000	1000	12000
Supplies	800	800	700	700	800	700	500	600	400	800	700	400	7900
Miscellaneous	100	200	400	300	200	800	200	200	600	200	700	200	4100
	21871	21971	22071	24880	28383	29183	28383	28483	28383	21971	22371	21571	299521

Fifth-Year Financial Forecast

Account	*Jan*	*Feb*	*Mar*	*Apr*	*May*	*Jun*	*Jul*	*Aug*	*Sep*	*Oct*	*Nov*	*Dec*	*Totals*
Income:													
New members					5000	5000	5000	5000					20000
Cottages	2475	2475	2475	3915	5355	18885	15315	18885	5355	3915	2475	2475	84000
Storage	59625	59625	59625										178875
Member storage	17335	17335	17335										52005
Member nonstorage			14000										14000
Sandwich shop	100	100	100	200	500	800	1000	800	500	200	100	100	4500
Boat charters				200	500	800	1000	800	500	200			4000
School	200	200	200	100	200	200	100	100	200	200	100	100	1900
Other income	150	100	100	200	300	400	500	500	400	200	100	100	3050
Gas and ice	100	100	200	300	600	600	600	600	600	300	200	100	4300
	79985	79935	94035	4915	12455	26685	23515	26685	7555	5015	2975	2875	366630
Expenses:													
Loan payout	10000	10000	10000	10000	10000	10000	10000	10000	10000	10000	10000	10000	120000
Payroll	6475	6475	6475	8860	11485	11485	11485	11485	11485	6475	6475	6475	105135
Taxes	1416	1416	1416	1940	2518	2518	2518	2518	2518	1416	1416	1416	23026
Insurance	1500	1500	1500	1500	1500	1500	1500	1500	1500	1500	1500	1500	18000
Electricity	600	600	600	600	800	1000	1000	1000	800	600	600	600	8800
Phone	100	100	100	100	200	300	300	300	200	100	100	100	2000
Equipment	1000	1000	1000	1000	1000	1000	1000	1000	1000	1000	1000	1000	12000
Supplies	800	800	700	700	800	700	500	600	400	800	700	400	7900
Miscellaneous	100	200	400	300	200	800	200	200	600	200	700	200	4100
	21991	22091	22191	25000	28503	29303	28503	28603	28503	22091	22491	21691	300961

Assumptions:

120 weeks/312 weeks occupancy of cottages (38% occupancy)
260 slips/270 slips occupancy of storage (96% occupancy)
40 new initiated members per year (12% turnover)
70 nonstorage members per year for clubhouse privileges (21% of membership)
45 member boats in water, 56 member families per weekend (17% of membership)

JOB ANALYSIS
FIFTEEN-YEAR PROJECTION

	Full Time			*Part Time*		
	Five-Year	*Ten-Year*	*Fifteen-Year*	*Five-Year*	*Ten-Year*	*Fifteen-Year*
Club manager:	1	1	1			
Schoolteacher		1	1	2	2	2
Charter crew			1	4	5	6
Clubhouse manager:	1	1	1			
Bookkeeping			1	2	3	3
Sandwich shop			1	5	6	7
Housekeeping		1	1	5	6	6
Lifeguards				2	2	2
Club boat storage manager:	1	1	1			
Boat handlers		1	1	6	7	8
Gas and ice island			1	4	5	6
Totals	3	6	10	30	36	40

CLUB INCOME PROJECTIONS

KINGDOM YACHTS SAILING CLUB
Income Estimates

Sailing club rental hours:

Memorial Day to Labor Day 8:00–10:00 A.M. then 4:00–6:00 P.M. Friday–Friday
Labor Day to Memorial Day 8:00–10:00 A.M. then 4:00–6:00 P.M. Friday–Monday

Sailboats:

Three 16-foot sailboats/motors/trailers (1,200 pounds boats; $15,000 cost)

Weekday rates:			*Scheduling Examples:*
$ 49	8 hours	= $6.125/hr	Out by 10:00 A.M. back by 6:00 P.M. (weekdays)
$ 99	24 hours	= $4.125/hr	Out by 10:00 A.M. back by 10:00 A.M. (overnight)
$196	96 hours	= $2.041/hr	Monday 6:00 P.M. to Friday 6:00 P.M. (weekdays)
$392	168 hours	= $2.333/hr	Friday 6:00 P.M. to Friday 6:00 P.M. (7-day week)

Weekend rates:			*Scheduling Examples:*
$ 69	8 hours	= $8.625/hr	Out by 10:00 A.M. back by 6:00 P.M. (weekend day)
$119	24 hours	= $4.958/hr	Out by 10:00 A.M. back by 10:00 A.M. (weekend overnight)
$196	48 hours	= $4.083/hr	Friday 6:00 P.M. to Sunday 6:00 P.M. (long weekend)
$ 25	2 hours	= $12.500/hr	Splash and dash and pickup

Bracketed return analysis: sailboat rental income

Assumptions: 52 weeks − 8 weeks (bad weather) = 44 weeks of operation per year

3 boats @ \$392/week (lowest rate) × 44 weeks = \$51,744 best of best rental income

Worst of the worst: No rentals = \$0 income (boats sold at cost)

Best of the worst: \$51,744 × 40% = \$20,697 rental income (see financial statements)

Worst of the best: \$51,744 × 60% = \$31,046 rental income

Assumptions: Clubhouse rental using lowest rate \$392/week (\$2.333/hr) for extra income

52 weeks − 8 weeks = 44 weeks × \$392 = \$17,248

Best of the worst:

Situation 1:	\$17,248 × 40% =	\$ 6,899
	Boat rental	+ \$20,697
		\$27,596 income

Worst of the best:

Situation 2:	\$17,248 × 60% =	\$10,348
	Boat rental	+ \$31,046
		\$41,394 income

Notes:

Special weeknight clubhouse would be \$25, with \$50 cleanup deposit.

Rental boats would be sold at the end of one year at cost (\$15,000).

Trailer hitches cost about \$75 installed. The club pays \$10 every time club members rent toward installation costs, up to \$50.

\$25/hr late fee, \$25 boat cleanup charge, \$100 deposit to hold boat.

KINGDOM YACHTS SAILING CLUB
(Startup Year)

	Jan	*Feb*	*Mar*	*Apr*	*May*	*Jun*	*Jul*	*Aug*	*Sep*	*Oct*	*Nov*	*Dec*	*Total*
Income	0	0	784	3136	3528	2744	2744	2367	1960	1568	1176	784	20791
Expenses:													
Bank loans	4000	515	515	515	515	515	515	515	515	515	515	515	9665
Building payment	2500	0	0	0	0	0	0	0	0	0	0	0	2500
Utilities	200	50	50	50	50	50	50	50	50	50	50	50	750
Copying and mailing	50	100	50	50	50	100	100	50	50	50	50	50	750
Car expenses	100	100	100	100	100	100	100	100	100	100	100	100	1200
Advertising	750	100	50	50	50	100	100	50	50	50	50	50	1450
Petty cash	20	20	20	20	20	20	20	20	20	20	20	20	240
Land lease	60	60	60	60	60	60	60	60	60	60	60	60	720
Supplies	25	25	25	25	25	25	25	25	25	25	25	25	300
Boat insurance	1878	0	0	0	0	0	0	0	0	0	0	0	1878
Total expenses	9583	970	870	870	870	970	970	870	870	870	870	870	19453
Balance	−9583	−970	−86	2266	2658	1774	1774	1497	1090	698	306	−86	1338

KINGDOM YACHTS SAILING CLUB
(Second Year)

	Jan	*Feb*	*Mar*	*Apr*	*May*	*Jun*	*Jul*	*Aug*	*Sep*	*Oct*	*Nov*	*Dec*	*Total*
Income	0	0	784	3136	3528	2744	2744	2367	1960	1568	1176	784	20791
Expenses:													
Bank loans	515	515	515	515	515	515	515	515	515	515	515	515	6180
Building payment	0	0	0	0	0	0	0	0	0	0	0	0	0
Utilities	200	50	50	50	50	50	50	50	50	50	50	50	750
Copying and mailing	50	100	50	50	50	100	100	50	50	50	50	50	750
Car expenses	100	100	100	100	100	100	100	100	100	100	100	100	1200
Advertising	750	100	50	50	50	100	100	50	50	50	50	50	1450
Petty cash	20	20	20	20	20	20	20	20	20	20	20	20	240
Land lease	60	60	60	60	60	60	60	60	60	60	60	60	720
Supplies	25	25	25	25	25	25	25	25	25	25	25	25	300
Boat insurance	1878	0	0	0	0	0	0	0	0	0	0	0	1878
Total expenses	3598	970	870	870	870	970	970	870	870	870	870	870	13468
Balance	−3598	−970	−86	2266	2658	1774	1774	1497	1090	698	306	−86	7323

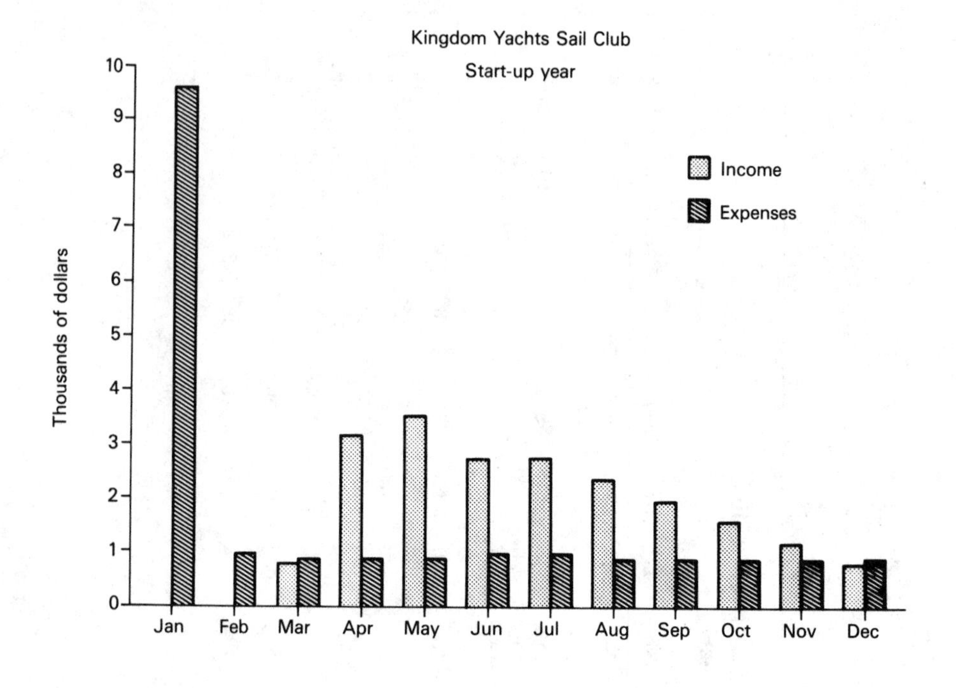
Kingdom Yachts Sail Club
Start-up year
Income
Expenses
Thousands of dollars
0
1
2
3
4
5
6
7
8
9
10
Jan
Feb
Mar
Apr
May
Jun
Jul
Aug
Sep
Oct
Nov
Dec

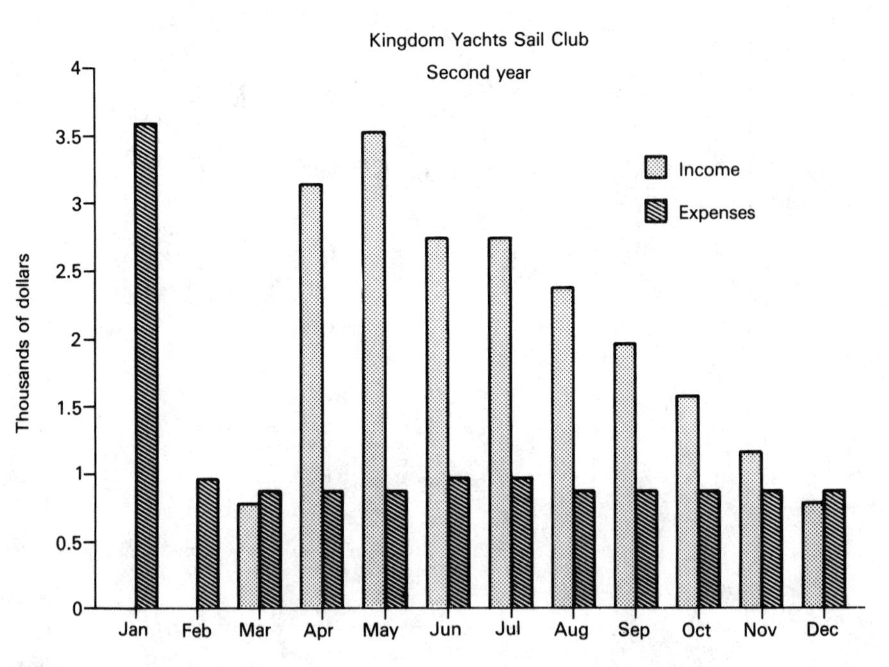
Kingdom Yachts Sail Club
Second year
Income
Expenses
Thousands of dollars
0
0.5
1
1.5
2
2.5
3
3.5
4
Jan
Feb
Mar
Apr
May
Jun
Jul
Aug
Sep
Oct
Nov
Dec

PROJECT FUNDING COSTS

Site acquisition	$440,000
Construction (see analysis)	$630,000
Professional fees:	
Architectural/engineering	$10,000
Developer/consultant	$30,000
Accounting/legal	$9,800
Closing costs	$12,500
Total costs	$1,132,300

CONSTRUCTION COSTS: ANALYSIS

	Each	*Total*
6 vacation cottages	$40,000	$240,000
1 tennis court	$10,000	$10,000
1 swimming pool	$20,000	$20,000
270 covered dry storage boat slips	$1,000	$270,000
10 acres cleared	$1,000	$10,000
Roads, storage slips, and parking areas	$20,000	$20,000
Clubhouse		$50,000
Landscaping	$10,000	$10,000
	Total	$630,000

CONSTRUCTION PERT CHARTS

M. J. Vallez Associates
Report Date 30OCT84 Run No. 12
SR0I Demo Sched Rep—Sorted by ES, TF

Primavera Project Planner
Master Control Scheduler

Kingdom Yachts Development
Start Date 15OCT84 Fin Date 1JUN85
Data Date 30OCT84 Page No. 1

Activity Number	*Orig Dur*	*Rem Dur*	*Pct*	*Code*	*Activity Description*	*Early Start*	*Early Finish*	*Late Start*	*Late Finish*	*Total Float*
1	20	0	100		Master schedule and business plan	16 OCT 84A	30 OCT 84A			
2	10	0	100		General arrangement and site plan	20 OCT 84A	28 OCT 84A			
6	20	20	0		Arch, Mech, Elec/club renovation design	22 OCT 84A	26 NOV 84		7 JAN 85	30
4	20	20	0		Grading, paving, and drainage design	22 OCT 84A	26 NOV 84		28 JAN 85	45
3	35	35	0		Obtain financing	30 OCT 84	17 DEC 84	8 NOV 84	26 DEC 84	7
5	20	20	0		Utilities design	30 OCT 84	26 NOV 84	8 JAN 85	4 FEB 85	50
12	60	60	0		Clubhouse construction	27 NOV 84	18 FEB 85	8 JAN 85	1 APR 85*	30
7	20	20	0		Clear and grub	18 DEC 84	14 JAN 85	27 DEC 84	23 JAN 85	7
15	80	80	0		Build cabins	1 JAN 85	22 APR 85	10 JAN 85	1 MAY 85*	7
9	20	20	0		Grade storage areas	1 JAN 85	28 JAN 85	29 JAN 85	25 FEB 85	20
8	40	40	0		Add boat shelters	8 JAN 85	4 MAR 85	5 FEB 85	1 APR 85*	20
11	15	15	0		Install utilities	15 JAN 85	4 FEB 85	5 FEB 85	25 FEB 85	15
10	15	15	0		Grade and surface roadways	29 JAN 85	18 FEB 85	29 APR 85	17 MAY 85	64
13	25	25	0		Construct pool and tennis courts	5 FEB 85	11 MAR 85	26 FEB 85	1 APR 85*	15
14	10	10	0		Landscaping	12 MAR 85	25 MAR 85	20 MAY 85	31 MAY 85	49

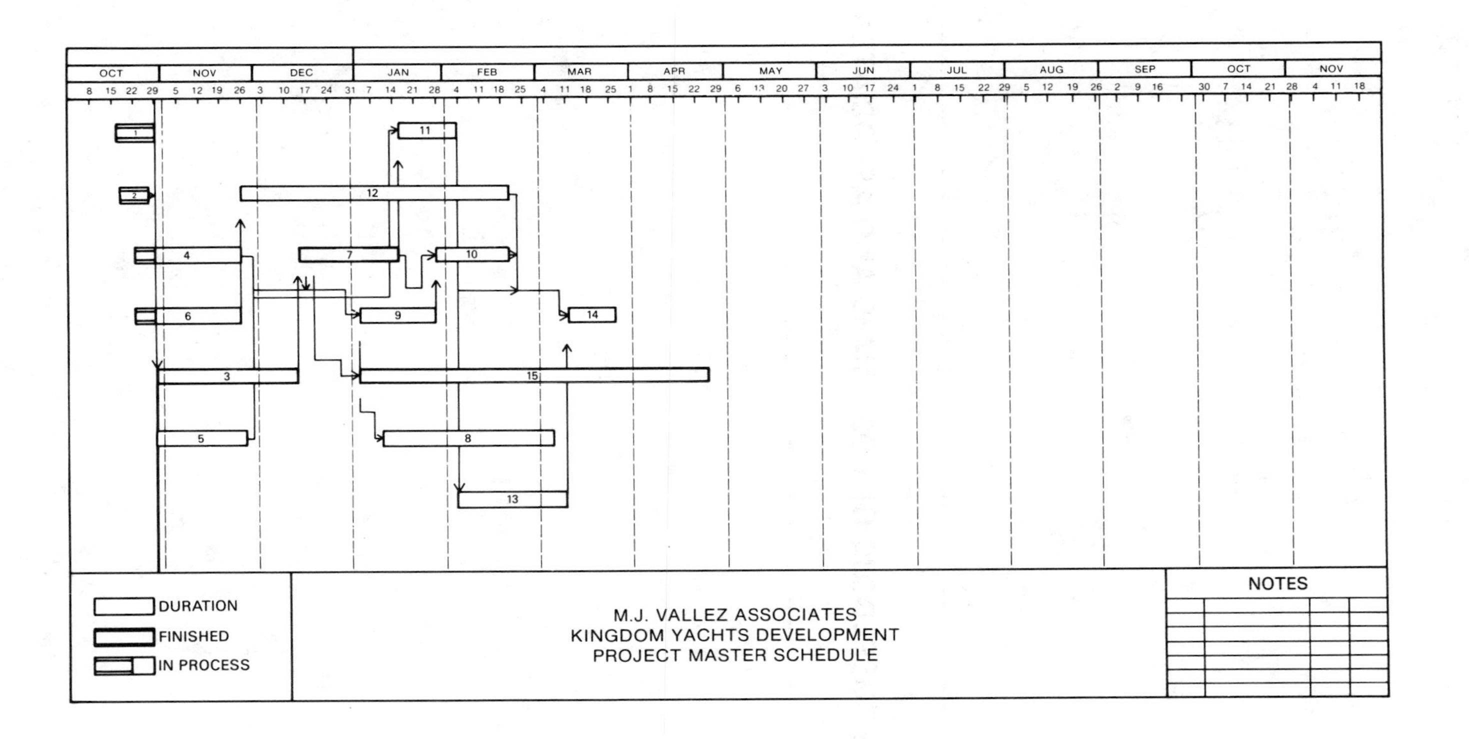
OCT
NOV
DEC
JAN
FEB
MAR
APR
MAY
JUN
JUL
AUG
SEP
OCT
NOV
DURATION
FINISHED
IN PROCESS
M.J. VALLEZ ASSOCIATES
KINGDOM YACHTS DEVELOPMENT
PROJECT MASTER SCHEDULE
NOTES

PROSPECTUS OF ROB JAMES AND ASSOCIATES

ROB JAMES & ASSOCIATES
P.O. Box 2132
Cumming, Georgia 30130
(404) 887-7966

PROSPECTUS

Purpose: To help small businesspeople plan and operate their business facilities, using our 25 years of business experience as a building block.

Education:

Georgia Institute of Technology	ENG.	1962–1965	3 years
Georgia State University	B.B.A.	1965–1968	3 years
Georgia State University	M.B.A.	1971–1974	3 years

Career:

C.B. Radio Sales and Service
Installation and repair of two-way radio systems.

Coats and Clark
Distribution center material handler.

Flight Test Center, Edwards Air Force Base
Electrical engineering research and development of high-technology aircraft.

S. A. Selby, Consulting Engineer
Drafting of mechanical systems for construction projects.

Fulton National Bank
Loan processing clerk and auditor.

American Oil Company
Marketing of oil products using retail sales outlets.

Univac Presentation Services
Computer sales proposal and presentation aid service.

U.S. Army Adjutant General Corps, First Logistics Command
Computer systems analyst for inventory and personnel files.

Sears Roebuck and Company
Production and operations management assignment areas: inventory control, merchandising, warehousing, training, personnel, security, credit, customer service, cashier, and store management. Computerized many of these areas.

Rob James and Associates, Facility Planners
Focusing on small businesses either starting or expanding in the marketplace. Client list on request.

PART TWO
LEGAL ISSUES

7
Real Estate Legal Concerns

The need for law stems from the need to create order and predictability of behavior among the members of some particular group. Laws are rules, set forth for specific reasons, among which are consistency of behavior, protection of individuals, establishment and maintenance of an orderly flow of activity, increased economic (and by extension personal) efficiency, and social stability (which is not the same thing as order).

Indeed, laws are contracts made among members of a group, under which those members mutually agree to give up some personal freedom for the sake of receiving the benefits that an orderly, efficient, and stable society can offer. The alternative to this, anarchy (the absence of law), would be viable if it were not for the fact that most people are not willing to voluntarily act in every case for the mutual benefit of all concerned. We restrict ourselves in various ways in our mutual dealings to ensure a happier, more productive, and safer existence. Herein lies the theoretical principle of law.

Because of the complex nature of our society, the legal process is quite complicated. Some laws are found to be restrictive. Some, though legally applicable, do not usefully apply in every case. Times change, and with the changing times comes the need for changes in laws as well. Because of these problems, our legal system includes not only rules (laws) but also court systems for administering and interpreting (judging) those laws, and a means by which we may change the laws to reflect changes in conditions under which we live (legislative bodies).

In commerce, there is a need for a particular type of law that can ensure the benefits of an orderly society. Many of these laws affect the process of securing and developing facilities. For this reason, it is important to investigate the chief legal concerns that one might encounter in the process of facilities development. We will do so by first discussing the legal environment of real estate in this chapter, and then by involving ourselves with the process of governmental regulation of the private sector in Chapter 8.

I. LAW SOURCES

A. Common Law

Common law, also known as English common law, is a body of laws developed over a long period of time beginning in the Middle Ages, consisting of court decisions based on custom and reason as applied to everyday disputes and events. These "customary" decisions were supplemented and in some cases altered by the small body of officially enacted laws that were decreed by feudalistic governmental units in the country. The principles of common law still stand as the basis of the legal system in the United States, as well as most present and traditional members of the British Commonwealth, all of whom share a common cultural ancestry. It acts as a framework for the more formal body of laws under which we are governed. Common law is contrasted with other elements in the English system of laws, including statute law, that body of law consisting of officially enacted statutes (as noted above) and equity law, as developed separately by the courts of equity. Equity law resulted from the differences between common law and common sense in some cases that pertained to property and rights thereto, and was designed to correct these differences. It dealt with problems such as monopolies, which are quite contrary to the principles of common law, yet certain licensed monopolies are allowed in cases where equity can be logically achieved only through the allowance of monopolies as in, for instance, the case for patents to inventions and the like.

In addition, the English system of laws, including common law, is in contrast with other law systems extant in the world, including the Roman law, used in most Mediterranean countries and South America, and law based on the Napoleonic Code, used in France and elsewhere. One of the main differences between law as practiced in the United States and other systems used elsewhere is the prime consideration that the burden of proof lies on the shoulders of the accuser, not that of the accused. This presumption of innocence is a major underpinning of the judicial system developed out of the English approach to law.

B. Statutory Law

Statutory law differs from common law in that, rather than being based on tradition and precedence, it is created through the legislative process and extensions thereof. Statutory laws are designed to cover areas that have no founding in tradition due to the more recent nature of the material covered. That is, statutes need to be supplied when events arise for which there is no precedence and therefore no tradition upon which the courts can call. The Constitution of the United States and constitutions of individual states are examples of statutory law, as is legislation passed by the Congress and state legislatures. The essential element here is that the laws are specific and written, as opposed to general and unwritten, as is the case with precendents set by court decisions and other forms of common law.

Among the statutory laws are those dealing with the conducting of business, both within a state and among states, and much effort has been given to the process of unifying these bodies of business law into a single unit. The effort has led to what we call the Uniform Commerical Code (UCC), which is accepted in all states with the exception of Louisiana (apparently clinging tenaciously to its roots in the Code of Napoleon, upon which so much of its commercial laws are based) and which regulates consistency from one location to another. Without this consistency, it would be difficult to know when one is or is not in violation of laws governing commercial endeavors, to the detriment of the continuation of commerce. Since the Uniform Commercial Code deals with such things as commercial paper and secured transactions, which are necessary for interstate business transactions, it is important in real estate and other activities connected with facilities planning and development.

II. GOVERNMENT RIGHTS

Under the Constitution of the United States, certain rights were guaranteed to the government to facilitate its operation and ensure its ability to effectively deal with the needs of the country as a whole. So cautious were the founding fathers of their individual freedom that these rights were specified so as to segregate them from all other activities, and reserved for administration by the individual states. Among the rights reserved for the federal government to administer under the Constitution are the power of taxation, the right to eminent domain, and police power.

A. Taxing Power

Under the Constitution, the government has the right to impose and collect taxes as part of the process of providing for the common good of the people. Likewise, state and municipal legislative bodies normally provide for taxation under the same principle. Limits are placed on the powers of the governmental bodies to impose taxes, as would be expected considering our social and political history, but the power is present, and rightfully so.

The collection and spending of taxes on public works, public programs, and operation of governmental bodies represents nothing more than a means of funding collective actions undertaken by the society as a whole, through those governing bodies. The principle of property taxes derives from the relative value of property and is considered appropriate under all of the theories of taxation, which are the ability to pay (resulting in progressive taxation systems), benefits received (where it is assumed that the person or institution which holds properties has more to lose without the benefit of government and therefore should be taxed on that property), or the principle of proportionality (where it is believed that taxes should be proportional to the relative wealth of the individual). Under any of these theories, landowners are taxed on their real property and on the basis of income received from that property. This has obvious consequences for the firm involved in an expansion of its real property holdings.

B. Eminent Domain

The principle of eminent domain enunciates the legal right of government to seize or authorize the seizure of private property for public use, providing that "just compensation" be provided to the private owner of that property. This right has to do with the rights of the entire society as opposed to individual rights and is a basic principle of law. Under eminent domain, for instance, local governments can condemn property that is required for public works, such as roads or municipal facilities, providing it can show that such seizure is necessary and proper. Without this stipulation in constitutional law, it would be nearly impossible to economically widen roads, control flooding in urban areas, and solve other community problems.

C. Police Power

Police owner is guaranteed to government under constitutional law as a means of enforcing the laws that the governmental body legislates. Without the right to institute police forces and to use them to protect citizens, enforce laws, and provide for order, society would become chaotic.

D. Escheat

The concept of escheat provides for the reversion of ownership of property to state government when there is no legal heir. Basically, this is part of the process of maintaining order. A distinct element of common law as well as capitalistic principle upholds the right of individuals to own private property and use it in whatever legal manner they may choose. Only under very specific circumstances can that right be impinged upon, one being in the case of eminent domain and another in the case of escheat. In both of these cases, the purpose is to maintain order and arrange for a distribution of property for the benefit of the society. In the case of eminent domain, as was cited above, the benefit to society derives from the joint use of property for public works. In the second case, that of escheat, the benefit to society is the means by which private ownership of the property can be reassigned in an orderly manner. Property without ownership under the law reverts to the government so that it may be reassigned through an auctioning or bidding process, or held by the state for the common use of the people, either as public land or as part of some public project.

Keeping in mind these rights of government under the Constitution and constitutional law, we now turn to some specific legalities that deal with real estate and its sale and purchase.

III. FREEHOLD ESTATES

By definition, freehold estates involve ownership interests in real property, whether land or other physical property, that will last for an indefinite period of time. Ownership interests (estates) of this sort are determined by a combination of common law elements dating back to feudal times and beyond, coupled with present statutory laws created for the purpose of establishing rights under conditions more recently encountered. Among the freehold estates issues are the concepts of fee simple estates, life estates, and qualified fee estates. In real estate dealings involving the disposition of land to be used for facilities, it is necessary to become aware of these various types of ownership interest. It should be noted that freehold estates may be either inheritable or noninheritable, and that on rare occasions this can create problems in land transfer.

A. Fee Simple

The term *fee simple* is a hangover from feudal times, when the transfer of land involved a straight exchange for a fee. Basically, the term still retains this meaning, though we recognize three possible forms of this type of ownership. The first of these three, the fee simple absolute estate, is the only complete form of personal land ownership. It is in this form that the owner of the land has the absolute right to transfer the ownership of the land or real property involved. A purchase of property that is held and transferred as a fee simple absolute estate is the "cleanest" and simplest of land transfers.

B. Life Estate

Another form of freehold estate is the life estate, in which the owner of the land retains ownership of the property only for the term of his or her lifetime. Under the provisions of this type of ownership, interests in the property are considered freehold estates for purposes of legality, but the designation may not be passed on to an heir because all interests are terminated at the death of the life tenant, or present owner. Another hangover from feudal times, the original purpose of this type of estate was to provide a tenant with land and real property without allowing for the passage of that real estate to heirs, as with a lord who wished to reward some subordinate, but did not wish to provide for inherited ownership of the property to following generations.

C. Qualified Estate

Still another possibility of freehold estate is the qualified estate, which in itself is of three types, fee simple determinable, fee simple on a condition subsequent, and fee simple subject to an executory limitation. Without going into a great deal of detail, a qualified estate is a fee simple estate whose status is determined by certain restrictions or qualifications. The three subdivisions mentioned above refer to the type of conditions to which the ownership of the property is subject.

1. Fee simple determinable. This arrangement can do such things as grant ownership upon the condition that the land be used for some particular purpose, or not used for some particular purpose, ownership reverting if the limitations are not followed.

2. Fee simple on a condition subsequent. In this case, the grantor (seller) reserves the right to reclaim ownership as specified in the contract. Unlike the case of determinable fee simple, the conditional does not automatically reverse ownership upon the occurrence of some specific event or events, but rather gives the grantor the right to take back the property as specified by contract. An example would be the reserved right of a grantor to enter the land in order to establish a right of way for a road to other property owned by that grantor.

3. Fee simple subject to an executory limitation. Here, a fee simple ownership can be terminated upon the occurrence or nonoccurrence of some event, as it can with other qualified estates, but in this case, termination results in the property's passing on to ownership by some specified third party rather than reverting to the original grantor. An example of this would be land granted to a business organization as long as it is used, for instance, for tree farming. When that purpose is no longer the use of the land, the land reverts to the state, or to a park, or to some conservation group.

All of these forms of land ownership complicate the process of securing land for facilities expansion and similar projects. It is important to understand the ramifications of such purchase agreements and the effects these arrangements may have on the future of the property. Temporary facilities that are designed to fulfill the firm's needs for a specific project that may last five to 10 years could logically be negotiated at a lower cost with one of the more restricted transfer covenants, much to everyone's benefit. As with most real estate law, the purpose of this myriad of possible ownership schemes is to fit action to need, which varies over time.

IV. STATUTORY ESTATES

Statutory estate property is property that is owned, as is indicated by its name, through certain statutory provisions not found in normal common law provisions. Under statutory law, estates can be created and maintained by virtue of condition or action that would not otherwise exist except for circumstance. Two examples of statutory estate are the cases of community property and homestead protection. As will be seen, these are estates that exist without benefit of real estate contract or the granting of freehold rights.

A. Community Property

The principle of community property allows for the joint (concurrent) ownership of property by husband and wife simply by virtue of their state of legal marriage. Under the provisions of community property laws, arrangement is made for the division of all property acquired during marriage on a fifty-fifty basis. Eight states hold the rule of community property to be law.

B. Homestead Protection

The right to homestead was first guaranteed under federal law in 1862. Under its provisions, anyone has the right to occupy and apply for ownership of public land in lots of 160 acres, providing that the land be occupied and cultivated for a five-year period. Since the first instigation of this law, a number of statutes have been added, including one provision that forbids the seizure of land in payment of debts incurred by the individual landholder prior to the time of the homestead patent. Although nearly one in three homesteaders failed to meet the five-year requirement, nearly 250,000,000 acres of public land have become privately owned through this procedure.

From the point of view of the individual firm securing land from private citizens, homestead ownership is no different from any other form. It is not an option for the firm to attempt homesteading, however, as the practice is limited to individual citizens for personal land procurement only.

V. LEASEHOLD ESTATES

A leasehold estate is exactly what the term implies, that is, a nonfreehold estate existing by virtue of a lease arrangement between the owner of the property (lessor) and the individual or firm seeking to occupy and have use of the property (lessee). We have already discussed leased property in various contexts. Here we deal with the legalities of the topic, particularly the main types of leasehold estates: estate for years, periodic estate, estate at will, and tenancy at sufferance.

A. Estate for Years

Under the provisions of an estate for years, property is conveyed to the leaseholder for a specific number of years, called the term of the lease. The term may be for any period of time from one month up to 999 years, though those in excess of 99 years are considered to be fee simple freeholds for all intents and purposes. At the end of the term of an estate for years, ownership and occupancy of the property revert to the lessor.

A firm developing a facility often has opportunities to secure the use of land on a rental basis using the estate for years approach. Under these provisions the firm is assured of a leasing period long enough to make investing in the development of the property worthwhile, yet the lessor is allowed to retain outright ownership of the property and is provided an income by the lessee for the occupancy of that property. This is the basic form of agreement encountered in long-term lease arrangements.

B. Periodic Estate

A periodic estate is similar to the estate for years, except that the period of the lease is unspecified. In this case, rather than a definite period of occupancy, the lease renews itself at the end of each periodic lease term automatically, unless one or the other of the parties entering into the agreement chooses to terminate the arrangement through procedures spelled out in the lease agreement.

In cases where the value of the property may be volatile or where the anticipated length of time for which the property is required by the lessee is uncertain, this type of lease offers a higher degree of freedom for both the lessor and lessee. Such lease arrangements often renew automatically on a month-to-month basis, which can be an advantageous arrangement for a firm. It may be dealing with test projects that may either result in long-term contracts for products produced or fall through almost immediately; demographics may be uncertain and the market location sensitive, or facilities may be required for a short period of time, as when businesses serve seasonal markets with their products.

C. Estate at Will

Also known as tenancy at will, this lease arrangement permits termination of the lease "at will" by either party, at any time and without cause, rather than under conditions specified in the lease agreement. Firms do not often encounter this type of arrangement and are not likely to seek it, due to the volatility and risk it adds to doing business. Apartment tenants may encounter this form, but they are usually protected by tenancy laws that require the landlord operating under a condition of estate at will to provide sufficient notice before terminating the lease, usually a period of 30 days.

D. Tenancy at Sufferance

This type of lease arrangement provides the least amount of protection under the law. Tenancy at sufferance occurs when an existing lease expires and the occupant continues to occupy the property without benefit of lease agreement. Tenants are technically allowed to occupy at the sufferance of the landlord, who has the right at any time prior to the signing of new lease agreements to terminate their occupancy. The purpose of tenancy at sufferance is to provide for a continuation of occupancy of a property while new lease arrangements are being negotiated, or to allow the parties to continue the lease arrangements by informal agreement until such time as formal arrangements can be made. As with tenancy at will, this is an agreement form usually encountered in the lease or rental of domiciles rather than commercial property, and the law normally provides for the protection of the lessee by requiring the lessor to provide 30 days' notice. It should be noted, however, that the tenancy at sufferance could occur in a business setting if a commercial lease expires before formal agreement on new lease arrangements have been made, thus avoiding encumbrances on either the lessor or lessee through a legal quirk requiring vacancy of the property.

VI. ENCUMBRANCES

Included in this term are any restrictions that may exist on a property and its ownership that would limit that ownership in some way. The only ownership considered free of encumbrances is the fee simple absolute form of freehold. The importance of encumbrances to the purchase and sale of property varies, but in general it should be obvious that each encumbrance will in some way reduce the desirability of the property as an investment, as it restricts use of that property by the purchaser. Depending on circumstances, a purchaser may find the encumbrances too restricting and simply lose interest in the property. As an example, a piece of property on which the mineral rights are reserved by the seller may be of little consequence to a purchaser who intends to erect a vacation cabin, whereas the same encumbrance would be crippling to a firm whose purpose is to exploit the natural resources of the property, or to sell displaced material from the construction process as a means of defraying costs. In this section, we consider some particularly worrisome encumbrances to the commercial purchaser.

A. Easements

Easements are nothing more than the right of a person to use property owned by another. Various types of easements exist, but the two main categories that concern us here are easement appurtenant (the right of some landholder to use part of some other person's land), and easement in gross (the right of a person, whether an owner of real estate or not, to use or have access to some part of another person's land). Easements can exist by expressed writing, as in a contract, by implication, as in general practice and custom, or by prescription. According to prescriptive easement, the easement exists under the law by virtue of the nonowner's using the property or a part of the property in question, for some specified period of time, serving to establish the right and necessity of that use by the nonowner. It is similar to property ownership by adverse possession, by which an individual may occupy and pay taxes on property owned by another and, if the owner fails to object, after a specified period of time, the "squatter" can apply for permanent ownership. In the case of prescriptive easement, however, ownership is not in question, merely the right to use of the land.

Easements are logical and valuable rights that serve to create equity in society that would otherwise not be present. As an example, the right to cross someone else's land in order to obtain access to one's own property is a common easement. Without it, a person could sell a parcel of land surrounded by his or her own land, and the new owner of the individual parcel would have no legal way to access the property. An easement is a simple way to allow access across the seller's property.

When purchasing and developing property for the purpose of facility development, easements can create undue hazards, limiting the choice of locations for structures and use of the land by the firm. Easements may render property useless due to restrictions; they are major encumbrances that must be investigated before securing land.

B. Deed Restrictions

This is a general category of encumbrances referring to circumstances under which the use of land is limited, those limitations being specified in the deed of sale. Property deeds normally contain guarantees and warranties, which may vary from contract to contract, but which include such items as a guarantee that the deed is free of outside encumbrances, that the purchaser has free and peaceful use of the land, and so forth. However, the deed may also include any restrictive covenants that are legal and agreed to by both parties. A seller may restrict the use of land in a deed, for instance, specifying that the land only be used for residential purposes, or that any improvements to the property, such as construction of outbuildings, will follow the same architectural theme of the present property or neighborhood.

Some restrictions may not legally be included in deeds, however, such as a restriction of resale specifying the new purchaser's race, color, religion, or national origin as a precondition to the sale of the property. Such covenant clauses are considered discriminatory under the law and therefore do not produce a legal contract.

C. Encroachments

An encroachment is an intrusion on one's property, or trespassing. Whereas the possibility of legal encroachment on one's commercial property is not a major issue, the purchase of land on which there is already an illegal encroachment may lead to difficulties, particularly in the case of adverse possession. Under conditions of this sort, ownership of property may actually transfer to another person without purchase, providing that individual occupies and fulfills all of the requirements for a petition of ownership under adverse possession laws. Specifically, in order to establish title under adverse possession, the party occupying the land must do it in such a way that is actual and exclusive; that is, the possession truly takes place and the petitioner is the only occupant of the land at the time. Occupation must also be open and notorious, meaning that the possession is obvious and in no way hidden from the owner or public, hostile, meaning possession is without prior consent or agreement with the owner, and continuous, indicating that the individual occupying the land does so without a break in that occupancy. It must also be done under a claim of right for a period specified by law; that is, the occupancy is reported, the person petitions for the right to title, and the petition is in good standing for a period of time legally determined to allow for sufficient time of rebuttal on the part of the present owner.

When commercial property is purchased from absentee owners who have nothing more than a casual or commercial interest in the land and have not kept up with the property's status, it is sometimes possible to make such a purchase only to find that the title to the land is not free and clear as expected but encumbered by an encroachment.

D. Air Rights

The purchase of air rights gives the purchaser the right to use the space above the land in question and nothing more. Under a condition of purchase of air rights, ownership of the land itself, all mineral deposits, and the surface of the land itself are retained by the seller. Whereas this may appear to be a useless purchase, it is often used in commercial property where, for instance, a building is constructed on a piece of property and the building itself is owned by one group of investors while the land upon which it sits is owned by another. In one rather unusual case, an entire complex of commercial buildings was constructed over a section of railroad yards in an urban area and built on pylons to allow for the continued use of the surface for tracks and switching equipment by the railroad, which retained ownership of the land involved. Such arrangements are often beneficial to all parties involved and offer an economical way to purchase a highly desirable location for a business concern without the high cost of purchasing the land itself. Obviously, such cases produce unusual encumbrances on all concerned.

E. Liens

A lien is a legal claim against a piece of property that has been used as collateral for a debt. When there is a lien on a piece of property, title is not free and clear to that property, as the person holding the lien can demand sale or possession of the property in payment for the debt should the borrower default on that debt. Purchasing a piece of property with such a lien extant upon it could result in either loss or further encumbrance on the property should a failure to pay the debt by the seller occur. One of the main purposes of a title search, under which the title to land is researched to discover all encumbrances, is to determine if there are any unknown existing liens against that property that would inhibit transfer of ownership. Under most lien agreements, the borrower's efforts to dispose of property put up as security are restricted, and new arrangements must be made among the lien holder, the seller, and the purchaser of the real estate before the sale can be legally finalized.

F. Subsurface Rights

Sometimes referred to as mineral rights, this type of encumbrance restricts the owner of real estate in the disposal of mineral deposits or use of subsurface space connected with the property. Sellers sometimes either retain or separately sell mineral rights to a property. Such restrictions can have serious effects on the use of the land if the holder of the subsurface rights desires, for instance, to mine the property or minerals, or construct tunnels for drainage of adjoining land. The purchaser who plans to establish a business facility must be aware of these restrictions and what effect they may have on use of the land for commercial purposes.

G. Leases

An existing lease can be an encumbrance, particularly if the purchaser desires to change the present use of the property in question. Depending on the nature of the lease covenant and purchase convenants, lessees may have the right to maintain their occupancy of the property for the duration of the lease, or even to extend the lease, if that is part of the several agreements involved. In such a case, an otherwise promising property may be rendered useless for development by the purchaser. This is a particularly knotty problem in the case of an estate for years.

VII. FORMS OF OWNERSHIP

Concurrent estates are those estates in which the interest in ownership rests with two or more people at the same time. There are a number of different types of concurrent ownership of property, the chief ones being joint tenancy, tenancy by the entirety, tenancies in common, undivided tenancy (community property), and estate in severalty.

A. Joint Tenancy

In the case of joint tenancy, the right of survivorship is extant. Through survivorship, upon the death of any owner, the surviving co-owners have the right to equally divide the share of ownership of the deceased owner among themselves. Such an arrangement allows for a smooth transition of ownership without involvement of heirs or outside interests.

B. Tenancy by the Entirety

A special case of joint tenancy, entirety is an agreement between spouses allowing for the right of survivorship, thus facilitating the legal transfer of property upon the death of a spouse and avoiding any disagreement as to inheritance of the property. Rather than being a matter of inheritance, tenancy by the entirety allows a direct and automatic transfer of property to the surviving spouse.

C. Tenancy in Common

This represents a concurrent ownership in a property that is undivided and proportional. That is, under the tenancy in common approach, if three children inherit the property of a parent who dies without will, it will be held in common, each of them having one-third interest in that property. If one of the three should sell half of their share in the property to an outside party, the property would be divided proportionally as one-third, one-third, one-sixth, and one-sixth among the four owners, respectively. At the present time, where ownership of land is uncertain, court systems appear to prefer this type of arrangement for concurrent tenancy.

D. Undivided Tenancy

In addition to these common forms of concurrent ownership is the concept of undivided tenancy, under which each concurrent owner has unrestricted rights to the use of the entire property in question. In addition to unrestricted use of the property, this type of tenancy includes survivorship, thus making it popular among married couples. Also a popular point is the fact that each owner is equally responsible for any liability incurred on the property.

E. Estate in Severalty

Though the term severalty seems to imply several if not many owners, it refers in reality to a single ownership without the involvement of others. The root of the term is the idea of a "severed" ownership, one in which ties with others have been broken and an individual holds all rights to a property exclusively. In cases of estate in severalty, an individual, whether single or married, enjoys all benefits of ownership to the property in question, but upon marriage may transfer those rights to form a concurrent ownership with his or her spouse. Estates in severalty, it should be noted, do not automatically convert to a concurrent estate upon marriage, and must normally be transferred by agreement.

SUMMARY

In conclusion, we remember that the laws involving real estate come from two main sources, common law, which takes a traditional view of ownership rights and forms, and statutory law, which is designed to deal with legal issues that did not arise in the traditional setting but have become important in the present era. By their nature they appear to be confusing, extensive, and detailed, with a wide number of possibilities for ownership and use of land available. Yet it must be remembered that their purpose is to create an orderly format for determination and transfer of ownership of real property in the society, and as such, they perform a vital function. Capitalism itself is based on the proposition that individuals hold the means of production, including land, and that they have the right to do with that property as they wish. This principle of private property gives us the flexibility to shift the use of resources as needed with a minimum of resistance from more traditionally oriented elements in the society. Clear right to use is absolutely imperative in such an economic environment, and the system of real estate laws has served to define those rights quite well.

For the firm in the process of developing new facilities, a wide range of choices are available to accommodate individual needs, and the laws governing those choices serve to ensure fairness and equity in the transfer and use of property for commercial purposes. The object is to provide a maximum of efficiency with a minimum of uncertainty. This is the reason that the body of law appears as it does, with all its intricacies and suspected redundancies. Like any other social institution, it will last only as long as it is sufficient to achieve the purposes for which it was designed. So far, it has achieved those purposes.

QUESTIONS

1. Give three examples of common law. How do you suppose these agreed-upon rules came into being?
2. Define *statutory law*. How does it differ from common law in its intent and purpose?
3. It has been argued that the concept of eminent domain is valid on the basis of the concept of the greatest good for the greatest number. Is this a valid argument? Why or why not?
4. Why is there a need for laws governing the purchase and sale of real estate and real property?
5. According to Adam Smith, capitalism works because of the free use of private property. Does this contradict the use of law in real estate?
6. Is the concept of private property in capitalism at odds with the concept of eminent domain?
7. What is escheat, and what is its impact on commercial real estate?
8. Name six forms of ownership.
9. How does eminent domain differ from police power under the law?
10. What are some of the ways in which real estate law can affect a firm's selection of a new facility site?
11. What is meant by freehold estate? Is its origin in common or statutory law?
12. Firms seldom operate under tenancy at sufferance. Why? When would it be to a firm's advantage to do so?
13. How might encroachment affect a firm, both as a property owner and as one guilty of encroachment?
14. Is homestead protection available to firms under the law? Why or why not?
15. In a large southern city, a municipal building is built over land belonging to a major railroad. The railroad's tracks pass beneath the building, which stands on reinforced pillars. What type of property ownership is probably held by the municipality?
16. A landowner purchases property and builds a house on that property. At a later date, valuable mineral deposits are discovered on the property. Under what conditions does the landowner have to allow mining on that property by other parties?
17. What is meant by the term *fee simple*? How does it differ from a life estate?
18. How do concurrent and joint ownership differ? How are they similar?

8
Regulatory Laws

As with real estate law, the purpose of industrial regulation is to attain and maintain a degree of equitableness in the development of business properties, to avoid conflict with other private and public use of land, and at the same time ensure that the rights of the producers of goods and services are protected so that they may carry out their business activities efficiently and safely. That is a tall order for any system, and in light of the degree of complexity of modern society, inevitable disagreement as to the effectiveness of regulatory legislation abounds.

Regulation is the result of two phenomena in modern commerce, one the advent and development of the modern economic system since the industrial revolution, and the other the extremely rapid rate at which events take place in the modern world.

To begin with, the Industrial Revolution served to alter the basic fabric of western capitalism by introducing the advantages of economies of scale inherent in industrial production. Prior to the advent of the steam engine and its offering of greatly increased power to do work, the western world was chiefly agricultural, its industrial component consisting of numerous small mills and production facilities, operated by water and wind power, and having little effect on the overall condition of economic life. Adam Smith, working in this preindustrialized world, created a concept of the ideal economy which he called capitalism. Under capitalism, many buyers and many sellers compete in the marketplace for resources and for sale of their goods and services to others. No one firm is dominant in the market; the system is self-regulating, self-governing, and always moving inexorably toward the best use of resources and most efficient production of goods and services. It was, in fact, a relatively close approximation of what was really going on. Then came the Industrial Revolution, heralded by the development of steam power, to join with the already extensive engineering technology of gears and cams associated with the more traditional waterwheel-operated mill. This new power source yielded such an increase in efficiency that soon factories and manufacturing facilities were springing up everywhere.

It was under this set of circumstances that a new phenomenon became apparent; that is, it seemed that by having larger production units and producing larger numbers of goods, efficiency of production could be dramatically raised. This has already been discussed as economies of scale, which became one of the chief components of the industrialization of the modern world. Couple with this the twentyfold increase in the productivity of workers, which, not surprisingly, led to an increase in factory employment and crowding of industrial centers (that is, urban areas), and the traditional version of capitalism was gone forever.

Smith said in his treatise on capitalism, *The Wealth of Nations*, that government should not involve itself in the affairs of commerce except to (1) protect property rights through courts of law and law enforcement and (2) produce certain "public works" which are necessary to the well-being of the entire society, but which cannot be produced at a profit in the private sector. He specifically decried governmental interference. But this was prior to the Industrial Revolution. In a traditionally capitalistic environment, no element of the economy is considered large enough or important enough to seriously influence the overall economic conditions of the economy as a whole. Not so with a reduction in the number of productive units and an increase in their size and power engendered in the industrialization process after 1800. Suddenly it was not only possible but relatively easy for individual productive units to influence the activities of the economy. And this is exactly what happened in the case of the trusts, from the Rockefeller interests in the oil industry to the steel magnates, like Andrew Carnegie. Suddenly, individual companies and their owners were controlling the behavior of the market, and the consumer was no longer in charge. Consumer sovereignty was a thing of the past and free enterprise in jeopardy. For these reasons, regulation of the private sector through the use of laws, executive decrees, and action by state and federally created agencies developed and continues to grow today. This development of powerful productive units is the first great cause for the necessity of regulation.

The second cause, the speed at which change takes place, is similarly the result of industrialization. We are now living in a postindustrial economic environment where services and information are the primary products, and the speed at which changes take place in those markets is greater than ever. As Buckminster Fuller has pointed out, the advent of the computer happened just in the nick of time to save us from destroying our own system by its own complexity. It is only through the use of this tool that we have been able, in any way, to keep up with and control the incredible rate of change that permeates the very fabric of our commercial system.

With this rapid rate of change superimposed over the topology of an incredibly complex and active economic structure, the system begins to break down in its ability to correct and adjust for changes as they occur. Lag time becomes more and more critical in the face of shorter and shorter cycles of change (or cycles of "commercial mutation," if you will), and the need for some outside agency to impress controls on the system is intensified.

Fortunately, such an external system was already in place as the need for it arose, since it had been functioning to control the inequities caused by industrial concentration that had been created by the industrialization process itself. The system is, of course, the regulatory mechanisms of federal, state, and local governments. These two causal sources—the increasing benefits of industrialization to create efficiency through economies of scale and the increasing rate at which change occurs—are responsible for the regulatory process now in use. This regulatory process has to be reckoned with in any attempt to expand facilities.

I. REGULATIONS

There are regulations governing the types of activities that a company will face in the process of facility expansion at virtually every level of societal organization. Each has its own unique raison d'etre, yet each has the same purpose as well, that is, to protect the rights of both the individuals and the companies affected by the process.

A. Federal

Federal regulatory activities usually center around protection of the entire population and the environmental assets of the country itself. Federal agencies and legislation are designed to regulate activities that affect commerce in general, particularly as it applies to interstate commerce. Beyond the specifically business-oriented regulations that may affect expansion of facilities, there are those dealing with the rights of individuals, such as civil rights legislation, and those affecting the environment, such as environmental protection and land use control regulation.

B. State

State regulatory efforts are similar to those of federal agencies, except that they deal with conditions within the state rather than the nation as a whole. The types of regulations and subjects covered can be expected to parallel those of the federal government for the most part, although special-interest groups and lobbies from influential regional and local industries or organizations can be expected to modify the type of laws one encounters. Protective regulatory pronouncements can be expected, sometimes to the detriment of the firm wishing to locate in areas where their particular type of industry is not traditional. For the most part, however, state regulatory agencies and efforts are simply designed to protect the state and ensure fairness in the operation of business in accordance with the views of the individual state.

C. County

At the county level, regulation tends to be directed toward local conditions and local interests, sometimes protective of county business interests, but seldom extensively so. Limitations on industry are normally designed to avoid interference with the already present economic structure, as in the case where industry is limited geographically to minimize interference with, say, farming or local commercial markets. Regulations may also be designed to encourage the development of industry of certain types or in certain locations within the county to increase the overall economic viability of the region. It should be noted that if there is a major metropolitan area within or adjacent to the county under consideration, regulatory legislation will probably be basically either encouraging or discouraging of development, depending on the attitude of the population. If the area is not highly commercialized, it may be due to the desires of the county population to maintain a suburban atmosphere adjacent to the more heavily industrialized urban center. Industrial location could be encouraged if the local population wishes economic expansion to come their way as an extension of the adjacent urban center. In general, however, regulation tends to be less severe in unincorporated, nonurban areas.

D. City and Municipality

The concerns of urban areas are different from those of a state or the federal government, in that the main considerations center in the best way to develop the whole urban area, and the most effective way to coordinate industrial, commercial, retail, and residential interests in a way that yields the best overall result for the good of the entire community. Thus there will probably be regulations involving zoning, taxation, use of utilities, size of facilities, use of local contractors, impact on sewage and waste disposal, and overall environmental impact to deal with. This is not to say that cities are against expansion into their confines. Quite the contrary; they wish development to take place, but they wish it to happen in a way that conforms to the terms set forth in the overall development strategy constructed by the local government. In addition, the limitations in large metropolitan areas are usually different (and more detailed) from those in smaller communities, and this too should be taken into account in seeking relocation. The primary difference between operating within a city and a true municipality is that the latter will often involve a plethora of regulating agencies and governmental bodies, often including city, state, and county regulations simultaneously. Metropolitan municipalities tend to spread out over larger areas and therefore entail more than one governmental body, understandable when we take into consideration expanded local markets, suburbs, and other closely affiliated urban areas.

E. Land Covenants

Lastly, there are regulations under the law not imposed by governmental bodies. These constitute the general class of agreements known as land covenants, representing contractual restrictions imposed in the sale of land at the time of that sale. Such restrictive covenants, as we discussed in Chapter 7, can limit the use of land or tie landowners to agreement with other surrounding landowners before proceeding with new or different uses. Under such circumstances, individual purchasers may find that a promising piece of property is in fact not suitable for their use, either because of direct land covenants or by implication through a reduction in flexibility for future changes in the use of that land. Although it may be possible to erect the desired facility at the present time, a firm may find restrictions imposed on it in the future when further expansion or alternative use of the land is contemplated.

II. FEDERAL AGENCIES

Federal regulation stems from two prime sources, the first, independent and duly constructed agencies, and the other, agencies created through the executive branch, generally for specific purposes. The rise of agencies as a means of regulation stems from the pattern of an increased rate of change within the society as a whole and within the business community in particular. As the speed of change increased, it became obvious that the slow-moving legislative process by which laws were enacted was not sufficient to keep up with the changes that were occurring. The solution was to create agencies with the power to promulgate rules, regulations, and standards of behavior, thus bypassing the need for a full legislative procedure. Since the agencies were created by the legislative branch, they hold legal status and can perform their functions relatively unimpeded.

Though some agencies have limited power, most regulatory agencies may carry out the following activities: to legislate by promulgation of regulations, standards, and rules; to enforce these laws by virtue of the powers vested in them by the legislative branch; to interpret (or adjudicate events surrounding the instigation of) the regulations, standards, and rules they have created through hearings and investigations; to investigate and, where appropriate, prosecute those found in violation of agency-produced laws; and to act as advisors to industry in how best to comply with the regulations of the agency.

A. Independent Agencies

Agencies have been a method of meeting the need for rapidly constituted law since 1887, when the Interstate Commerce Commission (ICC) was created. From the beginning, the powers given to agencies were widespread, as exemplified by the fact that not only was the ICC given the right to deal with cases of "noncompetitive practices in interstate commerce," but it was also given the right to define what constituted noncompetitive practices in interstate commerce. That gave the agency extraordinary powers to react to change in the system. A full list of independent agencies is given in Table 8.1, but among the agencies that particularly impact on the expansion of a firm's facilities are the Interstate Commerce Commission, the Federal Trade Commission, the U.S. Tariff Commission, the National Mediation Board, the Federal Power Commission, the Securities and Exchange Commission, the National Labor Relations Board, the Equal Opportunity Employment Commission, the Environmental Protection Agency, and the Federal Energy Administration. Each of these agencies has rules and regulations that can impact on the development of new sites or redevelopment of old ones, and this list should in no way be interpreted as exhaustive. Depending on the individual activities of a firm, others may play a key role in planning decisions as well.

B. Executive Branch Agencies

These agencies are in most ways similar to independent agencies, the primary difference being that they are created by executive proclamation rather than by legislative action. An executive branch agency is the result of decree by the President of the United States that creates the agency and determines its scope of operation. Since 1824, the executive branch has been a prime source of agency creation. It is, in fact, the more traditional approach, having been used for more than sixty years prior to the 1887 creation of the Interstate Commerce Commission. Such varied governmental functions as the U.S. Patent Office and the Social Security System were initiated

TABLE 8.1 Regulatory Agencies of the U.S. Government

Independent Agencies	
1887	Interstate Commerce Commission
1913	Federal Reserve Board
1914	Federal Trade Commission
1916	U.S. Tariff Commission
1924	U.S. Tax Court
1926	National Mediation Board
1930	Federal Power Commission
1932	Federal Home Loan Bank Board
1933	Federal Deposit Insurance Corporation
1934	Securities and Exchange Commission
1935	National Labor Relations Board
1936	Federal Maritime Commission
1938	Civil Aeronautics Board
1946	Atomic Energy Commission
1952	Federal Coal Mine Safety Board
1964	Equal Employment Opportunity Commission
1970	Environmental Protection Agency
1974	Federal Energy Administration
Executive Branch Agencies	
1824	Army Corps of Engineers
1836	Patent Office
1862	Internal Revenue Service
1863	Comptroller of the Currency
1903	Antitrust Division, Justice Department
1915	U.S. Coast Guard
1916	Packers and Stockyards Administration
1922	Commodity Exchange Authority
1931	Food and Drug Administration
1933	Bureau of Employment Security
1933	Commodity Credit Corporation
1933	Social Security Administration
1935	Rural Electrification Administration
1936	Maritime Administration
1951	Renegotiation Board
1953	Small Business Administration
1958	Federal Aviation Administration
1959	Oil Import Administration
1963	Labor Management Services Administration
1964	Office of Economic Opportunity
1966	Federal Highway Administration
1966	Federal Railroad Administration
1971	Occupational Safety and Health Administration
1972	Consumer Products Safety Commission

by this method of agency creation. A complete list of executive branch agencies is offered in Table 8.1.

Among the executive agencies that may be of concern to the firm attempting expansion of facilities are the Occupational Safety and Health Administration (OSHA), the Antitrust Division of the Department of Justice, the Army Corps of Engineers, the Rural Electrification Administration, the Small Business Administration, the Office of Economic Opportunity, and the Federal Highway Commission. Again, the list is suggestive rather than exhaustive.

Certain areas of agency regulation are so important that it behooves us to be more specific, since they will invariably affect changes in facilities when they occur. Firms already in operation are probably familiar with the aspects of these specific laws that directly impact them, yet when new facilities are created or old ones renovated or expanded, the changes can create new concerns. Specifically, the two areas to deal with include environmental concerns and social concerns, particularly equal opportunity issues.

1. Environmental concerns. Any change in a community, whether urban or rural, will have an impact on the environment and so on the ecological balance of the area. Changes to the ecology are of chief concern in land use and are controlled by agencies and specific laws designed to minimize the negative effects of industrial expansion and commercialization of the land. Of particular note are the following pieces of legislation.

a. Water Resources Planning Act of 1965. This law, along with the Water Quality Improvement Act of 1970 and the Clean Water Restoration Act of 1966, moved to set standards for acceptable levels of both air and water pollution, and required states to develop plans for controlling these problems. Such plans and standards may limit the availability of sites if the process to be carried out will produce either type of pollution, or if it will involve the cleaning of waste prior to its release into the ecology.

b. Solid Waste Disposal Act of 1965. This law moved to develop standards to control the disposal of solid waste, such as scrap metal, garbage, and trash.

c. Air Quality Act of 1967. Amended by various actions of the Environmental Protection Agency (EPA), this act was designed to establish and implement programs dealing with the problem of air pollution.

d. Noise Control Act of 1972. This act requires noise level standards to be developed for both products and equipment. Such standards could seriously affect the allowable proximity of industrial operations to residential areas.

e. Environmental Protection Agency (EPA). It is important to be familiar with the structure of the Environmental Protection Agency (EPA) because of its impact on commerce. With the founding of the EPA in 1970, more than a dozen separate government ecological programs were combined into a single agency. The EPA deals with all forms of pollution and hazards to the environment, including not only solid, air, and water waste pollution, but also radiation, noise, and such specifics as pesticide pollution control. The agency has four main responsibilities, including setting standards, enforcing regulations, conducting research, and offering financial and technical assistance to state and local environmental control efforts. Working closely with local and state governments, this agency operates to coordinate the entire range of programs dedicated to development and maintenance of an acceptable environment.

2. Social concerns. The increasing awareness of conditions for specific minorities has resulted in legislation designed to reduce or remove unfair discrimination among employees and contractors because of some minority status, such as race or sex. A number of laws have been enacted to further this cause; the following are among the ones of greatest importance to the firm in the process of expansion.

a. Civil Rights Act of 1964. This law prohibits discrimination pertaining to wage and job opportunity on the basis of race, color, sex, and national origin. This is the primary civil rights legislation of the twentieth century.

b. Equal Pay Act of 1963. This law prohibits wage discrimination on the basis of sex and gives the Wage and Hour Division of the Department of Labor the right to enforce this act. As of 1979, enforcement of this law was transferred to the EEOC.

c. Equal Employment Opportunity Act of 1972. This act guarantees equal rights for employment from all firms engaged in interstate commerce having at least fifteen employees. It impacts particularly in areas where there is a high percentage of minority labor, or in some cases, in areas where there is a shortage of qualified minority workers available to a company attempting to create overall parity in employment practices.

III. STATE AGENCIES

State activities parallel federal activities in the realm of regulation except that the emphasis is more locally and regionally oriented. Two areas that should be noted are those of master land use plans and state environmental agencies.

A. Master Land Use Plan

This is a comprehensive plan within a state or community designed to predetermine directions in which the area's growth will be allowed to move.

The master land use plan is based on an initial survey, which involves both a physical and geographical study of the region's assets and an economic use survey indicating how best to support the continued commercial growth and the "best use" scenarios that will accomplish these ends.

From the initial survey, a general plan is constructed that dictates what types of development will be allowed and what types disallowed. By using a planning commission to plan ahead, the state or local agency is able to ensure a cohesive, all-encompassing progression for the region in which growth in any one area is not allowed to outdistance growth in other areas. Through planning, housing can keep up with the rise in jobs from the growth of business concerns, sufficient retail and wholesale businesses can be provided to satisfy the needs of the increased population that accompanies economic growth, and rural and urban development can be coordinated to maintain proper balance for the overall welfare of the region. Such master plans can either encourage or frustrate companies in their attempts to locate on a given site. However, by working with the planning commissions, the firm can contribute to its own prosperity and that of the citizens of a state or region, and can also ensure continued long-run continuity by maintaining ratios of commercial and noncommercial land use and by increasing the flexibility of future resource use. The object is to find an area in which the needs and desires of the community match those of the firm. Only in this way may a mutually productive situation be created.

B. Environmental Agencies

Like federal agencies, states have environmental agencies charged with seeing to it that the regional environment is protected and that the ecology of the state is not seriously disturbed by commercial activities. The first line of defense for the ecology is the state agency, and if its efforts fail, the EPA steps in to deal with the problem. In actuality, the scope of operation and types of problems dealt with by state agencies are the same as those of federal agencies; they merely operate at a state rather than nationwide level. There are, however, several forms of state and local regulation that deal with land use which need be explored more closely. These include the various elements of the zoning process under which allowable utilizations of land are defined and controlled. This subject is discussed in detail in the next section.

IV. LAND USE

A. Zoning Laws

These laws represent the most common of land use control mechanisms at the local level. Under zoning laws, land is divided into zones designating legal uses of the land, regulating both the type of use allowable and the intensity of use allowable for the land within a given zone. There are three main classifications of zones: commercial, industrial, and residential. These zones are usually further subdivided into lesser classifications, such as single-family dwellings or multifamily dwellings (for residential zones), light industry or heavy industry (for industrial zones), and highway commercial district or light neighborhood shopping district (for commercial zones). In addition, some special-purpose zones are normally allowed for, such as agricultural, public use, or floodplain designations. When intensity of use, often referred to as density of use, is designated, such restrictions as the allowable height of buildings or number of buildings, houses, or other structures allowed per block or square mile are specified.

It should be remembered that the purpose of zoning is to maximally benefit the population of the region, to control land use as any other element of cooperative coexistence, and to ensure that the rights of the individual do not interfere with the rights of the community. The precedents for this type of control can be found as far back as feudal Europe, when it was quite common to restrict the types of activities allowable within and without the walls of cities and towns. This is a police function in principle. Enforcement of zoning ordinances and regulations is possible as a result of the requirement that an individual or corporation must obtain a number of documents in order to build or develop a piece of property. Such licenses and permits give the community the means to control land use by either issuing or denying these documents. In order to receive the permits, the landowner must be found to be in compliance with local ordinances. A landowner who builds or develops the property without the proper permits can be forced to tear down any structures and cease any activities not in compliance, under penalty of law. It is, however, at the discretion of the regulating body whether so severe a penalty is imposed. In any case, this is not the type of risk that a firm would consider taking, even ignoring the ethical question involved, for fear of losing its building. Furthermore, it is usually impossible to obtain financing without proper permits.

B. Zoning Variances

Zoning regulations, like other reasonable forms of legal restriction, are not without some capacity to be altered. The zoning variance gives the opportunity to deviate or vary from the zoning restrictions when compliance would create an undue hardship on the landowner. Such a case would exist where, because of the peculiar nature of a piece of property, slightly different specifications for buildings on their positioning would be necessary to efficiently utilize that property. In no case, however, should a variance serve to change the basic character or nature of a zone or neighborhood within that zone. In addition to the traditional variance, other alteration processes are possible, having the same effect as a variance, though more permanently changing the character of a zone. These include the amendment, which serves to change the definition of a zone, as in the case of an agricultural area's being changed to both agricultural and residential, which is useful as municipalities or suburban areas begin to spread out. Also included is the conditional permit, which allows specific use of land that is not within the present zone definition as long as that nonconformist use remains within the bounds set forth in the conditional permit. Another is spot zoning, which allows for the rezoning of a small parcel of land within a larger zone for the convenience of the population of that zone, as in the case of rezoning to allow for a laundry or small market within the neighborhood. These additional possibilities allow for the freedom of action necessary to appropriately use the land within the confines of the legally designated and constituted jurisdiction of the regulating commission. For the developer of facilities, this freedom increases the availability of feasible sites and allows for more flexibility in their planning as well.

C. Public Hearings

Whereas the use of a variance is an example of administrative relief under which the governing committee merely agrees to waive normal regulations for a specific case, the changing of a zoning designation through spot zoning, rezoning, or amendment requires an opportunity for public sentiment, of both those living in the affected area and those affected by changes in that area, to be heard. This normally means public advertisement of the proposed change, public hearings where concerned citizens and interest groups can present their case to the zoning board, and establishment of a reasonable and real justification for the zoning variance to be instituted in the first place. At these hearings, anyone who wishes to express a view on the proposal is allowed to do so, either pro or con, and the commission takes such testimony into account when rendering its decision on the motion of variance or rezoning.

D. Legal Action

Beyond the administrative and legislative forms of relief from zoning restrictions, there is the possibility of legal action, or judicial relief. This is available to the property owner if that individual is not in agreement with the decisions of the zoning commission involved. The individual landowner has the right of judicial appeal in the courts. In such a case, the property owner must prove that either the actions of the zoning commission were not constitutional, or that the decision of the commission was capricious, arbitrary, or unreasonable, thereby denying the owner right to use of the land without due process of law. Although this form of relief is provided by law and available at the discretion of the property owner, courts seldom overrule the decision of zoning committees unless the unfairness of that decision is obvious and grievous. The judicial body is simply reluctant to put aside the decisions of the legislative body in such cases.

V. COPING WITH GOVERNMENT REGULATIONS

The purpose of the government regulation process is to protect the rights and the property of the society as a whole. Unfortunately, this protection may cause difficulties for individuals within that society as personal or organizational goals come in conflict with the welfare of the particular community. Even so, it is necessary to operate within the confines of the law if one is to reasonably expect general freedom of action and safety in personal and business activities. One must cope with governmental policy and operate within the confines of the system in order to ensure such benefits. In planning a facility, a number of issues must be considered during the decision-making process that stem from the necessity for conformity and relate to the restrictive nature of some governmental regulations. Among these issues are the cost of compliance, the capital gains and losses resulting from necessary compliance, and the time delays that compliance will incur. It may be that if any one of these issues is excessive or that if the combination of them is excessive, a project that originally looked quite favorable may in fact be economically infeasible.

A. Cost of Compliance

Realizing that it is necessary to comply with agency regulations, the firm must make provision for the capital needed to do so. How much will it cost the firm to comply with government regulations? How much will it add to the cost of a project to, for instance, add scrubbers to smokestacks or filters to dumping bins to clean ejected waste flowing into local streams? How much will the cost of landscaping to soundproof surrounding neighborhoods from the noise of industrial production add to the overall cost of operations? These are the types of questions that must be answered in deciding how to go about the planning process. In the least costly scenario, additional cost of compliance may result in a specific site's being too expensive because of the additional costs inherent in rendering it tenable. In the worst case, governmental compliance may make expansion itself too expensive an alternative for the firm to consider. In any site selection process and any development of a plan for production, the cost of compliance with government regulation must be dealt with. It is a very real cost, and one that is not always immediately apparent. OSHA rules and regulations, for example, are massive and detailed. It is not always easy to determine on first inspection what types of actions are allowable and what types are not. In any event, the regulatory process will add to the cost of doing business. However, before the firm judges the regulatory process to be hostile or burdensome, it should be remembered that the purpose of regulation is to protect, and that that protection extends to the firm itself as well as to the environment within which the firm operates. In order to succeed, the firm must do its utmost in meeting the needs of that environment, through providing better products, more efficient production, lower prices, wider variety, and a more stable and equitable use of resources. Regulation may be cumbersome at times, yet it is necessary, in light of the complexity of the society within which we live.

B. Capital Gains and Losses

A more long-range problem for the firm is that of capital gains and losses from compliance with regulations. This is a double-edged sword, in that both compliance and noncompliance can create gain or loss. In the long run, a facility wil prove to be either economical or infeasible on the basis of its useful life. If a facility is not tenable for a long enough period of time, the investment will not be recovered sufficiently to warrant its development. If it has an extremely long life and the market is considered secure, then recovery of capital may become reasonable and profits thereafter will more than warrant the development of the site. Regulation will affect this process in two ways. On the one hand, compliance may be so expensive as to add excessively to the cost of developing the site so that recovery time creates a net loss of capital and therefore negates the benefits of the project. On the other hand, failure to comply not only could result in lawsuits, injunctions, and fines, but also could actually reduce the value of the property and equipment because of their negative environmental impact. In addition, the efficiency of the workforce may be reduced by noncompliance, which may result in, for example, increased cleanup costs and downtime on equipment, with noncompliance eventually leading to the closing of the facility. Capital gains must be sufficient to convince management that a project is worthwhile, yet the long-term versus short-term question of capital gains is often too heavily weighed in favor of the short-term, quick-fix attitude. One side says, "How can we afford to wait that long for recovery of capital just to keep the environment intact?" while the other side says, "How can we not?" Firms walk a constant tightrope between the desires of the stockholders to see higher and higher profits and the desires of the community to maintain a balance of development and protection of the environment. It is not an easy act. The firm must decide what a sufficient capital gains requirement will be and in the process, balance one element against the other. Regulation serves to ensure this process.

C. Time Delays

One other problem related to compliance with regulatory restrictions deals with the time delay encountered as a result of the additional efforts required by that compliance. Regulations may require inspection, additional construction, alteration of the basic site use plans, or other changes to the firm's expansion plans that are costly in terms of time delays. Schedules may become unreasonable, imposing unforeseen delays that result in unplanned and excessive costs to the firm. Such time delays may in themselves render the original plan untenable if they stretch development schedules beyond reasonable limits. To avoid this, it is usually best to choose excess over conservation in the planning stage and spend as much time as may be required to determine exactly what type of regulation compliance will be required when the project gets underway. The purpose of planning is, after all, to minimize future surprises, and this is one area of planning where it can be very costly to be wrong.

VI. ALTERNATIVES

Given the pitfalls of the regulatory process, how can the firm protect itself from the cost and legal vulnerability resulting from these restrictions? A number of strategies exist that can minimize the negative effects of the regulatory process. Among them are on-site expansion, multiple-division siting, land banking, site survey libraries, and parallel research and engineering projects. Each will be discussed below.

A. More On-Site Expansion

One of the advantages of a present site is that the facility is probably already in compliance with the regulations governing the firm's particular type of operation. Any regulations not covered by existing facilities but applying to new facilities can be readily determined. Thus an on-site expansion offers the opportunity to avoid many of the pitfalls inherent in new site locations. From the point of view of the regulation issue, the use of present sites to expand facilities or modernize them represents a minimum of hazard. It is for this reason that many companies opt for this alternative first, considering others only if present sites are not viable alternatives due to other reasons. In any case, on-site expansion should be considered for reasons of regulatory ease.

B. Multiple Divisions Sharing Site

Another possibility that minimizes the risks inherent in regulation is to allow more than one division to share a given site. Again, this is due to the greater ease with which an existing site can comply with regulations. Since the regulation process will occur with any new site, it is simpler and more efficient to go through that process, whether it be zoning or environmental regulation, as seldom as possible. If more than one division occupies the same site, the regulatory procedures can be handled for all of those operations simultaneously. If one division is already extant on the property in question, the addition of a new division will require only minimal effort in seeking regulatory approval as with the case of any other one-site expansion. This represents a more economical solution to the regulatory problem than would be encountered with separate sites for each division.

C. Land Banking

The process of land banking is a matter of saving land for future use; land is, so to speak, put in the bank. Companies that bank land buy parcels of property that have been researched and found to be suitable for their use, though not yet required for present production. If a firm finds five parcels of land suitable for a specific project, it may choose to purchase two or more parcels, keeping the unused properties in reserve for future expansion. Then, when a division of the firm is looking for an expansion site, that division is asked to first consider the sites in the land bank, which the firm already owns and has already researched, ensuring that regulatory compliance will not be a significant problem. Land banking is growing in popularity, particularly as the number of viable sites for industrial expansion drops in the face of increased urbanization and restriction.

D. Site Survey Library

Similar in philosophy to the land bank concept, the site survey library results from a company's instigating investigations to find suitable sites over a wide range of locations. Whereas this land is not purchased and kept in reserve, the information is kept in a site survey data bank that serves to minimize the need for additional research in order to find a suitable site. All important factors being determined for a large number of sites at one time reduces the cost of the research itself, and at the time of need, a much shorter time is required to determine what available sites are suitable to the purposes of the firm. Once this library, or data base, is established, it is rather simple and relatively inexpensive to maintain.

E. Parallel Search and Engineering Projects

This is another method of reducing cost and combating the time factor in the development of a new facility. With the parallel search and engineering approach, the traditional serial method of facilities development is bypassed. Historically, the development of a new facility proceeds in a serial fashion, from site determination to engineering to site development to construction to beginning of operation. With the parallel approach, both engineering and search projects proceed simultaneously, so that when the site is found, most of the engineering process is already complete and actual site development can begin without delay. The dangers inherent in this approach stem from matching the engineering design with the final site, and it can impose some restrictions on what will characterize an acceptable site. This approach does reduce flexibility in site selection, yet the cost advantages inherent in already having the engineering finished at time of site purchase more than compensate for those restrictions. On the engineering side, the engineers, not knowing the exact nature of the site, may be forced to "overengineer" the plant design to take into account as many contingencies as possible. However, although parallel projects are somewhat more costly and may require last-minute adjustments to fit plans to site exactly, the overall cost savings are more than sufficient to warrant the use of this approach.

QUESTIONS

1. In the development of facilities, government regulations are necessary for the protection of both producers and consumers in the marketplace. Why?
2. What are the sources of regulatory law?
3. What avenues are available to the public to bring about changes in regulatory laws?
4. Name five levels of government involved in issuing regulations concerning real estate and industrial development.
5. Name five executive branch agencies with the power to regulate industry.
6. Name five independent agencies of the federal government with the power to regulate industry.
7. Discuss an alternative to government regulation.
8. What are the primary differences between independent and executive-branch agencies?
9. What is a master land use plan? How does it serve to regulate industry?
10. Why, in your opinion, is the Environmental Protection Agency necessary? Whose interests does this agency protect?
11. Why is government regulation costly to the firm?
12. Define *zoning variance* and explain its use.
13. What remedies are available under the law when a firm is discovered to be in violation of regulatory law?
14. How does on-site expansion offer an alternative to dealing with excessive regulation by government agencies?
15. Imagine an environment in which there is no governmental regulation of industry. Describe what such a business environment would look like.
16. What is the role of OSHA (Occupational Safety and Health Administration) in regulating industry? Is it more of a hindrance or a benefit? What made its conception necessary?
17. What is a site survey library? How is the concept related to regulation?
18. Define *land banking*. What types of firms would be most likely to involve themselves in this strategy?

9
Insurance Concerns

In any human enterprise, there is risk. The very condition of being alive is inherently risky, and we accept risk every day of our lives. Risks stem from the fact that the systems within which we operate are so complicated and so convoluted in their structure and operation that it is virtually impossible to be aware of all the consequences of a given act. Unforeseen circumstances permeate our lives, and there is no logical way to penetrate the system's structure to eliminate uncertainty. Yet few of us allow this lack of foreknowledge to paralyze us in our day-to-day lives. Indeed, if there were total determination of the consequences of actions, this would be a very dull world.

In addition, most unforeseen consequences are minor in nature and negligible in their effects on our lives. We tend to ignore them and correct our actions accordingly when they do arise. Other consequences are of greater magnitude and represent the possibility of serious harm in the event of their happening. It is these unforeseen circumstances with which we tend to most concern ourselves.

The need to defend against risk stems from three central concerns: the probability of negative occurrences, the extent of the impact of their effect (how many individuals would be affected), and the magnitude of effect (how costly an event would be if it did arise). Primarily, we weigh the cost and probability of occurrence against the cost of defending ourselves against the event in deciding the means and extent of our precautionary actions.

Insurance represents a form of defense against the unforeseen that allows us to ensure that if some event comes about, we will be protected from loss due to that occurrence. How much we are willing to pay for such assurance of protection from harm is dependent on the degree to which we would suffer and the chances of the event's occurring, that is, how costly the event would be and how fearful we are that it will actually come to pass.

I. OCCURRENCES

Many kinds of hazards may be encountered during the process of developing a new facility or redeveloping an existing one. From the point of view of insurance, the way to protect against these hazards in the planning stage consists of first determining what the hazards are and then determining the degree of risk involved with each. Common mishaps are easily identified through experience and through analysis of the operations connected with the development of the facility in question.

There are three primary categories of occurrences with which to deal: those that are virtually certain to occur, those that are probable, and those that are possible though of relatively low probability. By analyzing what will go wrong and what could go wrong, the planners can begin to protect themselves against these contingencies.

Briefly, the occurrences which we refer to as certain are those that happen whenever a project of this type is undertaken. They are expected and allowed for in the planning stage so that they may be properly handled when they do occur. In the case of probable events, we find a group of events that will not always occur, yet they must be protected against because they can occur, and their occurrence has a serious if not catastrophic effect on the project if they should occur. These events can be insured against, and insurance limits the risk of damage from their occurrence. The third category, uncertain events, represents serious events that have a low level of probability yet would be disastrous if they were to occur. Although such events will most likely never happen, the firm may choose to insure against them because of the serious consequences. In order to understand this process better, each of these categories is dealt with separately below.

A. Certain

By virtue of the nature of development projects, some occurrences are considered virtually unavoidable. Although we would prefer not to have such negative occurrences, the fact that they are expected allows us to prepare for them in advance and guarantee with a reasonable degree of surety that they can be handled when they do occur. Among these types of occurrences are construction delays, construction changes, weather delays, moving expenses, loss of income due to delays, increased ongoing expenses connected with new facilities, and other factors. All of these can be anticipated and handled in advance of occurrence, since the fact that they will take place is a foregone conclusion in the development process.

These occurrences are not insurable, since they are certainties. The principle of insurance is to guard against mishaps that may or may not occur. The insuring agency is playing the odds by "betting" that the event will not occur, while the insured "hedges its bets" by paying a premium for the right to relief should the questionable event occur. With certainties, there is no possibility of beating the odds for the insurer, and insurers are therefore not willing to undertake the risk. This should not be confused with the situation in which one buys life insurance. In this case, the eventual death of the insured is understood to be a certainty. It it not the death that is being insured against but rather the premature death, that is, an early death that would cause hardship to those left behind. What the insuring body is betting in this instance is that the insured will live long enough to pay substantial premiums before he or she dies, thus creating the profit for the insurer. Although the same rationale could be applied to events that are bound to occur with facilities development, the premium needed to offset the losses would be so high due to the short time frame that cost would be prohibitive. For this reason, it is more economical for the firm to work to minimize the effects of the events and in the process to minimize costs connected with them.

1. Construction delays. Construction projects are very complicated affairs. A large number of elements are simultaneously occurring, and all have to fit together perfectly if the original plan is to be carried out. Unfortunately, this is not always possible. As Murphy's law states, if anything can go wrong, it will. Although this rather bleak view of the world is an exaggeration, it does point out a truism: the more complicated the system with which one works, the greater the opportunity for things to go wrong, and therefore, the higher probability that all pertinent events will not occur as expected. Things do go wrong, and in the process of construction, many result in delays. Too many elements are dependent upon the occurrence of other events to avoid construction delays. As we will see later, the entire point of Program Evaluation and Review Technique (PERT) and Critical Path Method (CPM) planning is to try to determine where delays will take place and how much time we have to adjust for them. Construction delays are simply a part of the process. It is a very rare case indeed when they do not occur. See the Kingdom Harbor Yacht Club case study in Chapter 6 for a PERT chart example.

2. Construction changes. Like delays, construction changes stem from the inability of the planners, no matter how good they are, to foresee all the elements that will in some way alter a construction project. There are too many changes in the environment within which they are working to account for all contingencies, so the construction process is fine tuned as it develops. A new method or material may suddently become available. An unknown factor may pop up, such as an unanticipated ground fault forcing an adjustment in the foundation, or the market may force the company to add new manufacturing steps that require rearrangement of the floor plan. A municipality may suddenly change its ordinances regarding allowable building dimensions, or an engineer may suddenly realize a more efficient way to perform some major function that was not apparent until after construction was underway. Numerous changes in the construction plan will take place. Insurance from loss in these cases is a matter of flexibility in the way the construction effort is planned and carried out.

3. Weather delays. As the saying goes, everyone talks about it but no one seems to do anything about it. Weather is capricious, and its capriciousness often seems to follow Murphy's law to the letter. One of the inevitabilities encountered in construction is that weather will periodically throw off schedules. Lead time must be embodied in the construction plans to allow for this contingency. It is normal, for instance, to begin construction projects, especially lengthy ones, during a period of warmth and, it is hoped, dry weather. Excavation and pouring of foundations require that the ground be soft enough to work (not frozen, or buried in snow) and that there be enough dry weather to allow for concrete and mortar to cure before the rains begin. In addition, those functions that need to be carried out inside can be accomplished while there is inclement weather, whereas the construction of the building's shell is mostly outdoor work. Yet we cannot guarantee nature's cooperation in our efforts. Rain or an unusually late spring can delay foundation work for long periods of time, whether it renders the ground too hard to work or creates a sea of mud that is unnavigable by heavy equipment. Very humid weather can affect curing time, and excessive heat can reduce the efficiency of the labor force, as well as forcing concerete to cure too quickly, requiring that it be kept wet lest it crack. These weather delays must be endured, as they are quite unavoidable.

4. Labor force changes. Like other factors in industry, changes in the labor force occur. Seasonal availability of labor for construction often varies considerably as, for instance, teenagers and college students return to school in the fall, or a major market opens up in some other part of the state or country, and workers move to it. In addition, the general availability of workers increases or decreases with shifts in the economy. As we approach a condition of full employment, workers become scarce, wage rates rise, and competition for workers increases. It should also be noted that since so many projects begin in the spring or require a large portion of the outside work to be done as soon as the weather turns warm, the shortages in workers tends to occur when a firm can least afford them. Again, the key is planning for and anticipating needs, allowing for the securing of a labor force to match those needs in advance of anticipated start dates.

5. Moving expenses. Moving to a new location will automatically result in moving expenses, including expenses for takedown, transportation, and setup of existing equipment, shipping of records, furniture, and office equipment, and expenses connected with moving key personnel. This includes the cost of moving households, living expenses during relocation, expenses connected with assisting the key personnel in selling old homes and procuring new ones, and a wide range of expenses connected with the general resettling process. It is not unusual, for instance, for key executives to incur rather substantial expenses in reestablishing themselves in a different community, which may include such costs as initiation fees at the local country club (a business expense connected with entertaining clients, officials, and so on) and the expense of joining local service and business groups.

6. Income losses. Also inherent in the shift to a new facility is the loss in income that comes from delays in production of goods for sale, delays in deliveries, and general downtime during the move process. Every day that the company must hold off on selling and production efforts results in loss of income. This loss must be anticipated and should be considered in the planning process. The timing of a move should be chosen to result in a minimum of lost income. Firms that close down at certain times of the year often choose those periods of time for the move process. Other companies may choose to move divisions one at a time, incurring expenses connected with the temporary maintenance of two separate facilities but reducing losses in income in the process. The key here is usually to accomplish the move as quickly as possible and with a minimum of adjustments to the schedule.

7. Expenses increased. The increase in general expenses incurred with a move to a new facility stems from two main sources. First, the equipment with which one has been operating has expended some of its useful life and therefore represents a low level of investment in the old plant, whereas the new facility will usually require, at least partially, new equipment. This new equipment will require adjustment, and personnel will need to gain experience in its use, which creates added expenses in training, initial inefficiency, and probably some turnover in personnel. Second, probably the more obvious source of expense is the added cost of purchasing the plant and equipment, servicing the debt connected with those purchases or suffering the reduction in income from money dedicated to the new facility, and a general increase in outlay as a result of expansion. Much of this increase in expense may be temporary; it is unavoidable and should be expected.

8. Transportation problems. In any move, large amounts of equipment and personnel are being shifted from one location to another and, as with any other extensive project, delays will occur. Unanticipated needs are certain to show themselves, schedules will be affected by outside influences, and poor judgment on the part of either the firm or the cartage contractor can result in the wrong equipment showing up in the wrong place at the wrong time. Such delays are expected and can be minimized by anticipating their occurrence, planning for them, and adding the necessary flexibility into the development process in order to handle them. Contingency plans should be developed and personnel made aware of them. In this way, when transportation problems do arise, the firm can handle them with a minimum of expense and time loss.

9. New equipment costs. The cost of new equipment will inevitably arise in the process of creating a new facility. Even if major productive elements are to be moved into the new facility from the old, often it is more economical to buy new equipment and install it in the new location rather than to move existing equipment from the old. The savings inherent in using the old equipment will often be more than offset by the greater efficiency of the new. This equipment, whether major productive elements or incidentals such as office equipment, employee amenities, and support facilities, represents a substantial outlay. It represents an unavoidable increase in cost to the firm that is expected to pay for itself in time. Such expenses need to be anticipated as well as the major production expenses.

B. Probable (Insurable)

A second type of event with which the firm must deal is that which involves occurrences with a definable probability of occurrence. Although these events are not certainties, they occur often enough to have an expectation of occurrence; that is, they are determined to occur often enough to be a possible serious problem, and how often they do occur is determinable. Since the events are determinate as to probability and their effect in monetary terms can be defined, insurers can statistically assign a risk to these events and calculate how much they would have to charge for insurance against the events to obtain a suitable profit. In other words, these are events that are insurable. They represent the type of event that insurance companies are willing to bet will not occur often enough to cause them a loss at the premium they are asking in return for the protection they are willing to offer. From the point of view of the insured, these events happen often enough to be of concern, yet not often enough to be certainties that can be planned against by normal means. The firm sees an opportunity to hedge against their occurrence by purchasing insurance protection high enough to meet its needs yet not so high that it becomes cheaper to absorb the cost of the event, should it occur. A number of these events or risks are noted below.

1. Employee accidents. Accidents involving employees occur. They are not inevitable and they are avoidable, yet they occur. The firm realizing that fact can insure itself against substantial loss in the event of employee accident by providing for the occurrence and then hoping that it never really happens. The production process is an inherently complex one and thus filled with unknown factors. These unknowns represent risks, in this case of the safety of workers, and the effect of these unknowns needs to be minimized. The only difference here is that the particular class of events connected with these unknowns are not certainties, but they do have a determinate probability of occurrence, so we protect against them in various ways.

There are a number of programs for employee safety that are designed to ensure against employee accident, such as company inspections, seminars to educate employees in safety, training in the use of equipment, and the use of rules, regulations, and safety equipment while on the job. Such safeguards are designed to reduce the probability of an accident's occurring and can be undertaken either by the company itself, through the use of a consultant firm, or under the auspices of the insuring body acting to minimize its own exposure. However they are done, such programs reduce premiums for insurance by reducing the risk that the insurance company is forced to take.

Beyond that, there is the general insurance against employee mishap represented by workman's compensation insurance, a form of insurance designed to compensate workers who are injured on the job, and required by virtually every state in the union. The risk to the firm is twofold: that is, the risk of loss due to the worker's medical needs and lost wages, and the risk inherent in the loss of the use of the worker and his or her expertise. Most accidents that occur on the job are minor and can be handled without much difficulty, yet some are catastrophic, both to the worker and to the employer. Protection against this sort of risk is necessary.

2. Customer accidents. Accidents involving customers can happen, although they are mercifully rare in most cases. Accidents resulting from the use of products or from events occurring at the firm's location can result in liability claims and high medical and compensation costs to the customer. Manufacturers are responsible for the performance of their products and, no matter how thoroughly the proper use of goods is indicated, occasionally someone will improperly use the item or some unforeseen circumstance will render it dangerous, causing accidental loss to the customer. The firm is at risk from this source, and claims by customers are costly enough when they do occur to warrant protection. Along with the insurance protection from loss in the event of a claim, the firm can also strive to make goods and services provided as safe and trouble free as possible, to properly label packaging, and to thoroughly test goods and services for any inherent hazards.

3. Construction accidents. This category is similar to the case of employee accident, except that individuals who are not employees may be involved. Construction projects are complex and filled with uncertainty. Accidents occur, and the firm must protect iteself as best it can from losses connected with injuries incurred on the job. The best means of achieving this is by education, training, and the use of safety rules and equipment such as hard hats and safety lines. Beyond this is the consideration of the employee's loss of wages and the firm's loss of time connected with construction accidents. Normally, the construction firm handling the project is covered for such losses and is expected to take reasonable precautions with its employees' safety. Yet it is not unheard of for the firm commissioning the project to be named as codefendant in legal proceedings related to construction accidents. If it can be proved that the structure itself offered unusual hazards that were not protected against, the firm can be at risk. In addition, there is the matter of company personnel visiting a construction site for various official reasons, and having an accident while on the site. Again, this is a risk that is insurable, since the occurrence is rare enough to not be common and yet happens often enough to warrant the seeking of protection against loss.

4. Land title problems. As we saw in the chapter on real estate law, not all land titles are free and clear, and in the event of a clouded title, it is not always apparent that there is some lien or restriction on the land at the time of purchase. Since construction of a facility represents such a large investment in terms of money and time, the firm may wish to protect itself against loss from unforeseen land title difficulties. It is common practice to insure against such losses, though they seldom occur, since the loss would be tremendous should it occur. Special policies exist for the protection of land purchasers against later claims stemming from unknown restrictions or liens on the property purchased. Think of the devastating effect of erecting a multimillion-dollar structure on a piece of property only to discover that you have no legal right to the use of the land! Obviously, the protection is well worth the effort.

5. Storm damage. Both during the construction process and thereafter, storm damage can be extreme. Depending on the part of the country in which the facility is located, there is danger from high winds, flooding, tornadoes, hurricanes, lightning, hail, and other phenomena connected with storms. The odds of such an event's having serious consequences is not high, although the loss of a roof or damage to equipment from rain and lightning is common enough to force the prudent firm to seek protection from loss in such an instance. Special insurance policies handle this type of risk, and premiums are adjusted according to the risk of loss for a particular facility or geographic area.

C. Uncertain

This third category deals with events that are rare, representing a small probability of occurrence, yet serious enough to warrant their being taken into account. The firm can presume that these events will occur at such rare intervals as to be of negligible consequence, yet some firms, by virtue of the nature of their operations, may choose to plan for them either through insurance or through contingency planning techniques. Rare but catastrophic events can overtake a firm or a firm's location, and when it happens, if the firm is not in some way prepared, it can incur substantial losses. Some such events are noted below.

1. Climatic conditions. Though variable weather can cause short-term delays in construction or even in plant operations, we expect general weather patterns to be fairly stable. Yet this may not always be the case. There are historical examples of extreme changes in weather patterns that have caused basic changes to the climate of an area or, indeed, the entire world. With these changes in climate come changes to the economic process in general and to industrial operations in particular. In the sixteenth century, for instance, there was a period known as "the little ice age" that resulted in nearly a century of unusually long, cold winters and short, dry summers. The effects on technology and commerce were striking. At one point during this period, the Thames River at London froze solid, halting all sea trade to that major city for months. This is also the period of time during which the chimney gained prominence in house building. Often structures were built with a separate fireplace in each room, more a matter of survival than of cozy convenience. In the last half of the twentieth century, climatic conditions have been affected by increased pollution, by a general warming trend (1985 and 1986 tied for the second hottest years on record, and 1987 took top honors), and extreme drought in several parts of the world. In addition, there are the mounting possibilities of earthquakes in California (this is a secondary climatic effect, as many geologists and meteorologists will attest to) and the possibility of the greenhouse effect on the rise, resulting in still hotter, drier years ahead. The center of tornado activity has been shifting eastward for years, as has the incidence of lightning strikes, both indications of general shifts in climatic conditions. Long-term changes might make a sizable difference in the risk a firm takes with the placement of its facilities.

2. Union activity. Historically, unions prosper in good times and decline in bad, though there is a temporal delay between the shift in one and the corresponding shift in the other. This is probably because during good times workers have little to complain about but want to get in on the prosperity of the country, and in bad times workers are too busy vying for available jobs to worry about unionization and the collective bargaining process. In specific industries, however, particularly those with a long-established tradition of unionization, unions react to good times by wanting a bigger piece of the pie and to bad times by protecting jobs. This is an oversimplification of the situation, not reflective of individual cases.

Unions can be as capricious at times as any other factor in the market. Changes in union positions, the sudden demise of a union, or the emergence of one in a given industry are not comon occurrences, but they do happen, and when they do they can have serious consequences for the firm. Unions exist for a purpose: to improve the overall economic welfare of the worker. Firms that have large labor forces and ignore that fact often find themselves faced with an influx of union activity when labor conditions are poor. Economics is a natural force, and it obeys the law that nature abhors a vacuum. If the firm will not provide for the well-being of the worker, there will come other ways to do so, sometimes through the unions. It should be noted that by treating the workforce with equity, unionism can often be avoided. Certainly changes in union attitudes can be forestalled.

3. Regulations changes. This has been discussed in Chapter 8 on regulatory laws. Industry is regulated externally by a variety of government agencies and through a variety of federal, state, and local laws. Regulating agencies are not within the control of the firm, and therefore the rate at which regulations change and the direction those regulations take are unknown and often unanticipated. Although consistency of behavior tends to be the rule, a sudden change in policy can result in regulation changes that are at once totally unexpected and completely devastating. Only by vigilance and constant reappraisal of position can the firm protect itself from these changes in the development of facilities or, for that matter, in any other activity.

4. Market changes. Changes in market structure can come from many sources, including changes in taste, technological displacement of products, demographic changes such as a reshuffling of the population geographically or economically, legal restriction, or changes in social values. These are not the kind of changes that can be anticipated, though for the most part shifts in the market are well understood, at least in the short run. When changes do occur, an entire industry may find itself in the midst of unexpected restructuring. Twenty years ago, for instance, the repair and reconditioning of typewriters was big business and largely mechanical in nature. Today, unless you understand electronics and electronic control theory, you are out of business. Check the yellow pages for typewriter repair shops in your local community, and you will see what we mean. There simply are different products and a different market to contend with today than there was 20 years ago. Since the further into the future one looks, the more difficult it becomes to be accurate with predictions, a firm involving itself in the construction of a facility with a life of 40 or 50 years runs the risk of owning a worthless productive unit before the life of the facility is completed. There is no way to anticipate these shifts with any degree of surety. Flexibility is virtually the only defense available to the firm, along with constant reappraisal of position in the market.

5. War. In this country, catastrophic events related to war are rare. Indeed, the continental United States has not suffered serious losses in productive capacity from the process of war in more than 120 years. This is a fine circumstance, if all of your facilities happen to be located within the confines of the continental United States. If war comes to America, it will probably be of such a devastating nature that no recovery is possible, at least, not in the short run. For these reasons, there is little need to plan against it. (It should be noted, however, that certain key industries appear to have such contingency plans, particularly as their operation affects national security.)

What of international operations? What of firms who have individual or joint ventures in locations such as the Middle East or southeast Asia? Here there can be considerable risks from nationalization of facilities, revolution, or general war damage. Much of that risk is unanticipated as well. Under such conditions, insurance may be available at a high premium (commensurate with the exposure of the facility to harm), but more commonly, the profits to be gained from locating the facility in the dangerous location more than offset the expected loss from war. At times it is necessary to simply take the risk. At other times, the payment of exorbitant premiums for insurance protection against extraordinary loss is justified.

II. INSURANCE RISKS

Given the types of events that a firm may encounter, whether certain, probable, or uncertain, protection from those events represents an important facet of company operations. This is particularly true of new ventures such as the development of new facilities, where investment and the degree of uncertainty is high. Insurance is the paying of a fee to a third party who will assume the risk of the project for the firm. Since the insuring body (insurance company or companies) is involved in the acceptance of such risks from a large number of firms for various projects, it is able to minimize its own losses while minimizing the risks of the firms for whom it is insuring. The premium itself represents a best guess on the part of the insurance company as to how much payment it must receive to generate profit in the face of expected losses from all events against which it is insuring. The company knows that a certain percentage of the firms it is insuring are going to experience events against which they have been insured. As long as the cost of those losses is smaller than the amount received by the insurance company from investing received premiums, the company can afford to pay off claims and still be profitable.

From the point of view of the insurance company, it is a matter of determining risk accurately and charging premiums accordingly. From the point of view of the firm being insured against loss, it is a matter of the risk of the event's occurring versus the cost of the premium paid weighed against the type of coverage received. Not all coverage is the same. Variance in coverage stems from two sources, one the type of policy the firm buys (all-risk policy, agreed-risk policy, or shared-risk policy), the other the types of deductibles involved, that is, limitations and exclusions attached to the individual policy.

A. All Risk

All-risk policies are by far the most inclusive form of policy the firm can acquire. Within the stated limits of the policy (that is, what type of policy it is, whether workman's compensation, fire, or earthquake), an all-risk policy specifies that the insurance company will accept all risks for negative events without restriction, limitation, or exclusion. In other words, it is a blanket policy covering all contingencies connected with the classification of events for which the policy is written. Since there is an often considerable possibility of unforeseen liability connected with such a policy, the insurance company is apt to charge a sizable premium for such protection. On the other hand, the firm gets more for its money. An example of this type of protection is an "umbrella" policy, one that is designed to supplement other insurance policies of a more specific nature. The umbrella policy is designed to cover only those items not specified elsewhere, thus filling any unanticipated gaps in coverage.

B. Agreed Risk

With an agreed-risk policy, the assumption of risk on the part of the insurance company is on the basis of agreed-upon factors. That is, both parties mutually agree on exactly what risks will be assumed and to what extent those risks will be assumed. The purpose of such policies is to protect the firm from the most devastating and obvious types of loss that can occur without forcing the insurer into unlimited liability for loss due to damage. This offers the firm all of the insurance protection it needs for probable events while keeping premiums reasonable through a limiting of the insurer's risks of financial loss should some totally unanticipated event occur. Fire insurance, for instance, could include coverage for smoke and fire damage, water damage, damage to adjacent and employee property, and so forth, yet not cover such unusual contingencies as repair to a motorist's car two blocks away from the fire, proven to have been damaged by skidding on a street wet with drainage of water from fighting the fire. This may sound ludicrously far-fetched, yet without limits on protection from damage related to fire, it is possible.

C. Shared Risk

Shared-risk policies are similar to the agreed-risk approach, except that the coverage is generally limited in terms of dollar liability on the part of the insurer, rather than the events covered. Shared risk involves an agreement by which the insurer agrees to cover losses for various events, but only up to a certain dollar amount, or the agreement in some other way restricts risk. Policies normally have a maximum dollar amount of coverage that limits the total amount of coverage involved, yet within that framework, there may be further restrictions as to how much will be awarded for certain services. Health insurance, for instance, often specifies maximum allowable payments for fees. For the firm creating new facilities, this limitation could be a splitting of all costs above a certain amount for specific losses or some other variable liability arrangement.

More commonly, shared-risk policies use the concept of deductibles to divide the risk between the insurer and the insured. Deductibles assume that the insured can handle small losses and that protection is needed only against larger, more catastrophic losses. Thus, up to a certain amount, the insured may be required to absorb its own losses.

Deductibles are quite common in auto insurance and in business insurance as well. They serve to discourage petty claims, thus lowering premiums to the insured while still providing the protection that the insured is seeking.

III. DEDUCTIBLES

There are three general classifications of deductibles, each designed to cover a specific type of risk problem and offer the most economical and direct solution to specific insurance needs. These three are fixed deductibles, variable deductibles, and percentage deductibles.

A. Fixed

Fixed deductibles are the simplest approach to shared risk and merely state a fixed amount of loss below which the insurance company will not accept responsibility. The purpose of this type of deductible is to separate losses that are considered certain from those that are considered probable. For example, a policy may insure a firm against roof damage from storms but include a deductible for $500. The deductible allows the insurer to avoid minor repair jobs that may be necessary after a heavy wind (as when a few sections of roofing are blown loose), and yet to offer the protection the firm needs against wind damage from high winds and tornados that destroy a substantial part of the building's roof. The first event, the occasional loss of shingles, is considered a relatively certain event (as any homeowner will testify), while the loss of the entire roof is considered probable, calculable, and therefore insurable.

B. Variable

Variable deductible is a hybrid form in which the amount of the deduction varies with conditions, circumstances, and time. Under this approach, the deductible is specified on a sliding scale that is designed to match the degree of loss. For instance, if a loss due to fire is small, the deductible may dictate that the firm handle all of the loss, while if the damage is great, the insurer may pick up virtually all of the loss. At points in between negligible and substantial loss levels, there is a variable deductible dependent on the level of loss incurred. In effect, there is a sliding scale that allows the deductible to change as the size of loss changes.

C. Percentage

The percentage approach to deductibles allows the insurer and the insured to share loss on a predetermined and specified percentage basis. At times this is accompanied by a fixed deductible clause as in the case of protection that specifies the first $500 of damage be absorbed by the insured and that the insurer will pay 80 percent of all losses over this amount. In this case, whatever the level of loss is, the insurer absorbs 80 percent after the first $500, up to the limit of the policy itself. This form of deductible keeps premiums down by virtue of protecting the insurer from some possible losses, but insures the firm against major loss due to catastrophic occurrence.

SUMMARY

Risk is inherent in any enterprise, and the more complex the enterprise, the greater the risk. Firms involved in the development or expansion of facilities will incur such risks. We can divide risks into those that are avoidable, as in the case of certainties for which we can plan and against which we can defend the firm, those that are transferable, as in probable and insurable risks whose statistical probability of occurrence can be determined and from which the firm can seek relief through insurance, and those that are assumable, as are those risks which are unlikely, yet catastrophic, and for which the firm may wish to simply assume any losses by a process known as self-insurance.

Various types of insurance are encountered in the operation of any firm and an expansion of insurance needs can be expected with the development of new facilities. Among these are life and health insurance for workers, key-executive or owner's insurance, retirement and pension plans, which are often in the form of annuity insurance, and government-induced insurance needs such as social security, unemployment insurance, and worker's compensation insurance. In addition, there are voluntary government insurance programs specific to industries, such as National Flood Insurance Association, Pension Benefit Guaranty Corporation, Federal Crop Insurance Corporation, Federal Housing Administration, Nuclear Energy Liability Insurance, Federal Deposit Insurance Corporation, Federal Savings and Loan Insurance Corporation, and crime insurance coverage. The degree to which these latter programs will have relevance to the firm is variable.

QUESTIONS

1. What are the three types of occurrences recognized by insurance companies?
2. Of the types of occurrences listed in Question 1, which are insurable, and which are not? Give examples of each.
3. Define *agreed-value insurance policy*. Is this a viable option for the firm constructing a new facility?
4. The alternative of self-insurance is attractive to many firms. What is it and why is it seen as desirable?
5. What is the effect of deductibles on insurance costs? Why is this so?
6. List four examples of certain occurrences.
7. Can a firm insure against loss due to war? Union activity? Storm damage?
8. Name three types of insurance risks.
9. How does agreed risk differ from shared risk?
10. Name three types of deductibles.
11. What is risk?
12. What characteristics define an acceptable risk?
13. A firm insures itself against loss due to construction delays from accident. Is this a legitmate form of insurance? What type of occurrence is represented here?
14. What are actuarial tables? How do they affect insurance premiums?
15. What is an insurance underwriter? How does it differ from an insurance carrier? An insurance company?

PART THREE
PERSONNEL ISSUES

10
Ergonomics

The study of ergonomics deals with the physical aspects of completing tasks, that is, with the way physical work is carried out by employees. For the facilities planner, this entails determining what jobs must be done by workers and studying the most efficient way to go about those tasks, given the limitations of technology, applicable capital, and the human body. People who are involved in ergonomics are sometimes accused of taking a "machine model" view of humanity, considering each person as just another input into the system that needs to be engineered as an integral part of the production machine as a whole. There is some truth in this, but only inasmuch as the ergonomist realizes the limits of human capacities and seeks to operate within these limits. Part of ergonomics deals with human comfort and the effects of psychological as well as physical conditions on the ability of the individual to work efficiently. Far from a purely mechanistic approach, it is the very human nature of the worker for which the facility is being designed.

Today there is a tendency to view the worker and the machine as parts of a whole, an approach known as man–machine integration. This approach, as we shall see, creates an integrated whole of the worker and the technology with which that worker is interfacing, matching the capabilities of each to the other. Such a synergistic, systemic approach to planning facilities and designing facility environments serves both to recognize the individual nature of the worker and the need for that worker's comfort in order to function well, and to improve the overall efficiency of resources, space, time and money. In addition, the proper ergonomic design of a plant can contribute immeasurably to the process of control, which is important in any type of manufacturing or on-line service industry.

Beyond the design of production workstations, any facility, no matter how far divorced from actual production processes, needs to be ergonomically designed to minimize the discomfort of the people who use the facility. An automobile is a facility of a sort, engineered for the comfort of passengers. It may be built, for example, to minimize difficulties in operation of the vehicle, to maximize visibility for the passengers, to avoid backaches and physical discomfort by proper seat design, or to allow for ease of entry and exit. The same is true of buildings; lighting may be designed to minimize eyestrain and maximize visibility in the workplace, door handles and elevator buttons may be set at a height easily reached from a sitting position for the handicapped, or rooms may be arranged such that acoustics either dampen sound where concentration is necessary or magnify it naturally, as would be needed in a classroom environment. Ergonomics is really a matter of human engineering. When facilities are designed which are to accommodate people, ergonomic factors must be kept in mind.

I. THE NATURE OF ERGONOMICS

A. The Study of Work

Ergonomics is not a new science. It has been with us since the Greeks first tried to determine how best to stack oarsmen in a three-banked trireme in order to increase the speed of the ship. But it has become particularly important in the twentieth century. The very foundations of the discipline we now call management originated with the attempts of early industrial pioneers to maximize the efficiency of workers through the study of better methods of work for the individual. For example, Frederick W. Taylor, the father of scientific management, used piecework pay schemes as a method of motivation and calculated fatigue factors into production estimates. Henry L. Gantt, a close associate of Taylor's, developed the concept of a bonus system of wage payment for output and also developed the still important Gantt production scheduling and coordination charts. Frank and Lillian Gilbreth were the first to seriously undertake the study of time and motion of workers by using advanced techniques such as motion picture film and micromotion study. Such people contributed heavily to the development of what today is called ergonomics.

The underlying concept of this science deals with the fact that in today's business world, humanity interacts with an environment that is meant to be productive, efficient, and economical in the way it carries out commercial activities. If this environment is to lend itself to the successful pursuit of commercial ends, it should encourage, rather than discourage, the process which it houses. Environmental factors that may affect people must be altered to suit the abilities and needs of the workers, serving to minimize psychological and physical discomfort. Ultimately, this results in a higher degree of efficiency in the individual worker and in the workforce as a whole, thus increasing the efficiency of the entire organization.

Areas studied in ergonomics include what motivates people to work, what the effects of various physical conditions will be on worker performance, what the limitations of the human body and mind are, and how best to engineer these elements into a cohesive whole. Main areas of concern can be divided into five categories: characteristics of the human body, environmental factors, special human factors, sociotechnical systems, and quality of life in the working environment. Each of these categories is discussed in detail later in the chapter.

B. Work Motivation

Despite our tendency to think of work as being for the sole purpose of earning a living, people actually work for a variety of reasons. Certainly, the desire to have money and to purchase the things that money will buy represents one motivation for working, but this is far from being the whole story. In the early days of scientific management, motivation was seen as essentially a matter of marrying productivity to salary to encourage the worker to be more productive. At first this worked well, partly because it had never been done before (the novelty itself encouraged higher levels of output) and partly because of economic conditions. When scientific management was taking root, most industrialists were pure capitalists operating in an oligarchic economy, where the few in power had control of the labor market and were able to keep labor costs to a minimum. Under such conditions, a raise in pay meant substantial improvements in lifestyle and thus served as an effective motivator. In time, however, it became clear that other motivators were also effective.

Abraham Maslow and others developed psychological theories of motivation that examined the effect of increased wages. Maslow found that people responded to a hierarchy of needs, beginning with basic survival, proceeding to social needs, and finally ending with higher-level needs such as self-actualization, which refers to the expression of a person's true individual talents and being. According to Maslow, most people, when a lower-level (more basic) need is threatened, will concentrate on fulfilling that need at the expense of higher-level needs. Thus, for instance, when one's physical safety is threatened (a basic survival need), the tendency is to forget about looking good to others (a social need) and to concentrate on getting out of danger. The relevance of this theory to worker motivation is clear. As long as workers are on the edge of poverty, a very real threat to their safety or physiological needs, they will be motivated by monetary gain. Yet higher-level needs still exist and can operate as motivators once lower-level needs are met.

Predictably, as workers' salaries rose with productivity due to incentive wages and bonus plans, those plans became less effective because the workers were less threatened and hence less motivated to protect their safety or physiological well-being. Obviously, if money alone was not the answer, then other motivators had to be found that would enable the job itself to fulfill higher-level needs of the worker.

People offer numerous reasons for working at their particular jobs and for doing them well. They may enjoy the companionship of fellow workers, they may find a sense of accomplishment in doing a job well, or they may do it just to have fun. They may even be doing it in order to support what they really enjoy doing (as does the professor who teaches in order to find time to pursue a career in writing, the true vocation of the individual). Motivation is a complicated, individual concept, one that does not lend itself to simple analysis.

There is, however, help in understanding the nature of motivation and in putting that knowledge to effective use in the design of jobs and of physical environments. It is generally true, according to Maslow, that the more needs that can be satisfied in the job and workplace, the greater the worker's motivation will be to do the job and to do it well. This is a large part of human engineering.

Maslow's theme has been expanded on by other researchers and industrial psychologists. Herzberg's research in the two-factor model led to the categorization of performance motivation into intrinsic motivators such as recognition, personal growth and achievement, and extrinsic motivators such as working conditions, job security, and salary levels. B.F. Skinner, one of the best-known of the behavioralist industrial psychologists, proposed a behavioralist reinforcement model that concentrated on the use of reinforcement and repetition to change the behavior of the worker.

There is one common thread throughout the motivational models: if the work and the work environment are designed to satisfy the needs of the worker, the worker will naturally be more productive, thus better providing for the needs of the firm. As we will see in the discussion of systems theory to follow, this is not at all a surprising revelation.

C. System Approach

Like scientific management, the systems approach is a product of the twentieth century with its roots in our industrial past. A systems approach is simply a rational way of organizing any information or body of knowledge so that it can be studied and comprehended. Used extensively during the second world war, the approach centers around the definition and structure of what is called a system.

A *system* is defined as a set of elements that operate together to carry out certain activities and/or achieve a specific goal. These elements are referred to as *subsystems* and represent systems within themselves, which are themselves all made up of numerous subsystems. Analysis proceeds by breaking down the phenomenon under study into its constituent subsystems and then breaking those subsystems down into further subsystems until the elements being studied are so simple that they can be totally understood as to content and behavior. How far one carries subdivision depends on the complexity of the subject of study and how far the elements must be broken down until they are totally understood. This process, called *reductionism*, is often said to be carried out "ad absurdum," although this is seldom done in practice.

The systems approach is particularly valuable in business because it allows the investigator to describe a process totally and to understand all the interactions that take place among the elements in striving to achieve the goal. By virtue of this knowledge, if something is not functioning as expected or if changes are desired, it is a simple matter to find what needs to be done in order to correct the problem.

An automobile is an example of a system, as is a corporation, a guided missile, the furnace in your house, or the human body. Each of these is made up of constituent parts that interact to achieve some goal, whether it be to stay alive and reproduce the species or to keep the house at a certain temperature, and each of the constituent parts is constructed in turn of constituent parts designed to cooperate in carrying out the purposes of the subsystem of which they are a part.

Systems have four basic elements: input, process, output, and feedback. The operation of the system is a matter of inputs entering the system (as land, labor, and capital are the inputs of the economic system), some form of process taking place involving those inputs (such as the production and distribution of goods and services using the inputs of the economic system), and a final output that is the desired end result or goal of the system (which is the goods and services actually created by the production process). Feedback is the communication element that completes the loop and compares the desired goal of the system with what actually took place so that further operations can be adjusted to control error. In the case of the economic system, feedback may be in the form of sales figures or profits and losses, which tell a company within that system whether or not it is achieving its stated goals by use of present production and sales techniques.

There are also three universal laws that apply to all systems, including you and me, and that govern the effectiveness of system operation. These are the laws of synergy, reciprocity, and balance. The first of these, synergy, is the action of two or more elements to achieve an effect that each is itself incapable of producing. It is the reason systems function in the first place.

To illustrate, consider the cases of specialization and cooperation in mass production. By combining the efforts of workers and machines and coordinating those efforts, total production is far greater than it would be if each person worked alone and then pooled individual goods. That is synergy in its simplest form.

The second law, reciprocity, is a basic fact of nature. This refers to the idea that what you put into a system is what you get out of it, and that one is a direct reflection of the other. There are many ways to express this concept, including Newton's observation that for every action there is an equal and opposite reaction. That's physical reciprocity. So are the laws of thermodynamics, of quantum mechanics, and, for that matter, orbital mechanics. Beyond the physical, there is social reciprocity (societies are systems, after all). Note, for instance, an individual case where out of fear, one may choose not to communicate one's true feelings to another and in return, the other person becomes suspicious (or fearful) of the reticence and decides to withhold his or her feelings as well. This is an obvious case of reciprocity. A saying in federal penitentiaries expresses the concept quite succinctly: "What goes around, comes around." In terms of ergonomics, this is apparent in the fact that by taking worker conditions and comfort into account, the resulting increase in output produces increased wealth for the firm. Such is reciprocity.

The third law apparent in systems is that of balance. In fact, it appears that the sole purpose of systems is to create and maintain balance, though that would be balance of a dynamic nature. Balance means that the system is performing perfectly at all times in its efforts to achieve the goal for which it is created, and that whenever there is a breakdown in the system (an imbalance), the system moves automatically to correct that balance and get back on the right track. Since the track may be changing constantly (thus creating a dynamic environment), remaining on track may need to be achieved by correcting for errors discovered through feedback.

This process may not be an obvious one. An example serves to illustrate the point. Consider the case of a person awakening in the morning and going to work. That person is a system with a rather complicated set of goals to achieve, goals that vary with circumstance, making them dynamic in nature. In the present example, we assume that the organism's goal is to go to work. When the alarm rings, a number of physiological reactions take place. The individual is initially asleep, with all of its systems functioning in what we can term "sleep mode." Alpha wave production is probably high and beta wave brain patterns low. Metabolism is low and body temperature down; muscle action is limited, and breathing is deep and steady. The system is in a state of balance. Then the alarm rings. Immediately, adrenaline enters the system, and the brain goes on full alert. Physical activity increases, even if only to shut off the alarm. Metabolism rises to supply more energy for active muscles. Brain wave patterns change as the individual starts to wake up.

Breathing increases and body temperature rises. All of these changes represent the system's attempting to reestablish balance and maintain it, albeit in a waking state this time. In short order, the person is up, has showered and dressed, grabbed a bite of breakfast (probably on the way out the door), and has finally gained full efficiency on the road while driving to work. This waking state represents a correction based on feedback from the environment and subsystems as to the state of the organism. The person is now fully awake and alert, all warmed up, refuelled, and ready to face the world. Balance is returned to the system.

In this example, it may appear that the system is not in balance because of the range of activities that are taking place. Actually, all these activities are for the purpose of constantly adjusting the system so that it does stay in balance. Although parts may be in a constant state of change, the overall system maintains its equilibrium and proceeds to complete its assigned tasks. Imagine a juggler standing on a board balanced on a moving cyclinder that swings back and forth like a seesaw. At the same time, this juggler is tossing apples into the air, swirling rings from the elbows, and precariously balancing five sets of cups and saucers from the nose. In the top teacup is a spoon on end. There is a great deal of activity going on in this scene, much of it chaotic. Yet all those actions taken together point to one spoon, perfectly balanced and unmoving, while wild gyrations continue below in an effort to maintain that spoon's stability. That's dynamic balance. Hearts pump, glands excrete, intestines digest, and blood circulates, but the sum total of those efforts yields a balanced organism that continues to survive and achieve its goal of living.

This is the way a system operates. Since we can define any physically organized activity as a system, providing it has a goal and is made up of a number of cooperating subsystems, we can use this model to create and analyze anything we might want to achieve. Ergonomics represents one way of analyzing a system in an effort to keep it functioning more efficiently.

II. ERGONOMIC FACTORS

A. Characteristics of the Human Body

The first element in ergonomic studies involves determining the characteristics of the human body. Since we are dealing with the worker as an element in our system, it is necessary to determine what the characteristics of that element are in order to know what the worker is going to be capable of doing. Such things as bodily dimensions, neuromuscular response to stimuli, and fatigue are studied to determine the capabilities and limitations to be expected when working with people.

1. Anthropometry. Literally, this means "measurement of the body." It is a term used to designate that part of anthropology that deals with measuring the physical body to determine what physical variations exist among humans regarding race, environment, sex and so forth. For our purposes, we use the term anthropometry to refer to the determinations of average bodily measurements and the application of those data to the design of work environments.

By studying the structure of human anatomy and the limits of such factors as reach, leverage, grasp, the pressure with which one squeezes, and agility (flexibility and speed of movements), the facilities designer is better able to develop work stations and general facilities layouts that are matched to the capacities of the people who must work there.

Frank and Lillian Gilbreth did much with this factor of ergonomics in their time and motion studies. By analyzing in detail the movements of workers and varying the placement of materials and equipment, they were able to design more efficient work spaces. They divided movements into minute characteristic actions (subsystemic elements) such as reach, grasp, turn, twist, and squeeze, and then rearranged these basic task elements into more efficient patterns. Today we see the results of this approach in the design of many devices, in the placement of instruments in automobiles and cockpits, and in the standardization of dimensions of such things as tables, desks, and countertops.

By first determining the characteristics of the human body and then testing the actions of limbs and muscles to find fatigue and stress points, ergonomists are better able to design facilities to maximize the strengths of the human body and compensate for its weaknesses.

2. Neuromuscular factors. These are the considerations relating to the interaction of the nervous system and the musculature of the human body. Neuromuscular factors include such elements as eye–hand coordination, tactility (sensitivity of touch sensations), agility (capacity to manipulate objects), reflexes, and response time. Such factors will define the limits of a worker's response to stimuli in the work environment and also that worker's capacity to manipulate machinery and material. By testing response time and manipulative abilities, the engineer can design jobs that are within the limits of the worker's capacities. Such elements as temperature, texture, size and shape of tools, and the number of operations to be carried out at a single location can be varied in accordance with what will create the most effective use of labor. Specialization is partly a result, although overspecialization has been found to be detrimental rather than constructive for both workers and their productive output. The proper level of response time and the proper types of stimuli must be incorporated into a job format to create optimal output. That is to say, in order for a worker to be efficient at a particular job, the content of the tasks should include a proper combination of variety and repetition. This enables the worker to take advantage of a repetitive job and get into the rhythm of what is being done, and yet have sufficient variety to keep from becoming bored. This subject is further discussed in the following section.

3. Physiological response factors. Response in this sense refers not only to response time but also to the overall response to the work process, complicated by the fact that different stimuli affect the physiology of different individuals in different ways. Maintaining a certain type of behavior over a long period of time contributes to such factors as fatigue, boredom and its accompanying lack of attention, muscle response capability, and permanent or temporary eyesight and hearing deterioration. Yet responses to repetitive behavior may be either negative or positive, depending on the content of the stimuli. Generally, however, since repetitive actions take less attention than do discrete actions, the more repetitiveness there is to a job, the less a person feels the need to pay attention. As a result, it takes more and more conscious effort just to pay attention to the job. This is a physiological response to repetitiveness that can result in a form of self-hypnotism. Also, various physiological stresses will occur with specific physical movements, and an unusually high level of stress or a continued recurrence of such stress can result in reduced overall efficiency. For this reason, workstations are designed to minimize such stress, to reduce the degree of stretch required for reaching and grasping (which also reduces time spent in these actions), and to allow for a sufficient variety of movement to minimize the possibility of repetitive routines becoming automatic.

4. Fatigue. This represents a major ergonomic factor. Fatigue, or exhaustion, is a physiological reduction in the ability of an organism (or parts thereof) to function properly due to the effects of extended periods of physical or mental exertion, the effect being a buildup of toxic waste products in the muscles and nerve cells. From the viewpoint of the ergonomist, fatigue is the chief cause of reduced efficiency in performing work.

When people are fatigued, the accuracy, speed, and scope of their efforts are reduced. Workers who perform specific types of motions or mental activity for too long a period of time begin to feel the effects of fatigue. It is not always immediately apparent that this is what is occurring. Like altitude sickness, fatigue can come upon a worker without the worker's knowing it until after it has already taken hold. At minimum, the worker experiences reduced productivity and at worst, fatigue can result in permanent physical damage through accident or the development of physical illness.

Fortunately, fatigue is easily combatted. The issue is not necessarily how much work an individual does as it is how long a specific type of exertion is continued. Thus, varying job content is an excellent way to avoid fatigue factors, as the variety allows specific muscle groups or the mind to rest while other activities are taking place. Periodic resting can be built into the job structure in such a way that only a few moments' respite offered over the course of the workday can greatly increase the productivity of the worker. The same is true for jobs of the nonphysical type, those requiring a great deal of mental activity. In either type, the degree of activity a job demands or the intensity of the required mental activity create stress for the body and the mind. Consider the position of an air traffic controller, or that of a trader on the stock exchange. These are high-stress jobs requiring mental agility and rapid decision-making abilities. Such activities, if carried on for extensive periods of time, can lead to extreme fatigue and sometimes permanent physical damage, such as ulcers, heart disease, and high blood pressure. Such high levels of activity cannot be sustained for long periods of time without taking their toll. Again, the fatigue factor can be controlled with surprisingly brief rests throughout the workday. Even a change in the color of the walls to a softer shade, or including in the job design requirements that necessitate the worker's standing up or moving around periodically can do the trick.

B. Environmental Factors

External factors play an important part in the way people are able to carry out their jobs. The human body exists not in a vacuum, but in an ecosystem comprised of conditions, other people, machines, activities, and social, political, and economic structures. This ecosystem can either enhance or detract from the individual's ability to perform properly. Chief among the physical environmental factors that must be considered in the ergonomic development of a position are temperature and humidity, lighting, noise, and vibration.

1. Temperature and humidity. The human body functions effectively within a narrow range of temperature and humidity gradations. If an individual is either too cold or too hot, function is more difficult, as the body suddenly has other problems to consider than just doing the worker's job. With low temperatures more energy is expended to keep the body warm, and this diversion of energy from the brain causes a sluggishness in movement and mental activity. The body begins to conserve energy and protect the vital organs of the body at the expense of those less vital. Thus when internal organs receive the heat, the body becomes agitated in an attempt to produce more heat, and efficiency declines. Although this is generally apparent only in cases of extreme cold (chattering teeth, obviously sluggish response time, numbness in the extremities, or unconsciousness due to hypothermia), even slight discomfort due to cool temperatures in the workplace can lead to physical and mental distraction from the job at hand. Too high or too low humidity can make the problem worse. Cold, moist environments feel clammy at best, which adds to physical discomfort, and humidity that is too low under cold temperature conditions tends to dry the skin, causing lesions, chafing, and general decline of skin sensitivity.

On the other hand, excessive heat increases the degree of fluid loss to the body which, when combined with physical exertion, can result in very rapid fatigue. When the temperature is too high, the body reacts by taking steps to dissipate the excess heat from the body through perspiration, increased breathing, and a reduction in physical activity, which itself creates heat. Bodily responses to heat can result in sluggishness, exhaustion, dehydration, and heatstroke (unconsciousness due to overheating) in extreme cases, but reactions to even mild overheating lead to slower physical and mental responses. Again, humidity magnifies the problem. In cases of low humidity, evaporation takes place quickly, cooling the skin and thus reducing internal temperatures. Granted, this is a positive benefit under hot conditions, but along with it comes the possibility of not realizing how much moisture has been lost. The unexpected result could be dehydration and/or heatstroke. With high humidity, high temperatures become even more unbearable as the body is unable to rapidly evaporate the moisture that serves to cool the system. Again, the result is general discomfort and possibly extreme fatigue or heatstroke.

The normal temperature of the body is 98.6°F, yet the most comfortable temperature for our environment falls between 68 and 72 degrees. It is generally believed that the slightly colder temperature of 68 degrees promotes alertness and physical activity while still being within the maximum comfort range for the individual. For the ergonomist this means designing physical facilities to maintain this range of temperature in order to create and maintain maximal efficiency from the workforce.

2. Lighting. The effects of lighting on productivity stem from two sources: one the physical ease with which the worker can see, and the other the degree of resulting effort that must be expended mentally to make up for any deficiency in available light. Poor lighting makes it difficult for the worker to function by increasing the possibility of accident, increasing the amount of time required to carry out a given function, and by promoting eye fatigue, which will further reduce the worker's capacity to function. Various forms of lighting problems exist, from bright lights causing a form of artificial snow blindness to computer screen output causing eyestrain due to long periods of focusing and harsh colors. Green or amber cathode ray tubes (or CRTs) are considered superior to black and white ones because eyestrain is decidedly less and the possibility of permanent eye damage is greatly reduced. Lighting must be individualized for the type of work being performed. In some environments, such as department stores and warehouses, soft yellow lighting may be preferable, whereas in environments where CRTs are used heavily, a soft white lighting may be better to reduce background glare. For general office work and plant operations, a light approximating sunlight seems to work best, unless the work involves fine details or high contrast, which requires specifically colored or intense lighting sources. It is also generally agreed that light should originate from the left of the worker if there is reading or calculating work to be done, as can be seen in any properly organized classroom, so that no shadow falls on the page from the person's hand. Of course, this does not take into account the left-handed student or worker. Individual differences can and, where possible, should be dealt with by adjusting the environment to accommodate special needs. If flexibility is built into the arrangement, this can be accomplished with a minimum of disruption to the overall design. One of the key elements in a proper design should indeed include such a capacity for flexibility while keeping in mind that project costs may be affected by such a philosophy.

3. Noise. Noise and noise pollution have become serious considerations in the past 20 years, particularly in environments where heavy machinery is used. Not only is it dangerous from the point of view of plant safety to operate without the ability to hear warning sounds and signals due to a noisy environment, it is also dangerous in terms of permanent hearing loss and in terms of physical fatigue and mental irritation resulting from such noise. The Occupational Safety and Health Act (OSHA) regulations firmly control noise levels in business settings. In addition, the facilities planner must consider the general level of noise created by the facility that leaks out into the outside environment. It should be noted that development of facility locations often includes actions designed to cut down on plant noise leaking out into the surrounding neighborhood or to protect the facility from the noise of nearby factories, highways, and/or other facilities. Table 10.1 shows a listing of sound levels for many familiar noises.

4. Vibration. Like noise, vibration can do temporary or permanent physical damage to the ear. Even at low levels, vibration can also result in decreased productivity by increasing general physical discomfort, irritability, fatigue, and inability to concentrate. Vibrations range in frequency and amplitude much as sound does. By minimizing the vibration encountered on the job, ergonomists increase the overall efficiency of workers. The primary methods by which this is accomplished are either to shield the worker from the vibration through the use of dampeners that absorb unwanted vibrations, or by eliminating the vibrations entirely through restructuring of the production methodology employed. The latter is more difficult when the vibration is the result of the actual process rather than proximity to other processes producing secondary or residual vibrational effects. In general, it is wise to isolate workers from sources of excessive vibration as much as possible. One exception to this rule is in the case of a tactile process involving vibration as a signal, or encoder. Skilled craftsmen, for instance, will sometimes use both sound and vibration to determine the effectiveness of their

TABLE 10.1 Sound Pressure of Common Sounds in Decibels

150	Jet aircraft
140	*Onset of pain*
130	Rock and roll band
120	Riveting
110	Car stereo/teenage driver
100	Boiler factory
90	*Risk of hearing damage*
80	Truck or bus
70	Class lecture
60	Conversation
50	Car cruising
40	Average quiet home
30	Small waves on beach
20	Whisper
10	Gentle breeze
0	Hearing threshold

Source: Reproduced from *Production and Operations Management* by Arthur C. Laufer with the permission of South-Western Publishing Co.

work, as in the use of manually operated power equipment such as drills, lathes, and saws. In such cases, the vibration becomes a useful means of obtaining information rather than a hindrance to efficiency.

C. Specialized Human Factors

Part of the job of the ergonomist is to consider the fact that people are individuals and to deal with their individuality as well as their similarity. Although in general the characteristics of the worker can be used to design a job or an environment, the facility should also include allowances for individual differences. Of particular note are differences in capacity among workers to perform specific types of tasks and reactions of the individual to workload, work scheduling, and temporal adjustment.

1. Organization of work. The manner in which work is organized needs to be structured, yet flexible enough to account for individual differences. For instance, the methods utilized by the Gilbreths to determine the most efficient sequence of events for completing a task were not always most efficient in practice because of differences in physical and neural capacities among workers. Something as simple as determining the optimum number of bricks to carry in a load becomes a complicated task upon realization that the number is going to vary from worker to worker, depending on stature, muscularity, height, and a host of other characteristics. Admittedly, the amount of brick-carrying going on in the modern production facility is not great, but the principle applies, no matter what the task. In your own workplace, wherever it may be, it should be obvious that different people work most effectively in different ways. The ergonomist will therefore create within the regimented framework of the job structure enough room to allow for these individual differences.

2. Shift work. Shift work becomes particularly troublesome when there is a variation in the shift that individuals are expected to work. Inherent in shift work is the concept of fully utilizing a facility by having different sets of workers operating at different times, usually in eight-hour segments called day shift, night shift, and graveyard shift. This works well, but it must be realized that such an approach will wreak havoc with an individual's internal clock if it is not kept reasonably consistent for that individual. To constantly move an individual from one shift to another reduces efficiency by destroying not only personal lifestyles and life schedules, but by insisting that the worker's internal autonomic system shift gears in response to the needs of the company. An individual used to working at peak levels in the morning cannot be expected to be at peak efficiency when suddenly moved to an afternoon or late-night schedule. Adjustment takes time, as anyone who has experienced jet lag can attest. Such changes in shift should be minimized whenever possible.

3. Talented workers. The exceptionally talented worker can be a problem, particularly when piecework is involved. Workers who are particularly talented in performing their assigned functions may outstrip their fellow workers in production rates, and if work is being compensated on a piecework basis, this can produce false estimates on the part of management of worker capacities, or at least can result in the talented worker's being looked upon unfavorably by slower fellow workers. Yet no individual should be penalized for being good at what he or she does. By necessity, jobs and workstations are designed for the average worker. What does one do with a talented worker who will either outstrip the production rate of others or become bored and subsequently less productive by trying to keep the slower pace of fellow workers? The key appears to be to allow for the higher rate of production, thus maintaining efficiency, while designing control methods to take such cases into account and avoid artificially inflating average production rates.

4. Aging workers. The aging worker poses a unique problem in that the worker may exhibit a slowing of pace through time, although overall efficiency may be maintained due to the worker's superior experience at the job. The firm must realize that it has a moral and ethical obligation, as well as a practical and economic one, toward the aging worker. It should be further noted that older workers represent a very real source of effective, experienced labor that offers benefits not found in younger and more agile workers. Aging workers may exhibit less physical strength or reduced reaction time, and often have reduced secondary abilities, such as poorer eyesight or poorer hearing than their younger counterparts, but many jobs do not require fast reaction time or acute hearing. Industry is becoming increasingly aware of the advantages of older, more settled and mature workers as the working population ages, and is beginning to take advantage of the benefits of this source of experience and maturity. Older workers tend to change jobs less, to be more reliable, and to require less training than younger workers. This saves the firm the cost of finding, hiring, and training workers, which can mean increased profits. By designing around the strengths and limitations of this sector of the population, the firm can increase the effective availability of labor and reduce costs in the process.

5. Handicapped workers. People in this category are an underdeveloped source of labor for the firm. Jobs that do not require a full range of movement or activity can be as easily done by a handicapped person as by any other worker, as long as the nature of the handicap does not interfere with the performance of duties. Use of hands is seldom diminished in someone who suffers reduced use of his or her legs. It takes no great strength or mobility to operate a computer, oversee the manufacture of micromachinery, or do many other manufacturing and service jobs. With adjustment, workstations and facility layouts can compensate for individual handicaps, and firms are in fact required to do so when it comes to individuals confined to wheelchairs, or in some cases functionally blind. This means that properly designed buildings include curbs that are formed into ramps for wheelchairs to easily negotiate, door handles that are lower to allow a seated person to reach them, elevator instructions in braille, hallways that are wider to allow for wheeled traffic, and bathroom facilities that are wider to allow chairs to enter the enclosure. By designing into a facility the capacity to utilize the talents of handicapped workers, the firm increases its available workforce to include a number of very qualified and efficient workers who would otherwise be blocked from employment in that environment.

D. Sociotechnical Systems

The modern work environment is highly technical and sophisticated in nature. It combines the full extent of our capacity to create and use technology to produce and distribute goods and services with a highly developed and broadly defined social system. Every facility containing two or more individuals represents a blending of machinery and humans into a cultural subunit with its own relationships, rules, mores, ethics, and beliefs. Some of these subcultural units become fixed within a given industry or technology, resulting in highly individualistic and peculiar behavior patterns, goals, and needs. To ignore these peculiarities in designing a facility could mean a fatal flaw in planning; they must be given due consideration.

1. Sociotechnical approach. This approach takes into account the fact that certain industries are centered in specific locations, either because of an abundance of raw material, availability of transportation, or possibly the expertise of workers, and that, as this centralization continues over time, there develops an entire social structure based on that industry. Cultural subgroups develop through this process, each with their own distinct features and social patterns. It is in reference to these patterns that the ergonomists must design the facilities within which the workers are expected to perform. This concept diverges from the old machine orientation that viewed the worker as just another element in the system, another cog in the wheel, as it were. Facility designs, in order to be efficient, must not only reflect the best arrangement of equipment but also the best arrangement of people, including the development of a feeling of space and of comfort for the workers, and reflections of the culture and attitudes of those workers. A compelling example of this is seen in the problem U.S. companies experienced in their early dealings with Japanese counterparts. A simple act, that of offering refreshments, has very different connotations in the two cultures. In the Western view, offering coffee or tea at a business meeting is an act of courtesy, stemming from the feudal tradition of offering strangers food and shelter as a matter of course. When one enters a home or begins a meeting, coffee or tea is offered as a sign of friendliness and hospitality. In the Japanese culture, however, it is seen as a mark of celebration. To the Japanese mode of thinking, the offering of coffee or tea at a business meeting would signal that an agreement has already been reached, that a deal has been struck. In Japan, it is only after the conclusion of business that such refreshments are offered. Imagine the consternation of Japanese executives at being offered refreshment at the outset of a meeting, only to find out that their hosts have not agreed to anything being negotiated. This creates a feeling of distrust, as if the American negotiators were going back on their word. More than one major international business arrangement has been destroyed or, at the least, hindered by this apparently minor faux pas.

2. Relationship of technology and the work system. Such relationships are not always clear, partially because preconceptions concerning behavior or technology are so ingrained in the culture that it becomes background and partially because technology itself is constantly changing. By generalizing our approach rather than developing our facilities in a manner consistent with subcultural social patterns, we run the risk of falling into the same trap as American and Japanese negotiators. A number of changes have been wrought by our increasing tendency to seek technological solutions, from the original separation of worker and product, through the technocracy fears of the 1920s and 1930s, up to the present total restructuring of our approach to production as a result of the computer and its handmaiden, industrial robotics, none of which has fully taken into account the effects of those changes on the culture. One of the chief failures in the attempt to use Japanese management techniques in the United States comes from the difference in culture between these two peoples. What is natural to a Japanese businessman is in many cases strange to an American and vice versa. Methodology must fit the culture or subculture, and as the tendency to develop manufacturing facilities overseas grows, this becomes an increasing problem.

3. Sociotechnical applications. Several examples of the development of specific cultures around specific industries should serve to illustrate this point. Large numbers of furniture manufacturers are centered in a small area of the southern Appalachian region, in no small part due to the large number of individuals with expertise in woodworking and carpentry. This expertise goes back as far as the original migration of artisans from Europe, many of whom were of Scottish and German origin and many of whom settled in Georgia, North and South Carolina, and Tennessee. The result is an industry that developed around a plentiful supply of raw materials (southern forests) and specialized labor. The skills of these artisans have been passed down from generation to generation, so that even as technology changes the methods by which the items are manufactured, those skilled in the use of the technology are still centered in the same group of people. The silver industry offers a similar example. Through the first half of the twentieth century the silver industry was centered in Connecticut, the home of International Silver Company, which was the country's largest manufacturer of sterling silver during that time. Other manufacturers also settled in the area, not because of the availability of silver bullion but because artisans and designers had long been centered in Connecticut, particularly around Meriden, the home of International Silver. To take a more recent example, close to 50 percent of all fiberglass boats produced in the United States are made in Tampa, Florida. That is where the industry began, and it has continued to flourish there unabated ever since.

4. Ergonomics and the sociotechnical system. From the point of view of ergonomics, the sociotechnical issue is one of building cultural and physical characteristics that carry positive psychological effects into the workplace. It is necessary to tailor the approach to the workers. If, for example, the workforce is of limited literacy, it would be necessary to use visual instructions wherever possible in order to minimize error. In dealing with a non-English speaking subculture, instructions must be bilingual. Populations who are, on the average, of smaller stature may require workstations of different dimensions, and care must be taken not to disturb differing cultural patterns with conflicting methodology. Some cultures are highly collective and would view separate workstations as unnecessarily isolated, while cultures that value individuality and territoriality might require partitioning of even the most private of work spaces. This applies to all types of facilities. For instance, schools that have substantial foreign populations need to accommodate those cultures in various ways. One school set aside an area for the use of Islamic students to use for prayer during the day. Other schools with large Hispanic populations are decorated with wall murals reflecting Hispanic culture. This is all a matter of fitting the work space to the social system operating within it to create harmony, efficiency, and productivity.

E. Quality of Working Life

People spend approximately one-third of their life working. With such a high percentage of their time spent in the work environment, the quality of life in that environment is as important as it is in the home. Effective facility design includes considerable effort to ensure that the experience of employees at the workplace is as enjoyable and rewarding as possible. This effort on the part of designers to create a pleasant environment is not merely an altruistic move. Productivity, turnover, error rates, cooperation, management–employee relations, creativity, communications, and product quality are all intimately tied to the quality of life experienced by all those who must work and spend time in a facility, whether as worker, manager, or customer. In other words, the basis for most economic activity is, in essence, a desire for a better quality of life for the individual, for the group, and for the community, and the workplace should be designed accordingly.

QUESTIONS

1. Define *ergonomics*. How did it originate?
2. What is a system approach? How does it help in the study of ergonomics?
3. Name three neuromuscular factors considered when designing work environments.
4. Name two physiological response factors considered when designing work environments.
5. How does fatigue affect productivity? How can the design of the work environment reduce or eliminate fatigue?
6. Name four environmental factors considered in designing work environments.
7. What is the purpose of work environment design?
8. Define the *sociotechnical approach*.
9. The combination of people and technology functioning together in the workplace is known as man–machine integration. How does this relationship between people and technology change different facets of the work environment?
10. What psychological effects are there to man–machine integration?
11. Give three examples of a sociotechnical approach to production.
12. Pick a job and describe how you would integrate ergonomic factors into the job.
13. Name three examples of positions for which handicapped workers are particularly well suited.

14. How do equal-access laws affect the design of facilities?
15. In what ways do nonhandicapped workers also benefit from designs under the equal-access laws?
16. From your personal experience, describe what fatigue is like and discuss its dangers.
17. What are some possible effects of shift work on the individual worker? How is a change in shift going to affect performance?
18. Define *sensory-motor task* and offer examples.
19. How can a facility planner's design account for the exceptional worker? For the aging worker?
20. In the case of noise control and abatement, list the following in descending order of danger to the health of the worker: (a) rock and roll band, (b) jet aircraft, (c) class lecture, (d) boiler factory, (e) riveting.

11 Psychological Impact of Facility Design

I. ELEMENTS OF HUMAN RESOURCES

As with any other subsystem, the nature of human resources—that is, workers—must be understood and taken into account when one is designing a facility. The job of the facility planner is similar to that of an engineer in that the planner is constructing a system into which all parts must fit in order to contribute to the successful functioning of that system and to the achievement of its goals. Each works with a number of subsystems that perform specific functions. It is the function of the facility planner to determine what parts to use, what characteristics (abilities, limitations, tolerances, and so on) they must have, and how best to arrange those parts to achieve the purpose of the overall system which that particular collection of subsystems comprises.

The facility planner has a list of priorities in designing facilities. Table 11.1 indicates the percentage of importance facility planners place on design factors. The primary difference in dealing with subsystems represented by people lies in the worker's characteristics and the way in which we view those characteristics. Whether we use a machine model or a humanistic/psycho-physiological model, the way we measure and characterize the individual

TABLE 11.1 Human Resources Utilization Design Factors

	Frequency of Use
Sales	62%
Quality of internal labor supplies	45%
Facilities expansion	36%
Workload	34%
External labor supplies	28%
Labor turnover	19%
New products	17%
Technological and administrative	17%
Budgets	11%

Source: Reproduced with the permission of South-Western Publishing Company from *Production and Operations Management* by Arthur C. Laufer. Copyright © 1984 by South-Western Publishing Company. All rights reserved.

worker will determine the way we believe that worker can be utilized in the productive process. Admittedly, the larger the number of ways in which we can describe the worker, the more expansive our understanding of the worker's possible contributions, and this is to be encouraged. Yet certain standard practices and perspectives have historically proven particularly fruitful. A number of these are presented here.

A. Role of the Workforce

The role of the workforce in the production process has changed with technology, yet it has not diminished. On the contrary, although the type of work done by individuals has generally shifted from a more physical orientation toward one of control and creativity, the worker has remained a necessary and integral part of the structure of producing goods and services. Currently there is a predominance of man–machine orientation in the production of both goods and services that demands a close, symbiotic relationship between people and the tools that they use. Computers do only what humans tell them to do. Industrial robots still have to be programed and controlled. Even automatic lathes must be initially programed and later adjusted by humans, who make the decisions determining the desired results. The effectiveness of this symbiosis lies in the proper use of any resource, whether mechanical or biological, so that it can contribute what it is best at contributing, whatever that may be. In this section we explore the psychology of using workers to their fullest potential.

B. Policy of Human Resource Management

Although the content of a firm's policy of human resource management may vary, the need for policies in this area are uniformly accepted. Policies are guidelines designed to channel thinking and action in productive directions, and in terms of human resource management this is simply a matter of determining what is needed, how to obtain it, and how to use it once obtained. In general, these goals are not different from those for any other resource. In the business of human resource management, they translate specifically into the functions of (1) determining what quantity of labor is required, (2) determining how to obtain that labor through a process of recruitment and selection of candidates, and (3) how best to train and develop that labor force once it is obtained. The specifics of those functions depends on the individual needs of the firm, but the overall process is fairly constant.

1. Quantity determination. Deciding on required labor is a process dependent on several factors, including the alternative production methods available, the spatial constraints of the facility itself, the cost of labor versus the cost of mechanization, expected downtime rates, required characteristics of workers, and many more. We are dealing here with numerous factors and may choose to mix them in different combinations of inputs that achieve the same end. Chief among the concerns just noted are the alternative production methods. The planner must determine the quantity of labor needed for any alternative production methods available to the firm and, in light of these alternatives, determine whether it is cheaper to pay labor or to expend funds on capital outlay. Labor-intensive production methods have the advantage of flexibility of inputs (one worker can theoretically perform a number of alternative duties), lower initial capital outlay, and low cost of adjustment in quantity of input utilized. Capital-intensive industries yield a greater consistency of quality and output through time, but they require a higher initial outlay of funds to pay for equipment (rather than a worker's salary). The question is which method yields the more economical way of getting the job done.

In addition to analyzing the alternative production methods available, spatial constraints may be imposed by budget, available land, or plant size that would artificially force a firm toward one or another approach. Again there is the matter of worker characteristics, particularly how often the stress of the job is expected to create downtime (absenteeism, illness, frequent breaks), the cost attached to this downtime, and the availability of workers with the characteristics needed to perform the required tasks. All these factors will affect the quantity of labor decided on for the firm.

These determinations appear more ponderous than they usually are. Normally it is simply a matter of past experience or a little research to determine ahead of time what the proper mix of labor is to be. By the time an actual hiring effort is underway, the methods of production to be used and the number of required workers, along with their characteristics, have already been determined. Also, in quantity of labor determination, the facility planner must be certain that the physical structure will accommodate the required number of workers, that the flow of traffic will not be congested, and that the psychological factors that contribute to worker efficiency are observed. Much of this may be quantity specific.

2. Recruitment and selection. This process naturally follows on the heels of determination of workforce quantity requirements. As soon as the firm has determined the number of workers required and the characteristics (or types) of workers needed, it can begin the task of recuitment and selection.

Recruitment is essentially a process of attracting candidates to the firm who have the desired skills and attracting them in sufficient numbers to ensure that workforce needs will be met. Generally, both internal and external sources of labor are tapped in the search for workers, current employees being used first, particularly in multifacility companies where promotions, transfers, and other incentives can be used to initially fill the ranks of the new facility. Actively seeking workers from inside the firm can be in the form of job postings that allow present employees to have the first shot at newly available positions and through a process known as skills inventories, in which an inventory of required skills for a position is developed and employee files are then searched to determine if any presently employed individuals have those required skills. Internal sources may also, by virtue of the technosocial nature of an industry, prove to be an excellent source of skilled workers from outside the company. This type of networking can yield surprisingly high numbers of workers at a relatively low cost if the process is nurtured and encouraged with bonuses and other incentives to workers who bring in new employees. Because of the high cost of procuring workers, many firms have found this a cost-efficient means of recruitment.

External sources may include procurement through professional agencies, educational institutions such as high school vocational counseling offices and college job procurement systems, and job-related organizations such as unions, military services, and professional associations offering job availability information. In a large facility expansion, it is probable that all of these sources that are appropriate will be utilized.

Selection is conceptually a matter of picking and choosing from the available candidates those whom the firm feels are best qualified for the positions available. It is a logical concept, yet not a simple matter since the determination of who is best qualified may be a complicated process at best.

The basic steps in the selection process consist of (1) screening applicants through processing and evaluating resumes and applications, (2) interviewing candidates to determine less quantifiable characteristics and to enhance feedback, and (3) testing and assessing candidates for suitability. This is a weeding process designed to more and more tightly define the group of applicants that will eventually be hired. The initial screening eliminates those candidates who obviously do not have the necessary characteristics to match with company criteria. This process may result in a reduction of possible hires from a few percent to as many as 90 percent, depending on economic conditions and available workforce.

In the interview phase, there is a more direct effort to determine possible fit between the firm and the applicant. The purpose of an interview is to allow both parties to take a closer look at each other, and skilled interviewers can find out more about nonquantifiable factors such as general attitude, work ethic, level of self-confidence, and adaptability than they could learn from all the employment records available. At the same time, interviewers are in a position to offer candidates a look at the way the firm operates and to answer specific questions or address specific concerns that the candidate might have. It should be remembered that the purpose of the interview process is to determine fit. Candidates are looking for positions that fulfill their needs and aspirations, and the firm is looking for workers who will do the same for them. Only if both are satisfied is there a match and only then will the "marriage" of worker and company be successful.

The third step, assessment and testing, is a process by which the specific characteristics of individual candidates are measured to ensure that they fit the physical, psychological, and aptitudinal characteristics of the firm's requirements and to find the best fit for the individual in the organization. A number of written and practical tests are conducted; those used are dependent upon the needs of the firm and on job content. Once this step is completed, final selection takes place and hiring occurs.

3. Training and development. This third phase of the human resource management process is designed to train workers in the specific tasks they will be required to perform. Herein are opportunities for the employees' development by way of increasing skills and knowledge, which in time will improve their productivity and provide for their advancement in the firm. Training is a process in which employees experience changes in their level of expertise in specific fields, are guided to mold their behavior and attitudes to those consistent with the firm's, and begin to enhance their capacity to achieve both the firm's goals and their own. It should be noted that the methods by which this training takes place vary not only with the nature of the task to be performed but also with the nature of the individual employee. Different people learn differently. Cognition, the process by which people assimilate, manipulate, transform, store, and use information, does not occur identically in all people. Some people tend to be verbally oriented and do much better listening than reading or examining diagrams. Others are symbolically oriented and can be more easily trained by studying schematics or text than by attending lectures. Still others are spatially oriented, and can best be trained through practical experience with the process to be learned. The particular technique used should fit the employee.

Development is similar to training in that it increases the capacity of the worker to be productive, but it differs in that its aim is to foster continuous growth in the employee rather than merely train that person to perform some function. Development is an ongoing process by which the employee continues to grow in the job and in the firm. Developmental requirements are dependent on the nature of the job itself, the potential of the employee for promotion and development, and the expected future needs of the firm. Since all of these elements differ from firm to firm, the type and level of development that takes place will differ as well.

II. LEARNING CURVE AND HUMAN RESOURCE UTILIZATION

All employees are not equally productive, and no one of them is consistently productive at all points in time. That is to say, not only must we deal with the fact that some employees are better workers than others, but also with the fact that a particular worker has varying degrees of efficiency at different times. The significance of this is that productivity and efficiency are variable, depending on what percentage of the labor force is new, what percentage is already trained, and what percentage is in the process of being trained. This has implications for the quantity of manpower required by the facility at different times. Essentially, we are confronted with the *learning curve*, a phenomenon observed in the American worker since the early twentieth century. It was initially noted at the Wright-Patterson Army Air Corps facility, when workers were noted to exhibit a peculiar learning pattern when taught to carry out mechanical processes. Since then, the phenomenon has been observed and studied by manufacturing companies to determine workload and productivity variations through time.

A. Time-Reduction Curve

Workers vary at their level of productivity depending on how long they have been at a job. That in itself should not be too surprising. It takes time to learn a job, as has been observed by anyone who has learned a new skill, whether it be mathematical manipulation or learning how to type. The surprising element that was noted in the 1920s at Wright-Patterson, however, is that this differential efficiency rate is predictable to a high degree of accuracy. It forms what is referred to as the *time-reduction* curve.

What was observed at Wright-Patterson was that the labor hours required for the production of an aircraft was reduced by 0.8 as the quantity of aircraft produced was doubled. In other words, if it required 150 hours to produce the first aircraft, the second would require only 80 percent as long, or 0.8 (150) = 120 hours. The fourth, which would be double the production rate of two, would be only 0.8 (120), which is equal to 96 hours. This relationship is an asymptotic curve, or series that approaches but never achieves a horizontal slope. The significance is that when workers first begin to produce goods or services, they are relatively unfamiliar with what they are doing, and production costs will be high with regard to labor costs, but as they learn their jobs and become more skilled, the rate of production per man-hour will rise until the number of hours required to produce the product will achieve a relatively constant level. This is one of the elements contributing to the increasing returns of scale noted in sinusoidal cost curves earlier. Armed with this information, the facility planner is able to separate the higher short-term labor requirements from the lower, more permanent labor requirements indicative of normal plant operations.

1. Logarithmic expression time-reduction curve. The difficulty with arithmetically analyzing this process is, of course, that the slope of the curve keeps changing as the number of units produced goes up. This is illustrated in Table 11.2, which assumes an initial production time of 200 hours and a learning rate of 90 percent.

TABLE 11.2 Sample 90 Percent Learning Rate Table

*n*th Unit Produced	Labor/Hours Required	Percentage of First Unit Time
1	200	100
2	180	90
4	162	81
8	145.8	72.9
16	131.2	65.6

Given a 90 percent learning rate, we can develop an analysis that allows us to determine the number of hours it requires to produce any doubled production rate once we know the time it takes to produce the first unit. The third column in Table 11.2 indicates the percentage of the time to produce the first unit that it takes to produce the *n*th unit. This is all very well if the number of units produced is a doubling number—1, 2, 4, 8, 16, and so forth, but what do we do to find the time to produce the seventh unit? or the fifteenth? For this we must use a logarithmic analysis.

If we approach the problem from a logarithmic point of view, the learning curve can be described mathematically by the formula:

$$T_n = T_1 * n^b$$

where T_n is the time necessary to produce the nth unit, T_1 is the time to produce the first unit, and b is the slope of the learning curve for a given learning curve rate (the index of learning exponent).

The value of b varies depending on learning rates. The value of b at the 90 percent learning rate is -0.1520, according to a learning-rate table, and is a logarithmic expression of the curve. By plugging it into the formula, we are able to obtain a table of learning-curve coefficients to be used in determining the production time of not only the nth unit, but also total production time of n units. This information will enable us to solve a host of engineering, planning, and cost problems. A small section of such a table is presented for a 90 percent learning curve in Table 11.3.

TABLE 11.3 Learning Curve Coefficients for 90 Percent Learning Curve

Units	Per Unit Time*	Total Time
1	1.000	1.000
2	0.900	1.900
3	0.846	2.746
4	0.810	3.556
5	0.783	4.339

*Solution key for Column 2 when $n^b = 1^{-0.1520} = 1.000$
when $n^b = 2^{-0.1520} = 0.900$
when $n^b = 3^{-0.1520} = 0.846$
when $n^b = 4^{-0.1520} = 0.810$
when $n^b = 5^{-0.1520} = 0.783$

B. Application of Learning Curves

To illustrate, if the time required for the production of the first unit is 135 hours and there is a 90 percent learning rate, then the fifth unit would require 135 * 0.783 (from column 2 of Table 11.3), or 105.7 hours. Similarly, the total time required to produce five units would be 135 * 4.339 (from Table 11.3), or 585.8 hours. See the Problems section for further examples.

Having this information available allows the facility planner to better determine labor requirements and project costs with an eye for developing the most efficient man–machine mix. This is particularly important in operations that experience small-batch job production, changing products, or where the complexity of the process means a rapid change in the learning curve as workers gain experience. Problems with the learning curve stem from the fact that workers learn at different rates and that the rate chosen is of necessity an average one. Also, projects may be so specific in nature that initial determination of learning-curve values or accurate prediction of production time for the first unit may be impossible. Finally, shifts from one product to another necessitate a constant reevaluation of learning-curve data.

With the exponent for a learning rate of 80 percent, an 80 percent time-reduction logarithmic curve was developed and is shown in Figure 11.1 for 1 to 20 units. A 90 percent curve could be graphed by using the data from Table 11.3. The two learning rates, 80 percent and 90 percent, should enable a facility planner to estimate most learning time requirements.

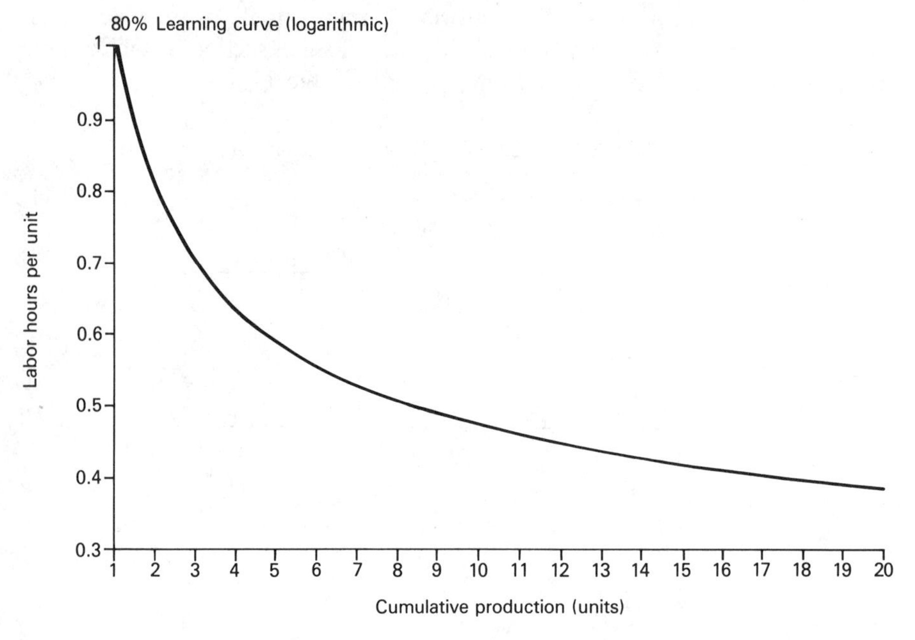

Figure 11-1

III. FACILITY WORK FORCE LOADING

A business facility is a dynamic system, constantly in motion and constantly adjusting itself to changing conditions. In terms of personnel, getting the most out of a facility means the most efficient use of the labor force. Loading is the process by which decisions are made as to how much and what type of labor must be where at any given time, and then scheduling the process to see to it that these goals are achieved. It is analogous to an inventory control system that tries to balance the cost of inventory versus the cost of not having that inventory when it is needed. In similar manner, loading factors attempt to optimize the use of available human resources in the process. In a food market, this could entail determining the proper quantity of labor to have on hand for properly stocking shelves, checking out customers, checking in shipments, and so forth, at any given time. In an academic setting, loading factors involve the availability of instructors and support personnel as well as the flow of students in and out of classrooms in a way that accomplishes the goal of producing an education without the high costs of wasted building space and manpower. Loading involves not only maintaining an inventory of suitable personnel but also scheduling that personnel so that it is where it is needed when it is needed.

For the facility planner, this is a major part of the production mix because through proper design of the facility, the process of labor-force loading can be done with a minimum of waste. In an educational institution, for instance, the way in which rooms are designed, their relationship to each other and to support facilities, and the general accessibility of travel routes through the buildings can contribute to the comfort and effectiveness of the personnel. In this case, the factors that must be considered include how many classes one professor is teaching, how likely it is that a student will be taking two or more of these, when the students are most likely to be on campus for other classes, when the professor is most likely to be on campus, the scheduling of office hours, what outside responsibilities and commitments individual professors have, the types of facilities needed for each class and how much competition there is likely to be for these facilities (such as lab space), and the time required for specific classes if it differs from the norm. This requires a tremendous amount of scheduling. Even less flexible scheduling situations involving office or production personnel require a determination of need and available workforce on a dynamic basis. Workforce loading is a key element in the design of a facility and of its operation.

IV. MOTIVATION

Business increasingly realizes the importance of properly motivating workers. Indeed, the main thrust of modern humanistic management theory centers in the ways that workers can be induced (motivated) to produce at top efficiency. There is a holistic approach being used presently that takes into account all aspects of the business process, including the motivation of the worker. This is not a new concept. There have been several theories put forth through the years as to what motivates a person to produce and be efficient. Do people work for money? Or for self-esteem? Or for enjoyment? Perhaps because it gives them a sense of power? Do they work better alone, or under group pressure? Are they better with a minimum of instruction or with close supervision? At this time, the answer to all of the above questions seems to be an unequivocal "Yes!" Motivation is not a simple matter, but a complex psychological puzzle that we continue to unravel. There are principles that tend to apply to the general population, but to apply any single rule to all people would be foolish. Some of the more accepted approaches to motivating workers (and people in general) are presented now.

A. Early Theories

Most early theories of motivation were centered around the concepts of fear and greed. It was generally held that what motivated people was fear of losing their jobs, of starving, of being judged a poor worker (or an incompetent person), or alternatively, a desire to accumulate wealth in greater and greater abundance. So went the arguments: if you can appeal to people's fear or to their greed, you can motivate them to perform the duties you want carried out. This concept can be traced back to before the time of Adam Smith, though it is clearly illustrated in his *Wealth of Nations*.

Smith and the classical economists felt that what motivated people to get involved in the productive process was a matter of self-interest, that is, avoiding the pain and misery of poverty and pushing for the benefits of success by maximizing profits. It should be noted, however, that Smith operated in a culture deeply imbued with the Protestant ethic of hard work, sacrifice, and pure living. Therein lies virtue, to be sure, but with a strict, puritanical mindset such as his it is little wonder that he viewed the early capitalists as being entirely profit motivated. Yet it is also to be remembered that this was a preindustrialized culture where farmers farmed for survival and merchants, only recently discovering the joys of international trade, reaped fortunes through their contributions to the redistribution of goods and resources, the likes of which the world had never seen. Perhaps in actuality, Smith was merely observing and reporting on the apparent nature of his world. Yet are desire for profit and fear of poverty the only motivations that people have? New light was shed on the subject when a study done in the United States in the early twentieth century yielded surprising results.

B. The Hawthorne Experiments

What started as a simple study in industrial engineering served to change the way management theorists look at the motivation of workers. The Hawthorne studies took place between 1924 and 1932 and began at the Hawthorne Works in Cicero, Illinois, which was a production facility of the Western Electric Company. There were four main parts to the study: the first, a test to determine the effects on productivity of changes in illumination in the work area; the second, a test in the relay assembly room to determine the effects of variations in work assignments and length of work week; the third, a series of interviews with workers; and the fourth, an observational experiment with a small number of workers in a controlled situation for the purpose of watching behavior patterns.

The researchers found that output was not just the result of changes in illumination or in working conditions, but that worker attitudes affect productivity as much as physical conditions and methodology do and that group norms were at least partially controlling output and productivity. With these revelations, theorists began to see the power of the group and the power of individual psychology in productivity. Such factors as group pressure, group rules, individual pride, identification, status, and a host of other factors came to be seen as important elements in production rates.

C. The Hierarchy of Needs

Another motivational theory, which we referred to in Chapter 10, was developed by Abraham Maslow, who saw motivation not only in the workplace but in all of life to be a matter of a hierarchical structure of need fulfillment. Maslow proposed that people are motivated in their behavior to fulfill certain universally felt needs, and that these needs exist in a hierarchy proceeding from the most powerful and basic needs to the more nebulous and less compelling needs. They range from survival needs, to social needs, to needs for growth and expression of the individual self. In order from most to least compelling, these needs are physiological needs, safety, belonging, self-esteem, and self-actualization. The first three of these are referred to as survival needs and the last two as psychological, though a good case can be made for "belonging" to have both survival and psychological elements, so it can be defined as transitional.

According to Maslow's theory, all these needs require fulfillment if an individual is to be fully healthy and balanced, but whenever a lower-level need is threatened, a higher-level need is set aside while the more powerfully motivating lower need is fulfilled. Thus a person who feels physically threatened is likely to deal with that rather than continue a conversation designed to make him or her feel part of some group (belonging). Actually, says Maslow, people need to fulfill all of their needs to some extent. According to him, individuals tend to "satisfice" their needs; that is, they fulfill each need to the point where it is no longer as compelling as other needs and then move on to partially fulfill others, rather than totally fulfilling each one in turn.

For the manager, the implication is obvious. The more needs that a job can fulfill, the more motivated a worker will be to perform that job and perform it well. Hence a job that just offers a good salary as an incentive is not as attractive as one that offers not only money, even a smaller though sufficient amount, but also offers an opportunity to be with people, to build toward job security, to feel a sense of pride in accomplishment, and to be creative (part of self-actualization) as well. Accordingly, firms that have strived to include as many elements of the hierarchy as practicable into a job can be expected to see increased loyalty, increased productivity, and decreased labor problems as a result.

D. Theories X, Y, and Z

The foundations of these theories lie in the work of Douglas McGregor, who first presented the concept in 1960. Dealing with two opposite leadership styles, McGregor noted that some managers tend to operate on the assumptions that workers are basically lazy, lack ambition, are interested only in security, and wish to be told what to do so that they may avoid the responsibilities of decisions. Such people, these managers believe, must be pushed, coerced, controlled, and threatened to induce them to perform their duties. This set of assumptions McGregor referred to as theory X. It is similar to a Freudian view of the universe in which people wish only to seek pleasure and, primarily, to avoid pain.

On the other hand, McGregor found other managers whose leadership style was based on the assumptions that workers are basically ambitious and creative, that they seek responsibility, are self-directed and self-controlled with processes they believe in, and that achieving goals is important to them. This approach McGregor referred to as theory Y.

According to McGregor, the leadership style that a supervisor or manager decides to employ in motivating workers depends on the motivational theory that that manager embraces. Actually, it could be noted that both types of workers exist, and that the same worker can exist in either state at a particular time. The McGregor findings are quite consistent with those of Maslow, if we assume that those exhibiting theory X behavior are caught up in lower-level need fulfillment, being basically interested in physical survival and safety, while those exhibiting theory Y behavior are actively involved with being part of the group, feeling good about themselves, and achieving a creative expression of their own self-ness.

More recently, a so-called theory Z has come into vogue. It is based on the Japanese style of management, which emphasizes making the worker feel a part of a family within the organization. In the Japanese approach, an employee experiences extensive screening and interaction before hiring, and once hired, advancement is slow. Yet once employees are hired, they are assured of the security of being a part of that company for the remainder of their working life. Also, they are involved with people from all levels of the firm in a cooperative effort to solve problems and achieve goals. The firm takes care of them as if they were family, and the firm expects and gets family loyalty in return.

Again there is a consistency with the Maslowian approach, this time the main thrust of the motivational effort being centered in the creation of a feeling of belonging (thus allowing group dynamics to take hold and influence the worker's behavior). Further, it is created in an atmosphere in which the individual feels secure and is offered an opportunity to gain self-esteem and in which creativity and productivity are encouraged and rewarded. One very important element in the success of this approach in Japan is the fact that the Japanese culture is highly oriented toward family, honor, and loyalty, all highly prized characteristics; thus the culture itself supports the approach. This is true only to a lesser degree in the United States, although modifications of the approach have been particularly successful.

None of these approaches are wrong, and none is the only "right" way to motivate. It depends on circumstance, on the nature of the job, and on the critical nature of such factors as time, money, creativity, independence, and cohesion of action in that circumstance.

E. The Motivation–Hygiene Theory

Another approach to motivation, proposed by Frederick Herzberg in 1959, divided job motivation into two categories, the first having to do with job content and experience, called *satisfiers*, and a second having to do with job context and environment, called *dissatisfiers*. The satisfiers were labeled motivators, and the dissatisfiers labeled hygiene factors. His reasoning for choosing the two terms stems from the fact that motivators actually make the job better if they are present, whereas the hygiene factors simply keep a job situation from becoming worse by means of controlling the environment in which one does the satisfying job. Hence the second is seen as a maintainer of good job health, much as personal hygiene reduces the possibility of disease. After all, negative job experiences are nothing more than an experiential dis-ease.

Among the motivators, Herzberg included the opportunity for achievement, acceptance of responsibility, recognition, advancement, and the satisfying nature of the job itself. As for hygiene factors, he included such environmental elements as interpersonal relations, working conditions, salary, and the degree and kinds of supervision present. Herzberg felt that by manipulating these two sets of factors a job could be structured to be maximally attractive and efficient, thus making it a positive experience for both the worker and the firm.

F. Patterns of Performance

Still another approach to the question of motivation deals with the relationship between performance of tasks and the perceived rewards that are dependent upon that performance. Notable among these approaches is the concept developed by Lyman Porter and Edward Lawler as an extension of the work developed by Victor Vroom. In this approach to motivation, it is assumed that the employee determines the reward connected with performing a task, as well as the probability of successfully accomplishing that task, and then proceeds toward the goal. Once achieved, the goal is again assessed in terms of received rewards and degree of satisfaction that the individual has experienced, and this information is used to reevaluate the desirability of again achieving the desired goal. The system has a built-in feedback loop that allows the employee to continually reassess satisfaction received versus effort expended and to adjust the amount of effort expended accordingly.

Of note is the fact that this approach emphasizes the changing nature of performance as a consequence of reevaluation on the part of the employee with each successive performance of a task. The reassessment is the key, being either supportive of or detrimental to future performance.

G. Achievement and Motivation

David McClelland proposed the idea that motivation is dependent on three specific needs: achievement, power, and affiliation. This is known as McClelland's achievement motivation theory. He defines achievement need as a drive to accomplish and reach goals. Certain individuals, McClelland found, have a high need for achievement, and they can be motivated to perform if they view that performance as supportive of achievement. Among the determining characteristics that high achievers exhibit are (1) a preoccupation with task completion, (2) a felt need to accomplish goals, (3) a high feeling of personal responsibility for the outcome of actions, (4) a willingness to take risks, and (5) a high need for feedback concerning performance.

Others are driven by the need for power, chiefly those who feel a need to be in control of their situation. Those motivated by power seem to move through a series of learning stages, moving from a manipulative, more paranoid type of power need toward a more altruistic power need that is lower key, less direct, and more interested in influencing the development of fellow employees than in actually controlling them.

The third need, the need for affiliation, represents a strong desire to belong to the group. As a motivator this pushes individuals to behave in patterns that they believe will more securely entrench them within the group, and is in effect no more than an expression of Maslow's belonging need. According to achievement motivation theory, personal motivation, as well as the manner in which a person will most effectively operate, is dependent upon his or her need orientation.

H. Attitudes and Environment

There are numerous ways in which the work environment can affect the attitude of the worker. In terms of physical environment, attitude can be affected by a host of factors already discussed, such as vibration and noise, but also by such factors as lighting, color, spatial relationships, design of physical equipment, and positioning of work space in the work area. Physical surroundings provide signals that workers internalize, often on a purely subliminal basis. Culturally, such things as how close an individual is to a door, how much space there is between workstations, and the predominant color scheme in the area all have content beyond the obvious. These environmental factors, when understood, can contribute to the motivation of the worker and help mold their attitudes. Hospitals, for instance, often use green color schemes to unconsciously support a feeling of growth and healing in patients. Prison systems sometimes use pink walls to calm inmates and reduce tension. Offices of supervisors are often placed near the corners of the work area to promote the feeling of power and control, and workstations are placed to give workers a feeling of personal space when possible.

Beyond the physical nature of the workplace there is the cultural environment created by the group within which the employee is working. Group dynamics have powerful influences on workers, whether we approach the subject from the point of view of Maslow's hierarchy or McClelland's achievement theory. The group can affect and in some cases control a worker's attitude as the latter reflects the opinion of the former. Group pressure affects absenteeism, feelings of satisfaction, and general attitude by offering a model for the individual to identify with, and as such creates a sizable portion of the environment. The attitude of workers can also determine productivity, creativity, willingness to excell, and motivational factors.

I. Motivation and Productivity

The purpose of studying and adjusting the environment of the workplace is to encourage employee productivity. Indeed, efficiency is defined as production with a minimum of waste, expense, and time, and how close we come to maximum efficiency is measured by the relative productivity of the worker. In order to be productive that worker must be motivated, by whatever means, to go through the processes involved in producing goods and services. So motivation becomes a necessity. It is not enough to motivate through fear and greed. These are effective only at the most rudimentary levels of operation, and they require constant reinforcement from supervisors to maintain their effectiveness. By using a wide variety of motivators and eliminating fear in the process, a firm is able to nurture self-starting, productive, high-energy workers who need little supervision and who will work towards the enhancement of the firm's position as an extension of enhancing their own. The most effective approach appears to be one involving workers in the decision-making process, making performance a matter of personal pride and accomplishment, and allowing employees the opportunity to accomplish while at the same time providing extrinsic and intrinsic rewards for that achievement. One method of doing this is through the Japanese management technique known as the quality circle.

J. Quality Circles

Another way of centering the production approach in the culture of the firm (sociotechnical approach), the quality circle is a method of problem solving and decision making that brings together individuals from all levels of the company and many different areas of function to collectively arrive at solutions. The real value of this approach is that quality circles not only produce solutions to problems, but also act as a method of developing employees within the company and within their jobs. This is a project approach that offers a specific goal for the group (providing focus), and if properly constructed, all of the members of the group, as well as the firm, will benefit from its operation. Certain characteristics distinguish successful quality circles from unsuccessful ones, one being the fact that they allow for employee development. Other positive characteristics include the following: it must be a voluntary operation in which the participants are there by choice, not by duress; management must be willing to recognize the efforts of the circle in its entirety, even for ideas and solutions not used; management should use at least some of the ideas set forth by the quality circle; the project being assessed and the job content of participants must be somehow connected in a real way; participants should be trained in the interactive techniques of quality control circles; and it must be realized that the quality circle is a group effort, not an individual one. Given that these criteria are met, the motivational benefits from the use of the quality control approach can be legion.

K. Significance of Management

In this motivational process, no matter how much autonomy is eventually handed to self-directed dedicated employees, management is still the key. By definition, the function of management is to get things done through people. In so doing, the decision still must be made as to what is to be done, who is to do it, and how. Control requires that the process be monitored, adjusted, and guided, all of which must be done by management. Motivating employees is an art as much as it is a science, and no matter how effectively the necessary institutions are constructed within an organization to create a highly positive, productive, and creative atmosphere, the art of maintaining high levels of efficiency is in itself a creative process requiring expertise and a subtle sense of what works. Management is responsible for providing this vital function.

V. APPLICATION TO FACILITY DESIGN

In designing facilities the planner is not dealing with physical space alone. That space has to be used by people, including employees, supervisors, customers, and outside vendors. The way in which they interact will be substantially affected by the way the facility is planned. Particularly with retail facilities and service organization facilities, the planner may find that the human element is not only significant but paramount in the way the building itself is designed. What must be remembered is that the design can either support or detract from the quality of life experienced by those who come in contact with the facility and that that quality of life has a very real impact on the productivity and efficiency, not to mention profits, of the plant. Design must be done with people in mind at all times. The building and its content are a tool by which people achieve results. The more that machine is designed with the people who must use it in mind, the better able that machine will be to accomplish its appointed purpose.

QUESTIONS

1. Define *human resource management*. What is its purpose?
2. Define *work force loading*. Is it necessary?
3. Define *time reduction theory*.
4. What is the concept on which time reduction theory is based?
5. How are quality circles used to improve efficiency?
6. Name three functions of the human resource management department.
7. What is the difference between training and employee development? How does each uniquely contribute to productivity?
8. Give two examples of learning curves and how they can be used to increase productivity.
9. What did the Hawthorne experiment conclude about motivation? Is it valid?
10. What is theory X? How does it differ from theory Y?
11. Theory Z is often called the family approach. Why?
12. Where was theory Z first developed? Where was it first used successfully?
13. How can motivational factors be built into the design of a facility?
14. Maslow's hierarchy of needs is used to explain individual motivational behavior. According to his theory, which of the need-satisfaction approaches is most effective? Why?
15. According to Herzberg, what is the difference between a motivational factor and a hygiene factor?
16. How is the need for achievement seen as a motivating force in the workplace? What opportunities exist for utilizing this force?
17. "A starving man will do anything for a loaf of bread." What does this say about motivation theory?
18. What actions can management take in promoting motivational methodology in facility design?
19. With the advent of the Industrial Revolution there began a separation of the worker from the produced product due to specialization and the division of labor. What do you believe to be the effect of this change on worker motivation?
20. List several psychological factors that might motivate some people to become artisans rather than industrial laborers.

PART FOUR
WORK FLOW ANALYSIS

12
The Work System

I. THE NATURE OF WORK

Work can be simply defined as the expenditure of energy over time. This is the definition that predominates in physical theory, and it has relevance to business in that work occurs whether productive results exist or not. Work does not have to be difficult, nor does it necessarily have to be productive. Yet when expenditure of energy is purposeful and has been designed to create products and services, it becomes productive work, and the more productive a given amount of work is, the more efficient it becomes.

All this is probably of no great surprise to the reader. In the production and distribution of goods and services, we expect work to take place. Indeed, each of us expends energy whether it is in mowing lawns, performing functions collectively referred to as a "job," or studying to master material and pass tests. All of this requires some level of work. Notice, however, that there is no inherently negative connotation attached to this understanding of the nature of work. Work can be as easily enjoyable and fulfilling as it can be painful, boring, or unfulfilling. The content of work is the work itself; the context in which we hold it is a personal matter. And individuals do not always hold the same work in the same context at all points in time. Sometimes mowing lawns is fun; at other times it is not. Sometimes downhill skiing can be a very rewarding experience, at other times it can be viewed as a necessary evil, as when it is done as a profession. Value judgments, then, are not the issue here. It is the content of work that we are interested in rather than the attitudes about that work.

When a facilities planner encounters the issue of work and job content, it is generally already set. At that point, the planner need only consider how best to build a facility that allows for the work to be performed while it is done in a manner that is both efficient and maximally appealing to those performing that work. The planner's purpose is to see to it that the work to be performed can be done with a minimum of waste and to design the facility in ways that will facilitate the completion of those tasks. Yet there are occasions on which the planner may be called upon to actually design the work itself as well as to facilitate the completion of that work. In either event, it is important to understand the nature of the work. In this chapter we consider the nature of work, how to classify and analyze its content, and finally how to design work functions for greatest productivity.

II. WORK AND JOB DESIGN

Jobs consist of a number of tasks, or units of work, that are strung together in a sequence that creates a desired end result. That end result can be anything from the completion of a computer program or production of a can opener (products) to cleaning a house or polishing an automobile (services). The purpose of the related tasks (that make up a job) is to reach the goal. Inasmuch as this is the case, we can formulate the necessary relationships and elements of a job into a system made up of subsystems strung together. Each of these subsystems can be divided into smaller subsystems that work to accomplish the individual tasks of a job's content. Since there is usually more than one way in which the sequence of events and actions (subsystems) can be organized and arranged in order to complete the task, we have the opportunity of designing the job to best achieve our purpose. Job design, then, is simply a matter of arranging work into a cohesive pattern that results in reaching a desired goal, and the more effectively we design that job, the more efficient we become. Efficiency, remember, is defined as accomplishing tasks with a minimum of time, energy, and effort; that is, accomplishing the task with the minimum amount of work.

Many job designs originally came about through a long and arduous process of trial and error whereby people changed their methodology through experience, eventually reducing the amount of time and effort necessary to complete a task. New methods would then be taught to others. In a world low on technology, slow-paced and without high levels of production, that was sufficient. It is not enough today. Changes take place rapidly in an industrialized economy. New technology and new products enter the market rapidly, and with their appearance it is necessary to create new job designs to accompany them. We no longer have the luxury of generations of tradition dictating the most efficient way to do things. Jobs are designed as a purposive effort and have been since the dawn of the Industrial Revolution.

III. MAN-MADE MACHINES

Technology consists of ideas, methodology, and man-made productive artifacts that assist us in carrying out the productive process. This third element, productive artifacts, are what we call man-made machines. The purpose of machinery is to facilitate the completion of tasks through mechanical means. Humankind has been using tools for an estimated one to two million years. Machinery is nothing more than a complex tool. As with any technology, the purpose of machinery is to make it easier to perform some task. It is a way of increasing efficiency through the application of natural law, and consists of a combination of many such applications which form a pattern of behavior that creates that efficiency. Even the most complicated machinery, such as a B-1 bomber or an industrial robot, is only a set of smaller machines that interact to perform a function. In today's approach, more than ever before, industry understands that the primary parts of this system of subunits are the workers themselves. Through man–machine integration we are able to practically and efficiently complete industrial tasks.

Mechanical devices cannot do it all. To be cost efficient there must be an involvement of people in the process, particularly if there is to be even the most rudimentary level of decision making in the process. The purpose of the job designer is to integrate the machinery used with the worker involved to develop a maximally efficient symbiotic relationship. Ideally, both the actions of the worker, with general-purpose abilities, and those of the man-made machinery, with specific-purpose abilities, should be developed at the same time to ensure a good match. Such a design would create a cohesive system.

IV. JOB DESIGN

Numerous rapid changes are always taking place in how we produce goods and services, a situation that forces us to continually design and redesign jobs. This rapid obsolescence would seem to frustrate any substantial efforts in job design. At best it appears that job design is a temporary adjustment to events. How can there be any experiential wisdom in a process that is continually being displaced by new technologies, new forms of man–machine integration and new needs?

What has developed traditionally through experience and remains relatively intact are the methods used to do job design. The jobs themselves change rapidly; the ways in which we go about designing them follow well-established principles. In other words, the process of job design follows a pattern of analysis and synthesis that has been in existence in one form or another for more than a century. Tools used in the process of design change have remained essentially the same. It is this pattern for analyzing job methods that lies at the heart of understanding job content.

A. A Pattern for Analyzing Job Methods

In analyzing job methods, the process is broken down into constituent parts and each part is analyzed, first from the top down by dividing jobs into steps, steps into activities, and activities into motions and movements, and then by analyzing each of the bottom-level motions and movements, building methods from the bottom back to the top. Particularly if the job content is to be created from scratch, this proves to be a most effective method of analysis. With job structures that already exist, it is sometimes possible to move directly to the bottom line and work toward the top, changing, shifting, and correcting elements as we go. In either case, the result is a detailed blueprint of the job, its contents, and its requirements, which the facility designer can use in creating a work environment.

B. Motion Study Tools

Three types of analysis take place when movements in the production process are examined. On the broadest level, a process analysis may be performed to get an overview of how the process of producing the product or service is carried out. Rather than concentrating on the details one by one, the entire structure is viewed as a series of operations at once independent and connected. In process analysis, the production function is studied as a series of operations to be performed in sequence. The main thrust of this form of analysis is to determine the proper sequencing of events for efficiency and productivity. By analyzing process content, it is sometimes possible to eliminate gross errors in plant layout and space usage that result in reduced efficiency.

Second, an activity analysis is performed in which the activities that make up each process are studied to create a more detailed view of each job. The purpose is to look at the number and nature of the activities that make up the processes, to investigate the subsystems within the system, and to determine if a more efficient arrangement can be found by combining, moving, or eliminating any of the activities involved.

Finally, an operational analysis is possible, which is a much more detailed approach. In operational analysis, the primary purpose is to look for small inefficiencies that individually do not appear to be serious, but which taken together and compounded by a number of stations all performing the same job, can be extremely costly. The changing of a single movement or the elimination of one part, such as a screw fastener, can translate into huge savings in time, energy, and money when multiplied by many repetitive performances of that process. This type of analysis is particularly useful for operations that involve a great number of repetitive movements or actions.

1. Process analysis. For an individual workstation, process analysis is a matter of studying the flow of material through the station, determining the different processes that take place at that station, and redefining the movements to increase production or reduce required time. Extensive work in this area was done by Frank Gilbreth early in the century. Known as the father of time and motion study, Gilbreth developed numerous ways of increasing efficiency through the analysis of process flows and individual motions. Using both intuitive trial-and-error techniques and scientific analysis of micromotion studies (which are discussed later in this chapter), Gilbreth was able to greatly reduce required time for completion of many tasks. He is considered the original efficiency expert, and much of the later work of motion analysis is based on his studies.

a. Symbols (Therbligs). Gilbreth classified simple hand movements and other worker activities into individual elements called therbligs. These elements could be combined to describe the flow of materials and actions taken on those materials in a sequential format which, combined with a diagram of the work station, offered a clear picture of a specific operation. Table 12.1 gives a description of the main therbligs used by Gilbreth.

b. Flow Diagram. By a combination of therblig analyses and diagrams of the work area, engineers are still able to completely describe and manipulate the content of the job process for any single station.

c. Assembly and Gang Process Charts. The logical extension of this process results in a much larger flow diagram that combines a number of processes into a single overview which shows how assemblies take place, including their individual steps. The term *gang process chart* is often applied to this type of diagram, since it represents an entire 'gang' of processes that interact to produce a single assembly map.

This type of detailed analysis allows the facility planner to become aware of all of the steps in a production process and to plan the flow of goods and materials through the system in a logical way.

2. Activity analysis. Activities are defined as actions that are required to complete a task. The analysis of activities involves defining each action taken within a process to determine its content. Similar in nature to the process analysis just discussed, the activity analysis simply involves a more tightly defined element. For instance, it may be necessary to carry out a series of activities in order to complete the process known as "assemble dimcadgen." To simply say that would probably not supply adequate infor-

TABLE 12.1 Symbols and Descriptions for Therbligs

Name	Symbol	Description
Search	SH	Trying to find item by using hands or eyes
Select	ST	Finding and choosing a particular item
Grasp	G	Contacting item and establishing control
Release load	RL	Letting go of item, loss of control
Position	P	Lining up
Preposition	PP	Lining up to prepare to use
Assemble	A	Putting items together
Disassemble	DA	Taking items apart
Use	U	Manipulation of item
Transport empty	TE	Hands move from one position to another empty
Transport loaded	TL	Hands move from one position to another loaded
Hold	H	Maintaining item in fixed position
Unavoidable delay	UD	Waiting
Avoidable delay	AD	Idleness caused by the operator
Inspect	I	Examination of item to ensure quality
Plan	PN	Thinking of the next step
Rest	R	Delay as part of the method

Source: Adapted with the permission of South-Western Publishing Company from *Production and Operations Management* by Arthur C. Laufer.

mation to the analyst, particularly if the analyst is not very familiar with the nature of a dimcadgen, whatever it is. The necessary activities may require the worker to pick up a casing, reach for a cylinder, grasp the cylinder, insert the cylinder in the casing, and place the assembly in a rack. Each of these activities is an integral part of the process of assembling the dimcadgen. Yet within each of these activities are a number of details such as where the parts lay, how large they are, how easily they can be grasped, and so forth. It is important to understand these elements if we are to achieve any degree of efficiency. If the casings are cumbersome to handle, for instance, we may opt for a process flow that places them in a moving rack of some sort rather than having the worker pick them up. Without an analysis of the individual steps or activities involved in the process, we would not be able to determine such a need with certainty.

3. Operational analysis. Here we have the most detailed level of job analysis, in which each action is analyzed and changed to maximize efficiency. In operational analysis, the elements of activities are broken down into what are called micromotions, each lasting for as little as a fraction of a second. Grasping is a general term in micromotion studies. Operational analysis would demand that the word be explained in terms of exactly what takes place when the object is grasped. This is where time and motion studies become most detailed. Motion picture footage of individual activities, a technique pioneered by Frank Gilbreth, is sometimes used, although videotaping is now more common. In large operations where many people are performing the same set of functions over and over again, the slightest difference in movement can increase production rates. Speed is not the only issue here. Micromotion studies deal with such issues as worker comfort and fatigue as well. The entire mechanics of the operations being performed are reduced to their smallest logical unit, and those units are studied for efficiency.

As we have already indicated, the purpose of all this analysis is to improve operations. Too often, the manner in which operations are carried out and processes are done is a matter of convention, tradition, or superficial logical sequencing. In many cases the errors inherent in the way an operation is set up, particularly a large and complicated operation, are hidden in the very activities creating them. Only through a thorough and systematic analysis of job content can the firm be assured of efficiency. There always seems to be room for improvement, either because of improvements in technology, improvements in methodology, or improvements in worker efficiency due to experience and the effects of the learning curve. By the very dynamic nature of doing business, systems are not maximally efficient except in the short run. Operations improvement, therefore, is an ongoing task.

C. Principles of Motion Economy

In developing flow diagrams and process charts, certain accepted principles are used to determine motion economy. These principles represent the current wisdom as to what works in creating efficiency of flow and motion in the production process. It makes no substantive difference as to whether we are investigating an industrial process, the flow of information through an office, the use of a computer terminal or, for that matter, how the boss's assistant sets up a conference. These principles are equally applicable to any activity involving flow and motion.

1. Use of the human body[a]

1. Both hands should begin their activities at the same time.
2. Both hands should never be idle at the same time, unless resting.
3. Whenever possible, arm motions should take place at the same time and in symmetrical patterns while moving in opposite directions.
4. Hand motions should be the simplest possible for completing the operation efficiently.

[a]Adapted by permission from *Motion and Time Study* by Ralph M. Barnes, © 1958 by John Wiley and Sons, New York, p. 214, Table 13.4.

5. Momentum should be used to promote the completion of the operation, or should be minimized if it takes muscular energy to overcome.
6. Arm motions should be smooth and flowing with a minimum of sharp directional changes and zigzagging.
7. Restricted and controlled movements are less efficient and require more energy than movements along natural arcs.
8. Rhythm should be used for pacing and to promote steady work progress.

2. Arrangement of physical elements

1. Everything involved in the process should have a definite and permanent place at the work space.
2. Tools and equipment should be placed for ease of use in the predetermined sequence of operations.
3. Physical elements should be located close to the operator and centered in front of the operator.
4. Wherever possible, gravity feed devices should be used to receive and transport raw materials and finished products.
5. Delivery of raw materials should be close to the point of use.
6. Effective operation requires adequate lighting.
7. The height of the workplace and the chair should be designed to facilitate ease of movement both while seated and while standing.
8. Workplace arrangement and furniture should be designed to encourage good posture.

3. Design of tools and equipment

1. Nothing should be done manually that could be done more efficiently by a mechanical device or other aid.
2. Whenever possible, tools should be multifunctional and specific to the operation.
3. Tools and materials should be prepositioned whenever possible.
4. When individual digits perform separate functions, the load requirements on those digits should be commensurate with their capacities.
5. Handles should allow for as much surface contact with the hand as is practicable, this being more important as the amount of pressure exerted on the tool rises. In general, screwdriver handles should be tapered toward the tip of the tool.
6. Levers, crossbars, and hand wheels should be positioned so that the operator can manipulate them with the greatest amount of mechanical advantage and the least amount of bodily movement.

QUESTIONS

1. Define *work*. How does this definition guide us in facility design?
2. Define *job design*. How does it affect efficiency and productivity?
3. What are some of the key considerations in developing job designs for the final phases of production?
4. How applicable is micromotion analysis in today's industrial environment? Where would it be most applicable? Where least applicable?
5. Name and define the three motion-study tools.
6. What is operational analysis? How is it helpful in work analysis?
7. What is a therblig? How is it used?
8. At what tasks would people be expected to outperform machines?
9. It is said that machines mimic human capabilities. Is this a true representation of the facts, or do machines perform in a fundamentally different manner?
10. What is a flow diagram? How is it useful in work systems study and analysis?

13
Product and Plant Design

I. OVERVIEW OF PRODUCTION AND OPERATIONS FACILITY DESIGN

The purpose of any facility is to economically produce the product or service that the firm intends to offer to the public. The key terms here are *economically* and *produce*. What every facility must do is create an environment in which the product can come into existence with all of the necessary capabilities and characteristics that it was intended to have, and to do this in a manner that is economical, that is, in a manner that achieves this end with a minimum of expense, time, and effort.

When any facility is designed, the way in which it is designed communicates to the workers what the desires of the firm are with regard to its approach to production. Quite literally, the design itself informs those working therein what is expected of them. Open spaces translate into encouragement of group activities. Closely defined work spaces with little room for movement translate into the worker's staying put. A lack of water fountains discourages taking breaks, while well-defined rest areas with tables and chairs encourage socialization. In this way, the medium becomes the message, to borrow a communications principle, and the individual behavior of the worker is influenced if not dictated by the facility's design.

It is possible, however, for the plant to be perfectly efficient in design and still not be economical to operate. There is always a degree of compromise between what is desirable from the point of view of physical efficiency and what is reasonable from the point of view of cost. In this chapter we discuss two aspects of this dichotomy, the first being the design of the product or service itself to ensure that it does what it is intended to do, and the second the analysis of investment and costs that must accompany any plant design in order to be certain that it is economical.

II. DESIGN OF PRODUCT OR SERVICE

A. The Nature of the Design Problem

The problem consists of determining how to design a product that is both functional and convenient while being aesthetically desirable to the customer. Much of this work takes place before the facility planner ever enters the picture, but once the product has been put into a final form, the facility design must coincide with the requirements of that design. Since form follows function, the facility design is dependent on what most efficiently and economically creates the product. Among the considerations in design are the characteristics of the product, the production design itself, tolerances, and economy.

1. Characteristics of the product. Every product is designed to have certain characteristics. The purpose of those characteristics is to create utility so that the item may be sold. Characteristics include not only the obvious factors, such as materials of construction, mechanical subassemblies employed, engineering principles applied, size and weight, weatherability, or power consumption, but also less obvious factors, such as ease of handling, aesthetics of the overall design, color, texture, ease of use, consumer appeal, and price. A physical product must have more than just practical utility to be sold. If aesthetic, psychological, and economic characteristics were not also important, everyone would drive the same type of car in the same color, live in identical houses, and wear identical garments. Utility implies much more than simple usefulness, and all of the factors that reflect individual needs must be taken into account.

Effectively, an infinite number of possible characteristics can be put into a product. The trick is to determine which are important and which are not, which is basically the function of the product concept. The product concept, a statement of the overall purpose of the product, including what market it will target and what niche it is to fill in the life of those who buy it, defines what is an important characteristic and what is not. If an item is being produced for the retail home-use market, it must be beautiful as well as functional, catch the eye of the consumer, and be convenient to use, even for the most obtuse of potential customers. Alternatively, if the product is for use in industry, its ability to perform its function economically is far more important than how pretty it is. The home consumer might be attracted to a screwdriver with a "space age" look, whereas the professional carpenter is looking for low price and utility. Thus there are different sets of characteristics for different markets. From the point of view of the facility designer, the inventory of characteristics considered significant to the product will dictate the production system design, since those are the characteristics that must be created in the finished product.

2. Production design. This is a reflection of the product content. Mechanically, a number of steps must be followed to create the finished good, and the manner in which these steps take place is the responsibility of the production process. In designing this process, it is necessary to (1) define the product in terms of its various characteristics and purposes; (2) define the subunits that make up the final product; (3) define the parts (further subunits) that combine to form these subunits; (4) determine the relationships that exist among these parts and subunits, such as what can be done simultaneously, what steps are dependent on prior completion of other steps, and which are time-critical steps; (5) determine what subassemblies and parts must be subjected to the same activities; and (6) determine how best to coordinate the simultaneous creation of all subunits, where possible, and final assembly of the finished good. This is a top-down, systemic design process whereby the individual elements of the system are defined and then their interrelationships are determined to create a model of the system in its entirety.

It should be noted that there is no right way to design a production process. There are simply designs that work and designs that do not work, and the criteria for defining what works is a matter of what is important to the individual firm. Again, there must be an acceptable balance between the economics of production and the efficient achievement of product ends.

3. Tolerances. This is an issue that is decided at the product design stage when desirable characteristics are determined. The issue here is one of how much error is considered acceptable in the manufacture of the product. As the word indicates, it is a matter of deciding how "tolerant" the firm intends to be of imperfections.

Tolerances are important for two reasons. Primarily, they are of importance in defining the level of quality of the product. One characteristic that a product invariably has is some level of quality. Whether the firm chooses to produce a relatively low-quality product for mass market appeal, or a higher-quality product with a higher price for a more prestigious market, the fact remains that the desired quality must be stated, and this statement dictates how rigid the tolerances of manufacturing will be. These manufacturing tolerances further dictate the type of machinery that must be used, with more exacting tolerances requiring machines capable of achieving those tolerances, as well as the nature of production controls to ensure quality, and the inherent cost of production that results from these tolerances. The facility planner must build into the facility the capacity to achieve the desired tolerances and the capacity to detect errors in order to ensure certain levels of quality.

4. Economic considerations. The economics of the production process permeates every state of the facility design. From overall budgetary constraints to individual savings in capital outlay and operating costs, there is a limit to the amount of capital that can be invested in a given facility, and the facility designer must be constantly aware of these limitations. With every step there is a trade-off that must be made between quality and economy. As mentioned earlier, the perfect production facility may be prohibitively expensive to construct and maintain, and what must be decided is what levels of cost and of production efficiency are acceptable rather than what is optimal. In this planning process, there must be a vision of both the long-run and short-run conditions under which the plant is to operate. It is not only a question of how much to spend, but also one of when to spend it. Is it more efficient to spend more now and avoid later maintenance costs, or to spend less now and absorb higher maintenance and alteration costs later? All these are considerations necessary in designing the product and the production facility.

B. Design of a Service

Functionally, the only difference between the design of a product and the design of a service is the nature of what is being produced. With services, there is still a product; it is simply less tangible and of a different structure than a physical product. All the elements of design that go into product design are still present. Depending on the nature of the service in question, there is still a need for maintenance, repair, processing of physical goods, transportation of goods or information, scheduling, and quality control.

For example, in the case of a service involving a physical good we could cite an industrial laundry. Although the essential element of the operation is the service rendered to the customer in cleaning the garments in question, there is a physical process going on. Clothing must be inspected and prepared for washing, then transported through a series of steps to achieve the final product (a clean garment), and throughout the process there must be quality control. Plant layout and machine tolerances are as important here for a successful and economical operation as they would be for the production of a physical product. Systemically, the process involves the following steps:

1. Determination of the service's characteristics and goals (to produce a clean and pressed garment, package it neatly, and have it ready for pickup or delivery at a certain time).
2. Definition of the subunits (receiving, identifying, cleaning, pressing, packaging, and distributing to pickup station or delivery).
3. Definition of the parts of each subunit (receiving involves writing orders, tagging garments, tendering receipts to customers, placing garments in proper receptacles, and so on).
4. Determination of the relationships among various steps (receiving is an ongoing function manned at all times, whereas the washing and cleaning function will sometimes simultaneously take place, and so forth).
5. Determination of what subunits must be subjected to the same activities (all garments must be received; all garments go through the press cycle, whether washed or dry cleaned, and so on).
6. Coordination of subunit activities (simultaneous operation of dry cleaning and washing equipment, simultaneous receiving and pickup counter operation, and so on).

Even when the service does not involve any physical goods at all, there is still a need for analysis and design of the facility. With an accounting service, for example, the product is the generation and manipulation of data to give useful information. Information may be nonphysical, but it is still subject to various levels of quality, tolerances, systemic determination of subsystems involved (balance sheet, profit-and-loss statements, inventory control reports, and the like) and characteristics (accuracy, punctuality, reliability, and so forth).

C. Standardization

Any process needs standards. Without them it is impossible to know whether or not what we intended to achieve and what we actually achieved are the same. Actually, standards are only a means of defining the important characteristics of a product in hard, concrete terms. By translating characteristics into a set of standards, the firm is able to use a process of communication to adjust its actions through time and achieve consistency in its production.

Standards are for the purpose of quality control. They define what the firm means by quality and operate as a gauge by which reality can be checked against the ideal. In any case, standards should be objective, measurable, and understandable. Whatever the criteria used for setting them, if they are not determinate by quantifiable means, they have little statistical value.

To borrow from communications theory, the creation of a product is a four-step process involving input (of raw materials, labor, capital, and information), process (by which there is a conversion of inputs into a finished product), output (which is the finished product, whether physical good or service), and feedback. It is this last element, feedback, that interests us in determining standards, since feedback is information about the nature of the finished product which can be used to determine if changes should be made in any of the other elements of production. It is only by establishing standards that we can make comparisons to determine what corrections should occur in order to maintain quality.

1. Methods of establishing standards. Several methods are used for setting standards, and in the more sophisticated production processes, all of them appear in some form. The purpose of standards, it is to be remembered, is to guarantee that the output of the operation is as close as possible to the desired output as defined. What is sought is adherence to the standards, uniformity of product, and a control system that stops serious divergence from the norm before it occurs.

a. Characteristics. These can be used as a basis for standards, and usually are. Whereas stated characteristics may be vague in nature, standards are as concrete and quantifiable as possible to allow for successful measurement of results. Hence a product characteristic may be "ease of handling," and in terms of product standards that may translate into a specific weight, a specific size, a specific surface texture, or a specific heat conductivity rating. The characteristic in this case is suggestive of an idea of what it means to be easily handled, and the standards that result are the concrete physical characteristics that embody that quality.

b. Industry Standards. This is an excellent source of standards, particularly minimal level determinations. Industry standards are usually a reflection of many man-hours of research and experience with the product or products in question, and represent the best estimate of what is important in producing the product. By starting with industry standards and then fine-tuning them to meet the specific product concept that the firm has, it becomes much easier to establish quality criteria.

c. Competition. This is another good source of standards, particularly minimum standards. By checking with the standards set by competitive manufacturers, the firm is able to determine what must be done, at a minimum, to successfully compete with the other firms in terms of quality. This approach also represents a first step rather than a final determination, and is useful in determining the general direction in which the final standards should go.

d. Parts Manufacturers. Suppliers are a ready source of advice on what standards should be set in order to achieve the desired results. Given the performance criteria of the product, subassembly suppliers may be in an excellent position to advise the final assembler on what standards are necessary to ensure achievement of those performance criteria. In some cases, the nature of the standards used in the subassemblies partially determine the standards for the whole product. This is not to say that the standards set by the firm have to coincide with those set by subassembly manufacturers for their average products. It is sometimes possible or even necessary to demand higher standards of performance of subassembly manufacturers than they are accustomed to encountering. The standards for government-purchased microchips, for instance, sometimes exceed the standards of microchip manufacturers for their products. As a result, only those microchips that "accidentally" achieve the higher government standards are made available for sale to this customer. Fortunately for the microchip manufacturers, the economies of the product allow for this.

e. Internal Testing. This is still another source of standards. Internally testing various parts of the prototype of a product yields a great deal of information about what is required of a successful production model. Through this process, the limits of the prototype are tested to determine if it will hold up to commercial use and customer requirements, and then the standards are set to coincide with these levels of performance. Today, much of the testing work done in engineering products is performed by sophisticated computer programs that simulate actual real-world conditions. Items as diverse as airfoils, automobile parts, and nuclear reactors can be tested through this method, and standards can be set. This is particularly useful where destructive testing would be too time-consuming or too expensive. It is far more reasonable to test an airfoil design in a simulator than it is to build it and drop it from the belly of a B-52, particularly considering the hazardous nature of a manned test.

2. Advantages and disadvantages of standards. As indicated, standards are necessary if we are to have any consistency in output. They are advantageous in that they give us something to measure reality against, they ensure that the product the public receives is consistent with the product the firm states it is producing, they allow for a control of costs and for unit pricing, and they increase the demand for the product in the market by instilling a feeling of confidence in the mind of the purchaser.

There are, however, some disadvantages to standardization as well. Standardization may lock us into a specific point of view that limits our freedom of action. When we set standards, we are defining the reality surrounding the product. We are in effect limiting the characteristics that the product is allowed to have and therefore limiting our potential for creatively serving the public. The purpose of standards is to help us control the production system, but standardization has the capacity to become tyrannical in its demands and thus control our actions as well. By constructing productive processes and machinery in accordance with tight standards, the firm may find that it has limited its capacity to change in the face of changing demands. The firm and the facility planner must remember that there is a balance to be maintained between the restrictions of standards and the freedom of flexibility. If either exists to the exclusion of the other, we limit our capabilities.

D. Metrication

Metrication is an issue that must be dealt with in both product and production facility design. It has become a major problem in international trade, but since a large percentage of a company's output is designed for foreign sales, the metric system has been established as the norm.

Because of the inherent benefits of using a ten-based number system, metrication is widespread over all of the industrial world with the notable exception of the United States. In this country, there has been a general reluctance to adopt the metric system, and metric conversion has been slow. This is due to a basic resistance toward change in the society, the fact that only a small percentage of U.S. production is targeted for the international market, and the cost and difficulty inherent in the conversion itself. Yet as we move more and more toward the realization that we exist in a true world market, the metrication process can be expected to gain strength. This is already evident in areas where even domestically targeted markets are supplied with goods containing parts of foreign origin.

There are basically two types of conversion that can be made in moving from the English system to a metric one. The first type, called *soft conversion*, is a matter of starting with measurements made in the English system and converting them to metric, offering data in both forms. The other type of conversion, called *hard conversion*, involves a straight conversion to metric with total elimination of the English system.

1. Soft conversion. This is considered by many to be the easiest and most palatable type of metric conversion. With soft conversion, machinery and tolerances are modified to present both English and metric scales so that they are simultaneously available. An obvious example of soft conversion is the speedometer on most newer automobiles, which gives speed in both miles per hour and kilometers per hour. With this type of conversion, the measurements may be originally made with either system and the data presented in all systems. The advantage here is that the engineer and the worker are allowed to continue working in the system with which they are most familiar, while results are offered in both systems. This type of conversion is least expensive, since it is only the terms used that are changed, not the dimensions in which things are built. As far as actual use of a product is concerned, the product has the same dimensions and tolerances whether they are expressed in inches, centimeters, or domgags (which can be defined as the span of the left central toenail of a fully mature Indian elephant).

2. Hard conversion. This is a more expensive and complete process, involving the changing of all machinery and designs to a totally metric system. To become involved in hard conversion requires a larger amount of outlay for such changes as converting digital readout displays and redesigning the physical dimensions of machinery and equipment, which would wreak havoc with domestic sales of products that will not interface with domestically produced English system products. Such an approach would also result in increased costs from revision of documents containing specifications and standards, retraining of personnel, and the necessity of maintaining two entire inventories of tools, parts, and casings.

3. Relationship of metrication to design and standardization. The relationship should be obvious. In designing the product and production systems and in setting standards, it must be decided what type of measurement will be employed. If it is necessary to use both an English and a metric system, it becomes necessary to account for size differences in the design and for how each will affect the availability of machinery and equipment, and to determine if there are to be separate production systems for each, as well as the cost of using either hard or soft metric conversion methodology. It is suggested that unless there are to be extensive foreign sales of the product in question or it is anticipated that there will be substantive advantages to metrication, the firm may find the cost of conversion more prohibitive than the advantages can justify. In this, too, the key to the standardization process is to balance the advantages of rigorous measurement against inherent flexibility in the system's design.

4. Problems of the conversion process. Though the physical difficulties of conversion have already been highlighted, it might be interesting to make note of some specific consequences of hard conversion. The United States uses the English system for measurement. Specifications for dresses designed in the United States and manufactured overseas must be converted to the metric system. Some of the fitting problems retailers face in the United States are from goods so manufactured. The same is true for automobiles that are manufactured overseas. Once they need repair, many a bolt is damaged because the wrong tool is used on the car. For the sake of efficiency, facilities should run on the same measurement system as other facilities with which they do business.

III. PLANT AND EQUIPMENT INVESTMENT ANALYSIS

Along with an analysis of the efficiency of production methodology, it is necessary to analyze the costs involved. In this section we look at some of the common methods used to determine the costs of production in an effort to determine the profitability of the proposed operation. A number of different financial techniques are used, some simple, others quite complicated. Whatever method is employed, its purpose is to shed light on the actual costs, both long-run and short-run, of developing the facility plan. Here we concentrate on determination of return on investment, the payback method of capital recovery, depreciation, the present value method of valuation, internal rate of return, the MAPI system and other issues.

A. Return on Investment

This is a method of determining the efficiency of capital outlays by the firm on any given investment through an analysis of the percentage of that investment returned to the firm in a given time period. That is, it is a measure of the percentage of the investment that is recovered through net revenues in a given time period, expressed as a ratio.

1. Simple rate of return. The simplest determination of the rate of return is the ratio of net income to total investment. In the case of the facility planner, this total investment would be the capital invested in the facility itself and net income would be projected net income from the sale of goods or services produced at that facility. The formula for the simple rate of return is

$$\text{Rate of return} = \frac{\text{Net income}}{\text{Total investment}}$$

For example, assume that there is to be an initial investment of \$15,000,000 in a major plant facility and that the project is expected to yield a net income of \$2,800,000 per year. The rate of return on the investment would be

$$\frac{\$2{,}800{,}000}{\$15{,}000{,}000} = 18.7\%$$

This is the rate at which the investment is returned to the firm, about 18.7 percent per year.

There are several relatively obvious reasons for determining the rate of return on invested capital. To begin with, it is necessary to recover the capital investment at a rate greater than the cost of the capital itself, or there is no profit to be made. If capital costs 10 percent, either in cost of loans or in opportunity costs because the firm chooses to invest the capital in the facility rather than some other investment, then the rate of return must exceed this 10 percent, or the firm experiences net losses on the project. The most advanced facility imaginable is useless if the firm goes broke attempting to run it. Second, one primary purpose of being in business is to make a profit, and firms set goals for the profit they should be realizing. If the rate of return on the facility is below the requirements set by the firm, then the project is not acceptable. In this case, the firm is not losing money; it is simply not making enough to warrant the venture. An acceptable level of financial return must be present to make it worth the firm's while to pursue the development of the installation.

2. Unadjusted rate of return on average investment. A slightly different approach to the determination of a simple rate of return on investment is to perform the same calculation with investment defined as average investment over the projected life of the facility. In this case, not only are initial investment costs taken into account, but ongoing costs as well. Beyond the initial investment, the firm can be expected to incur future capital outlay costs for machine replacement, expansion, and upgrading, and these costs of capital can be taken into account by calculating the total expected investment for the life of the asset and then determining the average expected investment for any time period. This can then be compared with the expected average net income for the time period. Thus if we were calculating the rate of return on a yearly basis, the formula would be

$$\text{Rate of return} = \frac{\text{Average net income}}{\text{Average investment}}$$

B. Payback Method

This approach to investment analysis bases its results on the length of time it takes to recover the capital initially invested. The importance of this figure lies in the fact that the faster the capital can be recovered, the faster the facility itself becomes profitable, allowing the capital to be reinvested elsewhere more quickly. Payback periods are susceptible to company policy, just as rates of return are. Many firms set upper limits on the length of time they are willing to wait for their investment to be recovered. It must be remembered that within the firm there are probably a number of other projects competing for the limited funds available. So, since money has a time value (as will be demonstrated shortly), the faster the investment can be recovered from a project, the more attractive that project will appear. It is generally accepted wisdom that faster paybacks are more desirable because (1) they are less risky since capital is not tied up as long, (2) they allow the firm to see profit sooner, and (3) they allow for a more rapid reinvestment of available capital, thereby making the use of the capital more efficient.

In most cases, the income that is realized from a project varies from year to year, and the rate at which it varies is itself variable from project to project. In the payback method approach, therefore, the schedule of positive cash flow for each project is compared to determine which has the shortest payback. To illustrate, note the cash flow table for three alternative investments given in Table 13.1. All three involve an initial investment of $15 million, yet the time required to recover (that is, pay back) the capital in-

TABLE 13.1 Comparison of Potential Investments Using Payback Method (Initial Investment for Each Project: $15,000,000)

Year	Project I	Project II	Project III
0	(15,000,000)	(15,000,000)	(15,000,000)
1	2,800,000	4,500,000	1,900,000
2	4,500,000	2,800,000	2,900,000
3	6,000,000	6,000,000	3,900,000
4	4,000,000	2,000,000	4,900,000
5	2,000,000	1,000,000	5,900,000
6	2,000,000	5,000,000	—
7	—	1,000,000	—
8	—	1,000,000	—
Totals	$21,300,000	$23,300,000	$19,500,000
Payback period	~3 yr 5 mo	~3 yr 10 mo	~4 yr 3 mo

Note: Initial investment for each project: $15,000,000.

vestment is shortest for the first alternative. All other factors being equal, this would be the one chosen as most desirable for the firm.

Note that though project I does not yield the greatest return on investment, the fact that it has the shortest payback period makes it the best bet for the company in this type of analysis. It is assumed that the lower yield is more than offset by the advantages of a quicker recovery of capital outlay.

The payback method is a useful approach because it has the advantage of being simple to compute, is considered a standard financial analysis approach that is easily understood, and is very applicable for firms with liquidity problems who may need to recover their cash outlays in a short period of time. On the negative side, payback analysis ignores both the time value of money and the returns experienced after the payback period. Were these considered in the example, an entirely different answer may have been derived. In addition, there is a tendency of the payback method to build into the firm a mind set that is predisposed toward shorter-range projects, which may cost the firm in the long run.

C. Depreciation

The principle behind the concept of depreciation is the fact that machinery and equipment, as well as other assets, have a limited useful life, and once that life is passed, the capital investment is no longer useful. This deterioration in usefulness does not occur all at once, however. It happens little by little over time. As such, valuing the investment at any point in time requires adjusting the value for the degree to which it has deteriorated, or depreciated. Assets depreciate at different rates and in different ways. A supply of inventory may depreciate at a relatively steady rate, a like percentage of the total supply deteriorating during equal periods of time. Machinery, on the other hand, may depreciate differentially, as does a yacht or airplane, which loses most of its value during the first year. It is not too surprising, then, that there are different schemes available for legally depreciating capital equipment, and since this item affects profitability in an indirect manner, it is important to determine which method of calculating this value is most advantageous to the company. The three approaches most commonly used by accountants are straight line, sum of the years digits, and declining balance. We will look at each of these in detail.

1. Straight-line depreciation. This is the simplest and most straightforward process of determining depreciation. In the straight-line method, it is assumed that the same percentage of value for an item is used up during each year of the life of the item. There are two factors to be considered in this analysis, those being the life of the asset in years and any salvage value that the item may have. Salvage value is the value that the item retains as junk, or the minimum value of the asset after the useful life of the item has expired. A machine that initially cost \$10,000, for instance, may have a salvage value of \$800. If the useful life of the item is assumed to be 10 years, the yearly depreciation will be given by the formula

$$\text{Depreciation} = \frac{\text{Cost} - \text{Salvage value}}{\text{Useful life}} = \frac{\$10{,}000 - \$800}{10\text{ yr}} = \$920 \text{ per year}$$

The primary characteristic of this type of depreciation scheduling is the uniformity with which changing values take place. Both the decline of value at which the asset is carried and the amount of total value depreciated change at a uniform rate. This means that any credit taken against taxable revenues through the depreciation of capital assets is taken at the same rate for every year of that asset.

2. Sum of years digits depreciation. Another approach to depreciation whose methods are collectively known as accelerated methods, operates on the basis of the assumption that assets depreciate more when they are new than when they are older. Accordingly, various schemes are used to depreciate assets heavily during the early years of their useful life and less in later years. One of these schemes that is generally accepted is the sum of the years digits.

With this method, the percentage of the asset's value that is depreciated varies greatly, the heaviest depreciation taking place in the first year, and a steadily decreasing amount taken each succeeding year. The mechanics of the approach involve determining the fraction to be depreciated each year from the ratio of the sum of the digits of the useful life of the item, and the total number of years left in the asset's useful life. Simply put, the formula is

$$\text{Depreciation} = \frac{\text{Remaining useful life}}{\text{Sum of years digits}} * (\text{Cost} - \text{Salvage value})$$

With this method, if an asset has a total useful life of seven years, an initial cost of $5000, and a salvage value of $300, then the depreciation value for the third year would be

$$\frac{4}{1 + 2 + 3 + 4 + 5 + 6 + 7} * (\$5000 - \$300) = \frac{4}{28} * \$4700 = \$671.43$$

With a short-lived asset, this type of analysis is not too time consuming, but with an asset whose useful life is 20 to 25 years, it can become cumbersome. Fortunately, the denominator in the fraction is easily obtained through the formula

$$\text{Denominator} = \frac{N(N + 1)}{2}$$

where N = number of years. In the example cited, this would be

$$\text{Denominator} = \frac{7(7 + 1)}{2} = 28$$

The primary advantage of this approach is the large amount for depreciation that takes place during the early years, allowing the firm to reduce taxes and net profits early, and free capital for other investments. The chief disadvantage is that if the income received from the new asset tends to rise through time, the higher levels of depreciation will not be there in later, more profitable years to counter those high returns. The trade-off could be unfavorable in terms of taxes if such a condition were to exist.

3. Declining-balance depreciation. Another accelerated method of depreciation is the declining-balance approach. The rationalization for this approach is the same as for the sum of the years digits, that is, that more depreciation should be taken in the early years of the asset's life than in later years. Here, there is simply a declining rate at which the depreciation is taken. Theoretically, any declining rate that can be defended can be used, although it is common to reduce the value at a rate equal to twice the percentage of a straight-line depreciation. This particular scheme is known as the double-declining-balance method. For instance, if the asset has a 10-year life, the double-declining-balance method would reduce the value of the asset by applying twice the straight-line rate to the remaining asset value balance each year. Hence the depreciation rate would be 20 percent of each year's value in a 10-year straight-line example (10% per year). Under this scheme, in the first year an asset originally valued at $10,000 would depreciate at 0.20(10,000), or $2000. In the second year, depreciation would be 0.20(8000), or $1600. In the third year, it would be 0.20(6400), or $1280, and so forth.

In this case, the rate at which depreciation takes place is the same each year, whereas the balance to which that rate of depreciation is applied varies. This creates a relatively heavy rate of depreciation during the early years and smaller rates later on. Table 13.2 shows a comparison for the depreciation of an asset originally valued at $15,000 with a salvage value of 0 (how convenient), using each of the three methods.

TABLE 13.2 Comparison of Methods of Depreciation

	Remaining Book Value		
Year	Straight Line	Sum of Year's Digits	Declining Balance
0	15,000	15,000	15,000
1	13,500	12,545	12,000
2	12,000	10,363	9,600
3	10,500	8,454	7,680
4	9,000	6,818	6,144
5	7,500	5,454	4,915
6	6,000	4,363	3,932
7	4,500	3,545	3,146
8	3,000	3,000	2,517
9	1,500	2,727	2,014
10	0	0	0

Original value of asset: $15,000. Useful life: 10 years.

D. Present Value

We discussed present-value analysis in Chapter 1; however, for the sake of clarity we reiterate briefly.

Which would you rather have, one dollar today or a dollar one year from now? If you're like most people, you'd rather have the dollar now. How about one dollar today or two dollars one year hence? Three dollars in a year? In fact, how much money would you have to receive in a year to convince you to forego one dollar now? The process at work here is the basis of what is called the present value concept. We can define present value as the dollar value at the present time of a future payment or series of payments which has been adjusted for the time value of money. Money has a variable value. The closer it is to the time that money is in our hands, the more valuable we believe that money to be. There are numerous reasons for this, some practical, some psychological. Whatever the reasons, we realize a difference in the value of money depending on when it is received. Facility planners can use this concept in determining the value of the return on investment received by a firm from the future income stream created by the production of a proposed facility. This type of analysis takes into account not only how much profit will be realized, but when it will be realized as well.

The determination of the value of money at a specific time is known as *discounting*, probably because the further into the future the realization of the money moves, the more we discount its importance. The rate of discount depends on the rate at which the value of money changes through time, usually determined by the interest rate available in the market. This interest rate represents the opportunity cost to the lender that must be incurred if that person has to wait for future payment of the money. In other words, how much more money will I have to be paid in the future to convince me not to keep my money now? There is a choice of money now or money in the future.

Let us assume that the opportunity cost (also called an opportunity rate or discount rate) is 7 percent. This means that if a person foregoes possession of the money today and waits a year, then that person will insist on receiving an additional 7 percent one year hence. Therefore, the relationship between money now and money one year from now is 1:1.07 (1/1.07 = 0.935, as shown in Table 13.3). Remember that if we use the going rate of interest as the discounting rate, we are operating with the average opportunity cost. Different projects lasting for different periods of time will have different discount rates due to the length of time involved, the degree of uncertainty and consequent risk, and the individual circumstances under which the investor is investing. Also, different individuals will have different attitudes about the time value of money, which may affect the rate as well.

Given some discount rate, we can determine by formula how much a future amount of money is worth in terms of today's dollar. This formula is given in the form:

$$PV = FV_t \left[\frac{1}{(1 + r)^t} \right]$$

where PV is the present value, FV_t is the future value in year t, t is the number of years into the future that the money is realized, and r is the opportunity rate or discount rate used.

Using this formula, one is able to find the present value of an amount of money received at any known time in the future. Fortunately, there are tables available for standard discount rates and numbers of years that can be used to avoid excess calculations. To illustrate, a reanalysis of the information presented in Table 13.1, with the added benefit of the present-value approach, shows us an entirely different picture of profits than the one offered in the payback method. Table 13.3 shows the result of this change in valuation.

As can be seen from the comparison of values before and after the time value of money is taken into consideration, what appeared to be the most profitable undertaking, project II, was in fact not the most profitable because

TABLE 13.3 Comparison of Total Returns with Present-Value Discounting (Rate of Discount = 7%)

		Project I		Project II		Project III	
Year	PV	Nominal Return (millions)	Real Return (millions)	Nominal Return (millions)	Real Return (millions)	Nominal Return (millions)	Real Return (millions)
1	0.935	2.8	2.62	4.5	4.21	1.9	1.78
2	0.873	4.5	3.93	2.8	2.44	2.9	2.53
3	0.816	6.0	4.90	6.0	4.90	3.9	3.18
4	0.763	4.0	3.05	2.0	1.53	4.9	3.74
5	0.713	2.0	1.43	1.0	0.71	5.9	4.21
6	0.666	2.0	1.33	5.0	3.33		
7	0.623			1.0	0.623		
8	0.582			1.0	0.582		
Totals		21.3	17.26	23.3	18.325	19.5	15.44

of when the income from the project was realized! In a complete analysis, the present-value method would be extended to include the determination of net present value, or NPV, which is just determining the difference between the initial outlay (in the example $15,000,000) and the total inflow adjusted for the time value of money. This would yield an adjusted value of income over outflow.

Present-value analysis is extremely important in determining the true profitability of any undertaking, particularly long-range projects, such as a facility construction program.

E. Internal Rate of Return

This is a short step beyond the present-value method of valuation. The difference with the internal rate of return method is that rather than deciding on an appropriate discount rate and calculating real returns, this method sets the value of the future income flow equal to the initial capital outlay and finds the discount rate that yields that parity. In other words, with the internal rate of return, the sum total of future income flows is assumed to be just as valuable in terms of present value as the initial outlay of funds. Internal rate of return (IRR) is designed to find the discount rate for which this is true.

For example, if we have a project that originally demands a capital outlay of $20,000 and will yield a return of $6000 for six years, we can determine the internal rate of return as that rate of return that will make $20,000 of present capital equal to the $6000 payments over a six-year period. The formula for this determination is given by:

$$\mathrm{IO} = \frac{\mathrm{ACF} * n}{(1 + \mathrm{IRR})^t}$$

where IO = the initial cash outlay, n = the expected life of the investment, IRR = the project's internal rate of return, t is the year for which the calculation is made, and ACF_t is the annual cash flow for the specific time period.

Said another way, how much initial cash would be needed to fund a project that yields $8000 per year for 14 years, when the company requires an 8% return on investment?

$$\mathrm{IO} = \frac{\mathrm{ACF} * n}{(1 + \mathrm{IRR})^t}$$

$$\mathrm{IO} = \frac{\$8000 \times 14 \text{ years}}{(1 + 0.08)^{14}} = \$38{,}132$$

In the next illustration given, it was necessary to calculate the values for the formula for each of six years with ACF equal to $6000, and IO equal to the initial $20,000 investment. Fortunately, this is a relatively easy task, as the only factor that changes in the calculation is the number of t, the year for which the calculation is made. Tables exist, in fact, for cash flows that are equal in each year, and this considerably helps in the calculations. For the example given, the internal rate of return is 10.29 percent.

To solve for IRR, using above formula, transpose below equation. Therefore:

$$\text{IRR} = \sqrt[t]{\frac{\text{ACF} * n}{\text{IO}}} - 1\ (100\%)$$

then substitute values

$$= \sqrt[6]{\frac{\$6000 \times 6 \text{ years}}{\$20{,}000}} - 1\ (100\%)$$

$$= (1.1029 - 1)(100\%) = 10.29\%$$

The resulting rate of return is then compared with either some cutoff criteria that the firm has for minimum acceptable rates of return, or with the internal rate of returns of other projects. If the project meets the company criteria, then the project is viewed as acceptable. For the example, as long as the firm's cutoff rate is less than 10 percent and no competing projects produce internal rates of return greater than the 10.29 level, then the project will be undertaken.

F. The MAPI System

This model was developed by the Machinery and Allied Products Institute in the mid-1960s, and offers a means of determining the rate of return based on a comparison of conditions with and without the anticipated asset. Whereas it is a relatively simple model from the theoretical point of view, an actual implementation requires the use of charts and tables available from the MAPI itself, and accordingly, it may have limited application unless one has access to these items. By way of introduction, the difference between absolute and relative rates of return is discussed below.

1. Absolute and relative rates of return. The difference between an absolute and a relative rate of return is that the calculation of the latter involves specifying the time period during which the cash inflow is received and relating that to the percentage of initial cost used up during that period of time. This differs from the absolute rate of return approach, which assumes that all initial cost takes place in year zero and that income over a period of time is compared with that initial cost by any of the various means already discussed. Suppose, however, that we choose to check the rate of return periodically by comparing the income of a given year with that part of the initial cost of the asset remaining in that year. Allocation of cost is done through a process called *deferment*, by which we spread out costs over time so that they are accounted for during the time period for which they are applicable. Depreciation is a recognition of this, deferring reductions in the value of capital equipment to the time when a given portion of the total original value is used up. With the deferment concept, we can measure rates of return by comparing periodic capital asset values with periodic levels of income. Such a rate of return is itself periodic rather than absolute for the entire life of the asset. For example, if a $10,000 capital asset has a 10-year life and straight-line depreciation is used, the value of the asset at the end of the fourth year would be $6000. Comparing this to a net income of $2000 in that same year, the periodic rate of return would be $2000/$6000, or 33.3 percent.

With this approach, it is still possible to use a net present-value approach, discounting the value of the income before calculating.

2. Next-year deferment concept. The MAPI approach draws on the theory behind relative rate of return calculation in that it determines a rate of return based on how well off the company will be next year if the project in question is undertaken. The basic idea is to compare the conditions that will exist in the following year if the project is begun with conditions next year if the project is forgone. A six-step process is used to accomplish this:

1. Determine the cost (net capital investment) of the proposed equipment purchase.
2. Determine how much net earnings have increased as a result of using the proposed equipment.
3. Determine the effect of the purchase of the new equipment on depreciation schedules, salvage values, and allowance for replacement purchases created by the project.
4. Determine the next-year depreciation (capital consumption allowance) by use of MAPI-supplied charts and tables. Here we are measuring the reduction in asset value resulting from the purchase and use of new equipment. This effectively offsets the reduction in the amount of allowance for old equipment depreciation, since that old equipment is no longer in use (see step 3).
5. Adjust the projected income tax allowances to reflect the increase in net income anticipated from the use of the new equipment or project.
6. Calculate the rate of return, known as the *urgency rating*.

3. Calculation of next-year relative rate of return. This process involves the use of a simple formula that is developed from the activities listed in the six-step process. The actual formula for the urgency return is

$$\text{After-tax net return} = \frac{\text{Net gain from project}}{\text{Net investment of project}} * 100\%$$

4. Comparisons of more than one year. These can be simply made by extrapolations of the above process. Two years into the future, or three, or any other number of years can be used in determining the return for a project. It should be noted, however, that in doing do, there is a reduction in the accuracy of the calculations that accompanies the new data, since the further away from the present a time period is, the less certain it is possible to be about the content of that analysis.

G. Other Equipment Investment Models

The range of models and theories behind them is enormous. Valuing various projects is a bit like analyzing history, in that choice of analysis methods depends on what you're looking for. What each firm considers important in determining project feasibility varies widely. Based on a firm's attitudes, different models will be chosen. Each has its own unique advantages and shortcomings.

In addition to those models already discussed, two others are well worth mentioning: the profitability index and risk analysis. These function as follows:

1. The Profitability Index. This is a very straightforward approach that calculates the expected profitability of the project through the use of the formula:

$$\text{PI} = \frac{\text{Original cost} + \text{Net present value}}{\text{Original cost}}$$

The value of this approach lies in the fact that it is possible to compare projects with different original costs by indicating the profitability of the net present value as a percentage of total costs. In this way, it is not the amount of income received but rather the percentage increase of expended funds that is found, which indicates how efficiently the capital is being used.

2. Risk Analysis. Such an approach computes the degree of risk of loss for a project based on present value, longevity, and other risk factors. Here the point is to determine which investment is the safest relative to the capital outlay. For conservative companies or firms who cannot afford to take risks, this may be the preferred choice of analysis. The primary difficulty with this approach is the determination of risk probabilities. Since the determination of probabilities may be highly subjective there is an inherent flaw in the approach.

H. Multinational Investment Decisions

In dealing with international markets, a number of problems and opportunities arise that tend to complicate the decision-making process. On the positive side, there is an increased opportunity to find alternative choices of location, equipment purchases, and operating costs because of the large number of potential locations and suppliers. The production methods and technological choices vary from country to country, and this creates a larger matrix of characteristic combinations to take into consideration. On the negative side, there are also increased uncertainties due to fluctuations in currency exchange rates and import–export costs, variability of labor and other inputs, and the necessity of dealing with different levels of risk. In the international arena, it is important to be aware of the differences that exist among cultures and among technologies.

QUESTIONS

1. A major component of modern facility design is standardization. Why is this so important to efficiency and productivity?
2. Equipment investment is a major consideration in the decision-making process of facility design. On what basis are such decisions made?
3. Define *internal rate of return*. How does it differ from return on investment?
4. What is the purpose of the MAPI system? Should it become the standard in plant design?
5. How does the design of a service operation differ from the design of a manufacturing operation? How are they similar?
6. Name and define three approaches to determining depreciation.
7. What are the primary advantages of standardization? What are the primary disadvantages?
8. In your opinion, is hard or soft metric conversion the most efficient method? Why?
9. Explain the concept of payback period and the payback method of plant cost analysis.
10. Name four considerations in the design of a new production facility.
11. How is production design dependent on the characteristics of a product? How is it independent of those characteristics?
12. How do multinational investment decisions differ from domestic investment decisions?
14. On what criteria should the tolerances of a product be based?
15. What are the steps in the MAPI system?

14
Design of Facility Processes

I. FACILITY PROCESSES

Within a productive facility there are a number of integral processes that must take place in order to create the product or service. These integral processes have elements in them as well, as does any system. We have defined a system as a set of subsystems that interact to perform some function or achieve some goal. And we have defined the way in which the systems operate, mainly that they involve input, process, output, and feedback, and that they operate according to certain universally applicable laws, basically the concepts of synergy, reciprocity, and balance. To further define a productive system, however, the following specific information should be reiterated.

Inputs. Inputs include not only machinery and equipment, physical plant, labor, and capital, but the customers and suppliers (the market), and the environment (political, economic, social, and physical), which includes technology, the state of international relations, general standards of living, and attitudes of the public.

Process. Processing of inputs involves all of the activities and internal elements necessary to convert the inputs into a viable finished product or service. This includes such considerations as the actual manufacture of the product, distribution (transportation, establishment and maintenance of wholesale and/or retail channels, and informational support), and support services for the manufacturing process.

Output. This consists of not only the goods or services produced, but also of the company image and returned support of the environment and community within which the firm is operating.

Collectively, this is the productive system. Note that in each of these subsystems there are a host of other subsystems that also have content, and that tying the entire package of subsystems together is a communications and corrective feedback system that is designed to control the activities of all the elements to insure that balance is maintained. In this chapter, a number of the subsystem elements in the productive process are discussed with exclusive emphasis given to the conversion part of the system that is known as processing.

A. Operations Systems

There are three primary operations systems used in the manufacture of physical goods, and these can be applied equally to the production of services with little difficulty. The particular type of operations system chosen is dependent on one major characteristic of the good: the number of units produced at any one time. Some goods are produced in a constant flow that for all practical purposes does not stop, such as the production of steel or the refining of oil. Other goods are produced in batches, usually to fill inventories rather than for the purpose of delivery to general customers. Such would be the production of replacement parts for machinery and equipment. Still other goods are produced intermittently, only on demand and in relatively small lots, such as production that takes place in small machine-shop operations. From these three different demand schedules for goods, we derive three types of processing known as continuous flow, batch, and intermittent.

1. Intermittent operations systems. Intermittent operations take place where small lots of products or nonrepeat order goods are being produced. Machine shops and cabinetmakers as well as print shops and general-contract prefabricates are all examples of this type of operation. With such manufacturers, the products are not standard items to be produced over a long period of time. They generally represent custom items that require custom runs. The machinery used is general purpose and can be used to produce a large number of different items, depending on the particular order. The workers are likewise specialized only in the operation of the machinery, not in the production of the particular item being produced. A cabinetmaker, for instance, can produce anything from the casing for a pipe organ to bookshelves to kitchen storage units. Although the products are different, they all require the peculiar skills of the worker and the equipment to produce them. Thus the production is intermittent; although reorders may occur, they are not on a regular basis, and they are not guaranteed. Intermittent operation is a very general approach to manufacturing. To make some cohesive sense of the facility, the plant is arranged to group machinery and personnel by process, with all milling work taking place in one area and all drilling or lathe work taking place in some other specified area. In this way, no matter what the product, the work in process can be trafficked from location to location in accordance with the steps in the production process.

2. Batch operations systems. Batch operations are similar to intermittent operations in that the production of the goods in question happens only periodically. Yet batch operations are dissimilar to intermittent operations in that there are periodic runs of the item, and these runs are always identical. They are done, however, in "batches" of either uniform or varying size, each batch being identical to the other in content. The item is standardized, as are the steps in its processing, but the frequency with which the processing takes place is variable. Since this is the case, there is still a tendency to use generalized machinery and generalized labor, although machinery and labor skills peculiar to the particular product may be employed if it is economical to do so. In simple cases, specialized jigs and setup equipment may be attached to general-purpose machines whenever the particular item is to be run. Batch operations often exist where there is a situation of production to inventory rather than production to sale. A firm producing a large variety of products may find it advantageous to produce the individual parts of one or more of their products periodically and store the run in inventory. When inventory runs low, another batch is produced to replenish stocks. A firm that services and repairs the products it produces may be expected to use this type of processing if their parts are nonstandard items and produced in-house.

3. Continuous operations systems. This approach to processing is used in situations where the product is being produced for sales and the demand for the product is relatively constant. Much mass production is done this way, with everything from automobiles to cigarettes being churned out in continuous-processing plants. The assembly line was designed to encourage economies of scale by use of the continuous-process approach in manufacturing. In this case, the firm is dealing with a product that is standardized, where only one product or a small number of versions of the same product are being manufactured, and where special-purpose machinery is used to produce it. Generally, the reason for special-purpose machinery is that, since only one product is being produced and is being produced in high volume, very specific functions can be carried out over and over again on the same machine without need of resetting tolerances or altering machine configuration, which with high volume serves to reduce downtime and allows for a maximally efficient arrangement within the plant. Economies of scale also tend to be high. It should further be noted that with this approach, flow of goods through the system can be highly specialized as well, thus reducing wasted time and effort. Also, the labor force tends to be highly specialized, thus taking full advantage of the learning rate factor in worker efficiency. Examples of this type of processing are so abundant that to cite further instances beyond those already mentioned would only serve to belabor the point. It is left to the reader to think of the numerous cases of manufacturing where the benefits of continuous processing are realized. The only caveat advisable is to remind the reader that continuous processing is equally applicable to small companies as to large ones, as long as the flow of raw materials and finished goods is continuous and the item itself is standardized.

4. Processes in operations systems. With all systems, there are basic processes or activities that go into forming the dynamic structure of the overall form with which one is working. In operations systems, these processes include facility processes, clerical processes, technical processes, and mechanical processes. Together, they form the total mechanism by which any operations system functions.

a. Facility Processes. This category involves the kinds of activities that must be carried out in order for the facility to function. Analyzing them is similar to examining the different systems within the human body, such as circulatory, respiratory, endocrine, and so forth. In the case of a facility, these processes involve such things as physical production, communication and information services, inventory, receiving and shipping, accounting and finance, personnel, maintenance, quality control, security, and community relations. All of these except the last involve internal processes. The last, community relations, involves processes that take place between the facility and the surrounding environment.

Note that, as with any other organism, all of these processes are operating simultaneously and continually, and that the system is designed to see to it that the processes support each other as well as contribute to the achievement of the firm's goals. Each process contributes to the overall success of the facility in its own unique way and deals with given types of changes as they occur, thus allowing the system to achieve and maintain dynamic balance.

Briefly, the processes noted involve the following:

1. *Physical production* refers to all of those activities actually involved in the production of the good or service in question. This is the heart and raison d'etre for the facility. All of the steps in the production process and the parts of the physical plant that are involved in the manufacture of the firm's product are included in this process.

2. *Communication and information services* include the mechanisms within the facility that are in place to gather, store, and provide information wherever it is needed. This is an increasingly important process in both manufacturing and service organization facilities due to the development of high-volume, inexpensive computer equipment. The firm's ability to accumulate, manipulate, and regurgitate huge amounts of data has enhanced the capacity of management to use data and methodology previously thought too complicated to be of use. It is now possible to analyze operations in detail, and to do so in a time frame that allows effective use of the information, not in retrospect, but in advance of the development of many problems.

3. *Inventory* is a process that has a different connotation in many businesses today than it did only a few short years ago. Since the advent of the JIT (just in time) approach to production, in which raw materials arrive on a continuous basis and finished goods are shipped as soon after completion as possible, the traditional inventory control systems that are designed to set economic order quantities and maintain control of on-hand inventory are less valuable than before. Inventory under JIT becomes a problem in logistics rather than storage; the main element to consider is the rate of flow of materials and finished goods in and out of the plant. Scheduling becomes a continuous rather than periodic function, with lead time, failure rates (in terms of probability of missed deliveries and shipments), and transportation methodology the main issues. For the more traditional type of production process, the inventory function amounts to control of periodic deliveries and shipments, storage capacity and control, and economic order quantities per unit of time.

4. *Receiving and shipping* are processes that are closely related to inventory in the case of JIT facilities. In more traditional facilities, these processes provide and maintain the necessary operations for the receipt and shipment of raw materials and finished goods respectively. Such factors as design of loading docks, types and numbers of materials-handling devices, location of receiving and shipping facilities in conjunction with other activities within the plant, and proximity of transportation and shipping modes are of primary importance. The specific purpose of operations here is to provide a process by which materials may be smoothly exchanged, accounted for, and properly delivered.

5. *Accounting and finance* are necessary functions of any operation, whether it be a production unit or any other firm activity. The economically rational firm attempts to maximize profits, and both the revenue and cost elements of achieving that goal must be controlled. Accounting and finance functions for the facility ensure that budgets are met, that

costs are kept to a minimum, and that economic efficiency is maintained. It is a matter of both planning and providing information for analysis. The accounting and finance process supports the efforts of all other processes in the system, providing data, control, and direction for the maintenance of balance.

6. *Personnel* is obviously a necessary element, considering that one of the major inputs into the productive process is labor. Although automation may reduce labor requirements, the need for control and supervision alone dictates a relatively extensive personnel need. The function of this process is first to obtain and maintain a labor force adequate to achieve company objectives, and second to handle all of the informational and regulatory records and compliance requirements connected with labor. Due to the extensive nature of these demands on the firm, it is normal to maintain a department within the facility to handle them.
7. *Maintenance* is a support function process by which the firm ensures that the equipment and physical plant will be available and capable of performing their assigned functions when needed. Maintenance is a continuous, scheduled process. Once a matter of machine repair, it is now seen as a preventive rather than corrective process by which repair is anticipated and becomes unnecessary. In highly mechanized facilities, maintenance is a first-line concern, since any downtime due to unanticipated equipment failure could jeopardize the entire productive process.
8. *Quality control* is similarly a continuous and preventive process in the manufacturing or service facility. The purpose of quality control is to ensure that the product produced is identical to the product the firm wishes to produce. By maintaining quality, the firm is better able to predetermine results of market actions and better able to discover problem areas in manufacturing. And, as noted, it is predictive when it is working most effectively. The trick is not to know when something is going amiss, but rather to know when something is about to go amiss. That is the goal of quality-control processes. This is a type of feedback mechanism that provides information to the system which the production process can use as a basis for making corrections that serve to reachieve and once more maintain dynamic systems balance.
9. *Security* is important to any operation. Admittedly, some businesses are more prone to security problems than others, but some sort of control must be imposed upon any system to avoid extraneous elements' entering and disrupting the normal flow of activities. Security can amount to anything from a set of rules for behavior (for example, no smoking or no parking areas due to insurance risks) to full-blown departments with funding, equipment, and manpower whose responsibility it is to protect the facility from outside elements. Whatever the extent and

nature of the security efforts, they must be present, thus making safeguarding the facility an integral part of its overall process.

10. *Community relations* is the final element to be discussed here. This deals with the interface between the facility and its environment, focusing on the role of the facility as a part of the community in which it operates and from which its labor force comes, and on the impact of the environment on the facility and the facility on the environment. The environment affects the facility in various ways, not only because of the general conditions of operations that it represents, but also because it acts as a source of change, and the facility system must be able to react to changes if it is to survive and thrive. Community relations could be a single individual performing the functions of this process or an entire department devoted to interactions with the community at large. The community itself is not necessarily a localized phenomenon. The term more aptly refers to the scope of the facility's influence than to those who share its geographic location. Community could be a neighborhood, a municipality, a country, or for that matter, the entire world economy.

b. Clerical Processes. This category includes those processes which perform the record-keeping function of the facility, and which deal with all of the various forms of paperwork that are required in order for the facility to operate effectively. Clerical processes inspect the data generated by the facility. In light of the large number of operations going on in even a small plant, that is a sizable job. This is an ancillary function, a support function that serves the main production function of the plant. Without some method of taking care of the myriad of secondary details inherent in creating a product, the whole system would come to a halt.

It is useful to think of the clerical processes as a form of overlay, always going on behind the scenes in every aspect of the plant operation. Clerical systems have their own infrastructure, complete with a nervous system for sending and receiving information, and a musculature in the form of functionaries who carry out the necessary activities that ensure control of the system. Record keeping, ordering, and controlling the flow of required material are all handled through some form of clerical position. And all this happens parallel to the actual production process, with each step of that process being accompanied by some form of clerical activity. To the facility planner, this means that it is necessary to allow for the activities and space requirements of this process when developing layouts and methodology and when allotting resources.

c. Technical Processes. These processes involve the state of the art in production as well as the whole body of technical expertise that goes into developing and carrying out the production process. From the standpoint of the technologist, technical methodology refers not only to the actual technical apparatuses related to the production of the good or service in question, but also to the whole range of technical activities involved in operating the facility. In some cases, ancillary activities require technology for efficient completion more than the actual production process. As a result, an entire network of technology is needed to handle operations within the plant, and this process is at once a system and subsystem in its own right.

In designing the overall functional process of the plant, it is necessary for the facility planner to coordinate and design into the system these processes so that they all function cooperatively to achieve the desired result, an efficiently and economically produced final product. Effective cooperative design in technical processes requires compatibility (the ability of one technical process to communicate and interact with other necessary technical processes), coordination (the operation of each technical process dovetailing effectively with every other technical process), appropriateness (technical processes that serve to fill real needs within the overall productive system), and reliability (freedom from unacceptable levels of downtime, confusion, or malfunction), particularly in cases where the technical systems are interdependent.

For instance, beyond the normal scope of technology connected with the production of the firm's product, there is another entire communications network tied into that process that is designed to supply information to such ancillary functions as accounting, inventory control, and management. It is necessary that that technology be able to easily and efficiently interface with the productive process in order to supply required data. The same is true of transportation systems, which must coordinate their distribution rates with plant output, shipping needs, availability of product, and so forth. Each technical process must be meshed with every other so that they can create the synergy necessary for maximum effectiveness in the facility.

d. Mechanical Processes. Such processes deal not so much with state of the art as with physical activity. Plants, whether they produce goods or services, tend to be filled with myriad legions of equipment, all performing functions that are integral parts of the overall effort. These mechanical processes may be similar or quite different, depending on their purpose, and they all require coordination, as does any other aspect of the plant. In many cases, it is necessary to design machinery from the ground up, particularly in the case of specialized assembly-line layouts, and these processes become a network that is both unique and a part of the overall scheme of plant operations. The technical process may require that a number of connections be made on some electronic part, whereas the mechanical process dictates how this is to be physically carried out, whether by robot or by fixed, nonelectronic machine. With the technical processes we deal with the highest understanding at our disposal of how something should be accomplished. The mechanical processes dictate the physical reality of those available techniques.

B. Production Processes

It is possible to categorize the actual processes involved in production into a small number of general categories, depending on the type of activities involved: forming, machining, joining and assembling, and finishing. A brief look at the major forms each of these take will serve to explain their nature.

1. Forming processes. These are processes that shape the work or the good into a more useable form. In the following step, basic raw materials are changed to conform to the predetermined specifications of the builder. To choose an exaggerated example, the only difference between a four-ounce slab of silver and a finely crafted silver teaspoon is the shaping that has taken place.

a. Casting Process. Casting is a major forming process. The material to be formed is liquified and poured or cast into a mold and then allowed to cool in order to conform to the desired shape of the part or product. Castings are often done in metal and concrete and to a lesser degree in plaster, plastic, and other artificial materials. The advantage of casting is that a complicated three-dimensional shape can be achieved in a single step with only general finishing necessary as a final step. The disadvantages of casting lie in the heat or chemical processes necessary to achieve liquefaction of the medium and in the limited size of the items that can be easily cast. Large castings sometimes suffer deformation of the medium as they cool. Though large castings are not uncommon, the problems connected to their size tend to add substantially to their cost.

b. Stamping and Forming Processes. These processes represent another category of shaping. In stamping processes, the medium (metal, plastic, and the like) begins as a blank, or raw slab, of standard dimensions that initially lacks any features of the final item. These blanks are in a solid rather than liquid form. As the name of the process implies, the blanks are stamped and then literally hammered, quickly and under tremendous pressure, to force them to take on the desired shape. The shape is translated onto the blank through the use of a die, which is a piece of metal or other material that has been engraved with a negative of the desired shape. The hammer used in the stamping process often incorporates a die as well, this one engraved with a negative of the opposite face of the final item. By stamping the blank between the two halves of the die, the material takes on the surface of the desired shape. An obvious and simple example of this is the manner in which blank circular disks are converted into coins. Another example is standard-issue military dog tags (identification tags), where the letters are stamped onto the surface of the metal blank.

Forming, on the other hand, is a type of deformation that takes place through the hammering process, but generally without the benefit of intricate dies. With forming, a more general-purpose hammering device is used, and the general shape is simply pressed or hammered out. This is the process by which a flat disk of silver can be turned into a beautifully sculptured goblet or a Revere bowl. Forming can also be done through oddly shaped rollers or other high-pressure devices that alter the general contours of the work.

c. Forging and Extrusion Processes. These are other methods of working with metal materials. With forging, the raw material begins as a standard shape of material, such as an iron bar or brass sheet, and the work is hammered into shape under pressure and at temperatures that are high enough to soften the material without causing liquefaction. At times, forging also involves dies, as in the case of drop-forged tools or other metal parts. It may also involve simply hammering under pressure to change the general shape, as in the forges that convert square ingots of metal into flat sheets.

The process known as extrusion can make use of either metal or plastic materials. In extrusion, the soft material is pushed through an opening that is shaped to match the desired final shape of the item being formed. I-beams and railroad rails can be produced in this way, as are square or round steel and plastic stock. The material is introduced in a softened, semi-liquid state and then allowed to cool once it has been extruded through the molding aperture. As it cools, it retains the shape induced by the extruder.

2. Machining processes. The fundamental distinguishing characteristic of machining is that the material being processed is cut rather than hammered. Because of this, materials are usually in a hardened state, which allows the technique of machining to be applicable not only to materials such as metal and plastic, but also to wood, stone, and artificial solids.

Whatever the nature of the cutting, machining involves the application of a sharp instrument to the material in order to remove mass and achieve a desired shape. Generally, we categorize machining into turning, drilling, milling, shaping and planing, and grinding. Each of these techniques is designed to perform a specific cutting function.

a. Turning. This technique is applicable to any work for which a cylindrical shape is desired. The piece to be shaped (referred to as the *work*) is put into a machine that rotates it about the axis of a cylinder, and a blade is then applied to the surface to cut away material until a uniform cylindrical surface of the desired diameter is achieved. This work is essentially done by the lathe, a machine designed to turn cylinders. Axles, pins, and rods are turned on lathes. Note that turning is extensively used where extrusion is not possible, as with wood and other permanently solid materials. The value of the turning process is that it achieves an absolutely uniform circular cross section and can be set to very close tolerances with a minimum of possible error.

b. Drilling. In this process the purpose is to create an opening in the material of some specific diameter and depth. Drilling can be achieved in one of two ways, either by rotating the work and cutting with a stationary drill bit, or by rotating the drill bit at high or low speed against the surface of the stationary work. The latter is most often used, though lathes that rotate the work can be set up to perform the function as well. The value of drilling is that the diameter of the hole and its depth can be precisely determined and controlled, allowing for tight tolerances, and that it is applicable to work that is solid, or work that cannot be cast or stamped with a sufficient degree of accuracy.

c. Milling. This is a grinding process in which the cutting is either done by a rotating blade or set of blades, or by an abrasive stone. The purpose of milling is to change surface contours minutely and to create a smooth edge or surface. Milling often accompanies forging or casting operations when the general shape of the item has been attained but further work is needed to smooth the surface or achieve the desired final shape. A very fine type of milling used to remove very small variations in surface contour and to create a polished finish is called *burnishing*. In this type of milling, it is often an abrasive such as sand or jeweler's rouge, consisting of gritty dust in an oily medium, that does the actual cutting.

d. Shaping and Planing. These techniques are closely related to milling. In the case of shaping and planing, however, the purpose is not to achieve a uniform surface but rather to cut specific shapes into the surface. Both employ cutting blades of some description which are either drawn across the work or are rotated as the work is drawn across the machine. Planer/joiners and routers are examples of machines that perform planing and shaping functions. In these cases, a specific surface contour is sought, and this contour can be anything from flat, as with a common door plane, to a complicated set of dips and curves, as with shapers or routers designed to put fancy borders on edges of wood. Again note that the process is applicable to a wide range of solid materials when cutting the excess material away is preferable to reshaping the entire material.

e. Grinding. This is a general-purpose technique designed simply to remove material. Grinding can be used to shape, smooth, or polish the work as needed. The cutting action takes place through the use of a circular wheel of abrasive material, sometimes natural stone and often artificially formed, that rotates at a relatively high rate of speed and cuts away material from the surface of the work. The coarser the surface of the grinder, the more material is removed with each pass. Very fine-grained grinders can be used to polish or burnish hard surfaces as well. The advantage to using grinders is that the work can be handled and shaped by the worker, or the work can be cut to size and shape in cases where cutting is not feasible due to the shape of the work or the conditions under which the work is being done. It is also applicable for sharpening as well as shaping and for removing material from large surfaces while creating a uniformly smooth surface.

3. Joining and assembling processes. This classification of manufacturing activity centers in the combining of a number of individually manufactured pieces into a single item. That single item may be either a subassembly or the final assembly of the product being produced. In either case, the purpose of this type of activity is to combine the pieces. This is a major source of synergy in the production process. It is through this joining or assembling that the added usefulness stemming from the combination of parts is achieved.

a. Welding and Brazing. These processes are used to permanently join metal parts into a single unit. The difference between the two approaches involves whether or not a joining substance is introduced. In the welding process, the two pieces to be joined are heated along the joining surfaces until they are hot enough to melt together or hot enough to be hammered together. The hammering technique was the only available technique of welding until the advent of butane and electric arc torches, which allowed for the local application of extremely high temperatures to the surfaces of metallic substances. Since these higher temperatures are now achievable, the hammering is no longer necessary or desirable in many cases, and the two pieces are simply allowed to melt into each other and then harden to form a single piece.

Brazing, on the other hand, uses a medium as a solder to form the join. With brazing, a bar of material with a melting temperature below that of the work to be joined is used. In this process, both pieces to be joined are heated to the melting point of the soldering substance, which is then allowed to touch the two pieces and heated until it melts. Upon cooling, the two pieces each become embedded along the joining edge in the soldering medium, which serves to hold the two together. With materials of extremely high melting point or situations where the strength of a true weld is not required, brazing offers an excellent alternative that is potentially faster and requires lower temperatures to achieve, thus saving energy.

b. Adhesive Joining. This process involves more than simply gluing two pieces together. Although that is basically what is taking place, adhesive joining has taken on a greatly expanded meaning in modern production. There are now a wide variety of adhesives available for joining, each with its own unique combination of characteristics, so that the medium can be tailored to the job. The important characteristics of an adhesive include solubility, permanence, workability for different types of material, setting time, volatility, and tensile strength. There are glues that hold materials more efficiently than welds. Some are specifically designed to adhere instantly, such as contact cement, whereas others take longer to set but form incredibly permanent bonds that actually leave the joint stronger than the material joined, as in the case of epoxy resins. The value of using adhesive joinery is that it requires little or no heat, no bolts, nails, or other fasteners, tends to be permanent, in many cases resists wear and stress better than alternative methods, leaves a finished edge, and is often cheaper to use.

c. Assembly Processes. These processes differ from joinery in that they represent the whole set of operations involved in putting the various parts of a system or subsystem together. Assembly is the final step by which parts become a whole. Along with more permanent joinery, such as welds and glues, assembly incorporates an entire engineering field having to do with how two or more items can be combined. Fasteners, such as nuts and bolts and screws, interlocking part technology, such as tongue and groove or dovetails, and step-by-step assembly methodology all go into this process. In designing a product, one of the elements that dictates the final design is, in fact, the assembly process expected to take place in its manufacture. It does little good to design a quality product if the cost of assembling it is prohibitive or too complicated to ensure quality control.

4. Finishing processes. These involve the last steps in the production process. In the finishing stage, the assembled goods are readied for final shipment. It is during this stage that much of the cosmetic attractiveness of the product is added, such as final polishing or painting, individual adjustments, cleaning, and packaging. By definition, all last-minute steps that need to take place after assembly to complete the job are performed in the finishing processes. It is in this stage that the product is completed; there are no other steps in the production process.

5. Processes in other industries. Some manufacturing industries have specific activities in their manufacturing process that exist due to the inherent characteristics of the products themselves or to the material of which they are constructed. Two particular industries where this is true are the plastics industry and the woodworking industry.

a. Plastic Processes. Plastic processing involves some specific opportunities and problems due to the nature of the material. On the positive side, the nature of plastic material is such that it is easily worked both in milling machines and in forming machines. Because many plastic substances melt at a low temperature, they are also particularly useful in molding parts through a process known as injection molding. With this form of casting, the plastic is injected into a breakaway mold at high pressure, and extremely fine detail can be achieved. Injection molding is a very common methodology in the plastics industry, used for everything from milk jugs to the bubble wrapping on convenience items. Large and small moldings can be made from plastic with this process. Extrusion is another common method of manufacture in plastics. With extrusion, the material can be brought to a thick, flowing texture at relatively low temperatures, and due to the high cohesion of plastics, particularly polystyrenes and other plastics, can be easily extruded and formed.

On the negative side, many plastics are volatile, which eliminates high-temperature processing, and the low melting point of the material that is so advantageous in injection molding becomes a hazard when cutting technologies are used. Turning and milling produces heat, and if the rotation of the cutter or work is too high, melting or unwanted deformation will occur during the cutting process. Most plastics must be worked at a relatively slow rotation rate because of this. For the same reason, (heat caused by friction), plastic is difficult to cut, and its hardness as a solid tends to dull cutting equipment quickly. Fortunately, it also exhibits many of the crystalline characteristics of glass, since it is a lattice of carbon compounds, and it shears nicely. Many plastics in particular can be cut by scoring, much as one would a piece of glass. Finally on the negative side, plastics can be quite heavy per unit mass, causing difficulties in handling and in finishing.

b. Woodworking Processes. Woodworking processing has its own peculiarities. Because wood is a substance that, unless mixed with polymer resins as a particular substance, will not melt, most processing is in the form of machining. It is also subject to combustion at fairly low temperatures, and to deformation from humidity and heat. Although it has an excellent structural tensile strength, it can be easily cut, dented, or splintered. These characteristics create both opportunities and hazards.

Opportunities arise from the soft, easily workable nature of the material and the fact that it maintains its integrity well when cut. Light, variably shaped parts can be made from the material where excessive friction and wear are not a problem. It turns, mills, and shapes easily and holds that shape once established.

On the negative side, wood cannot be molded or extruded without extensive treatment, generates a tremendous amount of waste in the form of sawdust and wood chips, and is difficult to shape. The waste presents a problem in disposal and in machinery and equipment maintenance, as well as potential fire and health hazards. Storage and transportation also present unique problems because of the effects of heat and moisture on the final product and because of the softness of the material, making extra care in handling a necessity.

*c. **Processing Industries.*** These are industries that are capable of handling many different types of materials. They are flexible processing systems that allow for a wide range of materials to be processed. Canning plants, where corn or other vegetables may be canned one day and varieties of meats the next, are an example of this type of industry. Their value lies in their ability to change readily as market dictates without encountering excessive costs. On the negative side, they often operate more slowly than a facility that specializes in only certain products because of different handling rates necessary for different types of products.

II. NUMERICAL CONTROL OF MACHINE TOOLS

Numerical control is not a particularly new idea, but it has developed into the backbone of heavy-production methodology, particularly where processing machinery is used. It is unfortunate that a misunderstanding of its nature has led to a belief that *automation* refers to the replacement of people with machines in the production process. This is actually quite untrue. It is not the replacement of people, but rather the replacement of human work with machine work that creates automation. Numerical control (NC) was the first step in this process in the last half of the twentieth century.

A. Automation of Process

The concept of automation of process uses technology to transfer responsibility for the effort that goes into producing the goods from humans to machines. This is an idea far older than numerical control. Indeed, the automatic looms of the nineteenth century were technically numerically controlled devices, depending on punch cards in long strips to give specific instructions to the looms. In general manufacture, the use of automation provides for the actual work to be done by the machines rather than workers, thus taking the actual work performance out of human hands and relegating to the humans the job of initiating and stopping the machine process. The value of automation lies in the increased speed, accuracy, invariability, and lower unit cost inherent in automatic machinery. It is an example of practical technology at its best.

B. Nature of Numerical Control

Numerical controls came into their own during the period between 1955 and 1980. They were seen as the new technology in production. Numerically controlled machines operate by data stored on punch tape, on magnetic tape, or in computer programs that are fed to the machine, which enable it to automatically perform the operations. This eliminates the need for a worker to tell the machine what to do or to perform any direct work. In actuality, the numerical control approach offers automatic changing of machine settings usually done by workers. Thus nonskilled workers can be used to perform jobs that required skilled workers before the advent of numerical control devices. Under such circumstances, labor costs dropped, production rose with the increase in machines used and their inherently higher speed, and the quality of the production improved with reductions in variation due to human error. With numerical controls, a machine can be constructed and programmed to impose milling and cutting activities on a piece of raw material in three dimensions simultaneously, often with finishing and in-and-out feeding taking place unaided by anyone.

C. Economics of Numerical Control

The beauty of numerically controlled devices is in the reduction in operating cost, lower labor cost, increased productive output, and higher quality. Disadvantages are the increase in initial capital outlay for the purchase of the equipment itself, initial training costs to familiarize unskilled workers with the operation of the machinery, and higher maintenance costs because of the dependence on machinery as opposed to human labor.

D. Potential of Numerical Control

The continued use of numerically controlled equipment is fairly well assured, in spite of the increased use of newer production concepts such as robotics and computer-aided manufacturing (CAM) systems. It is a natural adjunct to these newer concepts, and in fact, they can be seen as a logical progression of technology and an outgrowth of the NC approach. Inclusion of numerically controlled machines in a CAM system is logical and natural. Although some see NC as nothing more than an important step in the development of robotics, there are still a large number of applications in which its use is more economical and efficient than the use of the robots themselves. By way of analogy, when pneumatic tools were developed, they didn't destory the market for monkey wrenches. Numerically controlled machines are still the most efficient means of mass production in situations where the work is repetitive and each piece is to be identical to every other. Although robots can achieve this as well, it is not their forte. The power of robotics lies elsewhere, as we shall see.

III. ROBOTICS

Robots are computer-controlled machines that can be programmed to perform a series of complex operations in a manner similar to that of humans. They often interact with human operators in a symbiotic fashion or as part of a man–machine integration. Through the use of a computer, the robot can be programmed to do a number of different complex tasks and to do them in a human-like fashion. The robot represents a kind of "missing link" between man and machine in that it has a degree of human-like flexibility while having the capacity to repeatedly carry out the same set of instructions in exactly the same manner, as if it were a specialized machine.

On the negative side, robots work at only one speed and are capable of exhibiting (obviously) mindless behavior. For example, it is possible to have a robot welding seams for car bodies out in space. If something should prevent a car body from reaching the exact location necessary for welding to take place, the robot would continue to weld, totally undaunted by the fact that the procedure was not completed. The robot will never know it has done anything wrong; it simply performs what it has been programmed to do. Without a human being to oversee the robot, it could produce a great deal of junk.

Although robots may perform as specialized machines, they are actually general-purpose machines; their specific design may vary by category of work to be done. Humans also are general-purpose, but those with an advantage in sports are used in sports, and those with an advantage in calculating numbers are used in fields like actuary sciences, accounting, or physics. This is not to equate humans and robots or to insinuate that robots are in any way on a par with humans when it comes to performing activities. Robots can, however, perform a limited number of functions as well as or better than a human counterpart, and do it more consistently and more economically.

A single-armed general-purpose robot, for instance, can be "taught" to stand on an assembly line and spray-paint auto parts. It learns by programs, and once the lesson is learned, it never forgets and never changes the program. It will repeat that learned set of steps again and again without the slightest variation until it is taught to do something else. Another robot of the same manufacture could be taught to weld a given set of joints on an auto frame, or to drill holes, or to fasten door panels, or any other set of functions of which it is physically capable. And it performs accurately. Robots programmed to work in electronics can perform delicate soldering tasks without making mistakes and without variation in quality, as long as they are adequately supervised.

There is, of course, a great deal of opposition to robotics from labor, particularly from labor unions. This is unfortunate, since what the robot is doing is nothing more than what numerically controlled machinery did 30 years earlier by taking the physical part of the work out of the hands of workers and giving it to machines. In the process, more jobs are actually created through the robotic approach, in spite of an initial drop in the demand for labor in the performance of specific jobs. Further, as labor costs continue to rise and the production and use of robots continues in other countries, this opposition to the use of robotics has the potential to cause greater loss of jobs to stronger, more efficient foreign competitors who are not so reluctant to embrace the new technology. In some industries, retraining efforts are helping to teach workers how to function in technological positions by offering opportunities in the production and maintenance of the robots themselves and in ancillary positions elsewhere within the company.

With process design, the need for robotics is self-evident, particularly in situations where a high degree of automation is economical. New facilities should keep in mind that it is easier to include robotics in the process planning than to add them later, as there is no initial labor force to contend with. In either case, robotics may prove to be the answer to operations efficiency problems. The costs of robotic technology at the present time are high, and will be so for the near future, but as the use of these tools becomes more common and their rate of production grows, their costs of purchase and operation are likely to diminish significantly, rendering them an essential part of many new facility plans.

IV. LAYOUT OF FACILITIES

Once the production process has been defined and designed, it becomes necessary to finally lay out the physical plant in such a way that the activities of the production process are carried out with a minimum of cost and time expenditure. This physical layout stage is critical, as it will dictate the final size and shape of the facility itself to a large extent. As with all other areas of facilities planning, there are different approaches to choose from, and picking one is a matter of matching the characteristics of the approach with the needs of the specific production process.

A. Layout Types

Layouts are classified into three basic types: the process layout, the product layout, and the fixed-position layout. Each has its strengths and weaknesses; the characteristics of the manufacturing to be done determine which type is most appropriate.

1. Process layout. The key in developing process layouts is the process to be performed. This type of layout is usually used where a number of products are to be produced in batches, as in the case of a cabinetmaking company or machine shop that produces nonstandard items to order. Since there are different steps being taken in a different order depending on the particular job run, the logical way to physically organize the shop is to divide it into segments according to the type of operation (process) being done. Therefore, all the lathes are located in one area, all milling equipment in another, all drilling in still another, and finishing is located generally at one end of the shop, since finishing is generally the last step taken no matter what is being produced. In the process layout, the machinery employed is normally of a general-purpose type, and the settings of the equipment can be changed to meet the needs of the individual job. Routing of work in progress is done from department to department and is scheduled according to the steps necessary to create the product in question.

Some specific characteristics in addition to the general nature of the machinery include the following:

1. The workers tend to be highly skilled, since they must be able to produce a large number of different items on the machines they are operating.
2. Materials- and product-handling equipment is generally mobile in nature because of the changing order of departmental function inherent in handling a number of different products.
3. Planning and scheduling must be done continually, as does inventory ordering and storage because of the changing nature of the work being done.
4. There is a tendency to use automatic retrieval and storage systems for inventory.

2. Product layout. Product layouts are centered in specific products and are used where large or continuous runs of the same products or product groups are to be performed. In this case, the constant factor is not the processes to be performed, but rather the product to be produced. This means that it is possible to plan the design of the facility around the steps to be taken in producing that product. Since the steps to be performed and the settings on the machinery are constant, economics of operation and design can be achieved through the use of specialized machinery. This increased efficiency stems from the fact that there is no need to change settings for other jobs, only to adjust tolerances within a narrow range. There is a savings in labor in this type of layout, since the work force is required to perform a very narrow range of tasks. Less training is necessary, less skilled workers can be used, and the opportunity for automation and use of NC equipment rises. Finally, materials-handling and storage equipment can also be specialized, since goods always take the same route through the facility and the goods being stored always have the same characteristics. Ramps, belts, and other assembly-line delivery systems can be used effectively in this type of operation.

3. Fixed-position layout. This type of layout exists where the product being constructed cannot easily be moved about during its manufacture. In such cases, the firm may choose to transport the production equipment and personnel to the work rather than the work to the equipment. The production of ships, railroad stock, and large airplanes are all examples of fixed-position layouts. Of course, for such really large projects many subassemblies may undergo production in a process or product layout, but the final layout is centered in the fixed-position approach. It would be exceedingly difficult, for instance, to transport a Lockheed C-5A, the world's largest airplane, from one production point to another. The parts are brought to the airplane, rather than the airplane to the parts. It should be noted that there are other circumstances where this type of production method is used. Examples would include on-site production, as in the case of swimming pool production for in-ground pools, or where special environmental conditions must prevail, as in the production of satellites in "clean rooms" necessary for avoiding contamination by dust and other particles of matter.

B. Approaches to Layout Problems

The more complicated the production process, the more complicated the layout of the facility. In individual cases, even relatively simple production processes may present special problems because of the specifics of the product design. This means that choosing a layout that works and is efficient is not a simple matter. Problems arise and must be analyzed. Multiple choices have to be studied and ranked. A single change in one step of a layout design can call for numerous other minor or major adjustments elsewhere in that same design. Even after a specific type of layout has been chosen, the problems to be solved are legion. In this section, some of the more common methods of working with these problems are discussed.

1. Process analysis. The first attemps at layout problem solving were in the form of process analysis, and this has continued as a basis for many modern approaches as well. In process analysis, the process by which the product comes into being is analyzed and structured into a system consisting of a number of chronological steps. These chronological steps are then analyzed in terms of the sequence in which they occur, and this is used as the basis for the layout analysis. Layout is then organized in accordance with this sequencing.

2. Quantitative techniques. A number of quantitative techniques are available for use in developing the most efficient layout of a facility. As with other aspects of layout, the choice of which method to use depends on the individual characteristics of the process. Two general categories that create different needs are the process layout and the product or assembly line layout. Different considerations must be dealt with in each case due to the differences in the individual layouts; in the case of the process layout there is a constant change in routing, depending on what product is being produced, as opposed to the product layout, where the same sequence occurs each time.

Process layouts require that the designer pay particular attention to the distances between various process locations and the size of the batch being transported. Since the material is moving from station to station, the arrangement of stations in a manner that minimizes distance and maximizes batch size is required in order to maximize efficiency.

As a simple example, suppose the product being produced required milling, assembly, painting, finishing, packing, and storing, in that order, and that the material had to be moved one piece at a time, except when it was finished, boxed, and stored. It would be foolish under those circumstances to place the stations requiring movement of items one at a time at opposite ends of the building from each other while connecting painting, packaging, and storage. This would entail a great deal of movement of small batches over long distances, and short hauls of large batches. It would be more economical (since moving materials requires time and therefore costs money) to have one-at-a-time steps, particularly if they are consecutive, close to each other, and then have the materials sent in large batches to painting. Once assembled in large batches after painting, the items can more efficiently be carried longer distances to packaging and again to storage, since the movement of many items is carried out all at once.

This leads to the load–distance model that is used in determining plant layouts for process-oriented facilities. The basic formula for the combination of distance and number of trips that must be made is:

$$F\ (\textit{Load factor}) = \sum_{i=1}^{N} \sum_{j=1}^{N} \mathrm{L}_{ij}\mathrm{D}_{ij}$$

where N = the number of stations, L_{ij} = the number of transfers or loads of material between station i and station j, D_{ij} = the distance between stations i and j, and F is the loading factor. By determining the value for F between each two sets of stations and adding them together, the total amount of distance times the total number of transfers can be determined. If we build a matrix of the work centers, we can determine the minimum load factor, F, that will exist for that arrangement. The same can be done for other matrix arrangements, and then the minimum cost matrix can be chosen. This is similar to the linear programming matrix technique known as the transportation model often used in goods distribution. The difference here is that there are a number of transports to be made, and they must be made in a given sequence. Rather than attempting to send goods from point to various point, depending on what is most efficient, the designer here strives to arrange the points in such a way as to minimize the distance traveled and time expended in order to minimize transportation costs. The traditional transportation model does not have the luxury of being able to move the geographic locations around to suit their needs.

The model is really much simpler to use than it first appears. In reality, all that the planner does is (1) determine the number of loads in a given time period that must be transported from one station to another, (2) determine the distance between the two stations, and (3) arrange all stations in a manner that minimizes the sum of the $L_{ij}D_{ij}$ combinations. In the example previously cited, the process involved milling, assembly, painting, finishing, packaging, and storage in that order. If we label these stations 1 through 6, respectively, we can build a matrix that indicates the number of loads between stations as indicated in Table 14.1. For example, there are 150 loads moved from milling (1) to assembly (2), and from finishing (4) to packing (5), there were 10 loads. This chart was put together by actually observing the loads moved from operation to operation over a period of time.

TABLE 14.1 Total Loads

Departments	From \ To	1	2	3	4	5	6
1. Milling			150				
2. Assembly				150			
3. Painting					150		
4. Finishing						10	
5. Packing							10
6. Storage							

Once this information is obtained, it is simply a matter of configuring the arrangement of stations (the layout) so that it will minimize transportation distance and time costs. Note the two arrangements given in Table 14.2. It is assumed that each station is one unit distant from the station next to it in all three directions (Note: in both arrangements, station 3 is two distance units from station 4). Using the flow path for both examples of $1 \rightarrow 2 \rightarrow 3 \rightarrow 4 \rightarrow 5 \rightarrow 6$, in arrangement A, Table 14.2, the sum of the number of loads from Table 14.1 times the distances (load * distance) traveled from Table 14.2 (load * distance) is as follows:

$$1 \rightarrow 2 + 2 \rightarrow 3 + 3 \rightarrow 4 + 4 \rightarrow 5 + 5 \rightarrow 6 = \text{flow path}$$

$$(150 * 1) + (150 * 1) + (150 * 2) + (10 * 1) + (10 * 1) = 620 \text{ load factor}$$

For arrangement B, Table 14.2, using the same flow path but a different facility layout, the sum will be

$$1 \rightarrow 2 + 2 \rightarrow 3 + 3 \rightarrow 4 + 4 \rightarrow 5 + 5 \rightarrow 6 = \text{flow path}$$

$$(150 * 2) + (150 * 2) + (150 * 2) + (10 * 1) + (10 * 1) = 920 \text{ load factor}$$

TABLE 14.2 Process Flow Distances (Follow the Arrow)

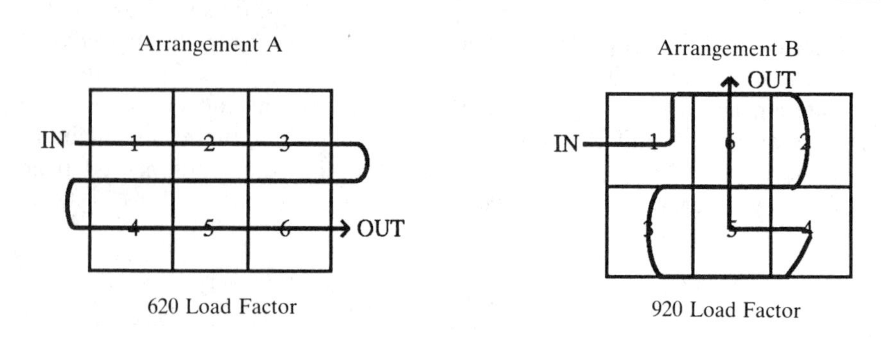

In this simple example, the first arrangment is far superior to the second, particularly if the cost of moving goods over a one unit/load distance is \$5.00. Opting for the first arrangement would mean an ongoing saving of \$5.00 × 300 (subtracting 620 from 920 equals 300 unit/load difference), or \$1500.00! Of course, there are few layout problems as simple as this one. Many processes require a large number of steps, with repeat movements back and forth among the various stations. Painting may have to take place, for instance, several different times during the processing, with returns from different stations. A more realistic analysis using the load–distance model would probably require a computer, either utilizing matrix algebra to minimize the sum of rows and columns, or a systematic number crunching approach that merely runs through all the various combinations of station locations and finds the minimum values. This is a particularly useful approach for batch processing situations, where a number of different products requiring different orders of subprocesses exist in the same plant. By extension, after determining the unit/load values for all of the different products produced, the task then remains to add up the cost for all different products using a specific arrangement of machinery and equipment, and finally to choose the configuration with the minimum total value.

Note what this means. Suppose there are three possible configurations and three different products. Each product's unit/load cost is determined for each configuration (layout), and they are totalled. The configuration that offers the lowest average total cost for the combination of all three products is the one to choose, even though it may mean a higher cost for one product than some other configuration. This can be further weighted in the analysis by the probability of a certain product being ordered.

As an example, assume the three products produced are produced in the ratio of 40:40:20. That is, 40 percent of the work done by the firm is in building product A, 40 percent is in product B, and 20 percent is in product C. These relative weights are attached to the unit/load values for each configuration to figure total costs. Note the matrix in Table 14.3. In this matrix,

TABLE 14.3 Load Factor Process Mix Table

Products	Configurations I	II	III
A	1,000 (*0.40) = 400	1,500 (*0.40) = 600	300 (*0.40) = 120
B	850 (*0.40) = 340	150 (*0.40) = 60	400 (*0.40) = 160
C	900 (*0.20) = 180	600 (*0.20) = 120	2,000 (*0.20) = 400
Unweighted average	917	750	900
Weighted total	920	780	680

the unit/load value for each of the nine possible configurations is shown, weighted for the total amount of firm output wrapped up in each.

The solution for column A, Table 14.3, is as follows:

Unweighted average solution: 1,000 + 850 + 900 = 2,750/3 = 917

Weighted total solution: 400 + 340 + 180 = 920

Note that if we do not weight the configurations for the expected percentage of total output that each product will represent—that is, if we assume there is an equal chance of any one product's being produced, and in equal numbers—the average expected unit/load cost is most favorably represented by configuration II, and the expected cost at $5.00 per unit/load is (750 * $5.00) = $3,750. However, if we weight the results, the minimum expected unit/load factor is represented by configuration III, at 680 unit/loads, and the expected cost is (680 * $5.00) = $3,400. Assuming our ratios are accurate, weighting gives us a much more accurate answer and a lower expected cost. This increases our efficiency and, further, allows us to budget costs more accurately in the process.

An alternative mathematical model, which is based on the PERT concept, is effective for assembly-line models where the sequence of events is always identical, and where a number of subassemblies are being produced simultaneously prior to final assembly. Two different criteria may be used in this approach, one based on maximizing output per day or other unit of time, and the other based on minimizing costs or time of production, each to be charted over some period of time. Actual analysis usually involves both of these elements.

Assembly lines offer unique opportunities, since they can be defined once, and all activities continue to take place in the same sequence over time. This means that an optimal arrangement can be sought without consideration of the number of products being produced. By definition, there is only one process taking place, and only one product being produced in an assembly-line approach. Further, since the same process is carried out each time, there is an opportunity to carry out a number of steps at the same station through the use of specialized machines. This can greatly reduce transportation costs and stabilize load size and scheduling. Add to this the ability to carry out activities simultaneously, and the efficiency can be greatly enhanced at the layout stage.

For an acceptable design, there are several criteria that must be met. First, the required level of output must be maintained so that production rates meet firm needs in terms of units produced in a given time period. If the facility is to produce parts to be incorporated into final products at some other location, production rates must be commensurate with production of the items for which the subassembly is an element. If it is a finished good, the plant must be able to produce at a rate commensurate with expected demand. Second, the arrangement of steps in the process must be workable. That is, the building of the process itself must be technically possible, and it must accomplish the desired task. Third, the total process must maintain overall efficiency, being constructed in a configuration that minimizes total production costs for a given process design. The layout analyst has the job of seeing to it that these criteria are met in the final layout design.

We are, therefore, dealing with plant output capacity, subprocess sequence design, and economic efficiency. This is where the multitasking model becomes so valuable. This approach, which is heuristic in nature (meaning that it is a general guideline), consists of six steps leading to the discovery of the layout combination most suited to satisfying the multiple criteria noted. The six steps in this analysis, called *balancing the line*, are as follows:

1. Define the basic tasks to be performed.
2. Identify any precedence requirements among these tasks.
3. Calculate the minimum number of workstations needed.
4. Determine the combination of tasks to be performed at each station.
5. Calculate the efficiency of the layout.
6. Continue to adjust and improve the layout for optimization.

The calculation that most interests us here is the determination of the LOT, or longest operation time. This is a method by which the particular path of work in the process leading to the finished good that is the longest path is determined. This longest path, in terms of time, will determine the minimum amount of time in which the project can be completed. As an example, if there are five separate subassemblies that are being constructed simultaneously during the process, and the times required to complete each are (a) 10 minutes, (b) 5 minutes, (c) 7 minutes, (d) 12 minutes, and (e) 3 minutes, respectively, the critical subassembly operation is the (d) at 12 minutes (the longest amount of time). It doesn't matter how long the others take to complete; the item can not be totally assembled until after the completion of the 12-minute process. This knowledge is useful in assigning tasks to a given station in the line, since those subassemblies with less than 12 minutes' duration have excess time; if more than one operation can be assigned to the same station, this can reduce idle time considerably while reducing the number of stations (and accompanying travel time) used in the operation.

In the example, subassemblies (b) and (c), requiring 5 minutes and 7 minutes, respectively, could be done one after the other at the same station without increasing production time. Further, it would be possible to reduce the number of workstations by one if the firm is willing to raise the longest operation time to 13 minutes, by combining steps (a) and (e). If there is no dependency of any of these subassemblies on prior completion of other subassemblies (see item 2 in the balancing the line format), this arrangement reduces the number of required stations to a minimum. The "before and after" illustration of this concept is presented in Figure 14.1.

Note that by carrying out a number of operations and groups of operations simultaneously, the time of production is reduced from 37 to 13 minutes. Also, the number of workstations necessary is reduced from five to three. It should be noted that it is not always possible to maximize the time relationships by performing a combination of steps at any given station, either because of prior completion requirements or because the expense of outfitting the station to do certain combinations of tasks is prohibitive compared to the expense of adding an additional station. What is sought is the proper combination of stations and LOTs for the operation to minimize overall costs.

Throughout the mathematical approaches discussed in this section, there have been specific steps taken in the layout process. Among these are reducing travel distances, analyzing relationships, and improving the flow of materials. Although these have been discussed as an integral part of the two models presented, a special word on each is appropriate.

3. Reducing travel distances. It is often the case, even with professional design engineers, that the person responsible for plant layout is hampered by self-induced limitations in design possibility. Layouts tend to follow traditional patterns, because of the proven reliability of certain designs and because of the tendency on the part of designers to stick with what they know works. However, some novel layouts may be used that can decidedly reduce travel time between locations. Nontraditional shapes that go beyond the U-shape, the H-shape, the S-shape, and the straight-line shape can create a more efficient use of space while reducing travel distances, as can thinking

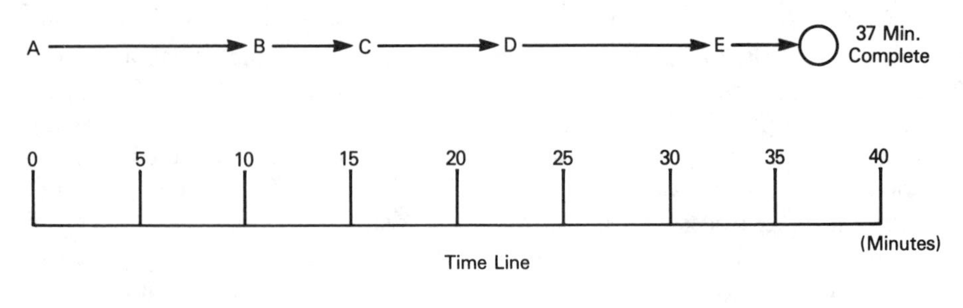

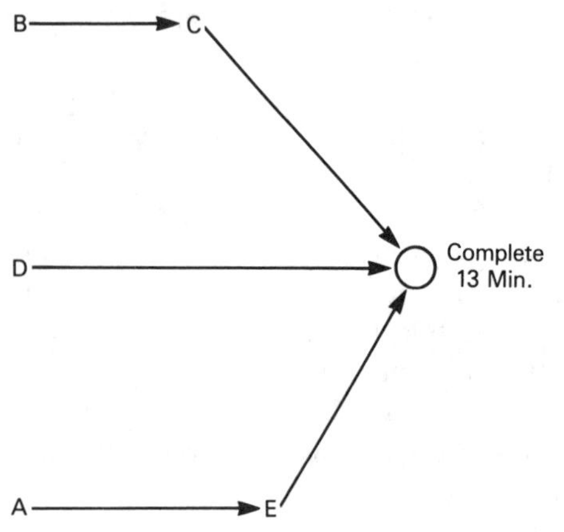

Figure 14.1

in terms of three-dimensional space. Gravity shoots are economical ways of delivering parts, especially bulk materials and small parts, and by placing earlier assembly stations above later ones, or storage above work areas, multistory stationing is able to significantly reduce not only travel distance, but travel time and cost as well. Spirals are also a novel possibility, as are lazy-susan and ferris-wheel conveyor approaches. The imagination with which the designer attacks the problem of plant layout is often the measure of how closely the operation approaches maximum efficiency.

4. Relationship analysis. This part of the modeling process is a matter of determining how each of the steps and operations in the production process relate to every other step. It is a form of systems analysis; the subsystems are the elements of manufacture, and the analysis is of their interconnectedness. During this phase, attention must be given to the totality of the operation, not simply the steps in the manufacture. That is, not only should the processes be analyzed, but the ancillary activities as well. Support facilities such as maintenance, supervision, production control, and quality control must also be taken into account in the total design, since these functions require space and integration into the production process. Developing a layout that follows the process of manufacturing the product is not enough, particularly if it does not include the inspection stations, the access of supervisors to controls, and the needs of personnel. And all of this must take place within the limits of the physical space available for the layout, something that is not always at the discretion of the designer. If, for instance, existing facilities are to be converted, and the building itself is already extant, the possibility of radical spatial designs without extreme property alteration is eliminated. In other words, designs are often pragmatic, not ideal.

5. Improving flow of material. A caveat is necessary here. Improving material flow is a major factor in achieving and maintaining efficiency of operation, but it is not the only item of importance. There are a number of ways of improving the flow of material, both by reducing travel time or travel distance, or by combining operations in a single station. Yet there is a cost to this as well, especially in developing specialized operational stations designed to combine operations that are not naturally compatible. Improving the flow of material by increasing total facility cost may not improve overall profitability. It is sometimes more efficient to have higher transportation costs or to stick to accepted layout designs than to move to an inherently more efficient design that is too costly to construct. As an example of a megaproject that illustrates this, there was a period of time in the late nineteenth century when all of the post offices of Paris were connected by pneumatic tubes laid under the streets through the sewer system. It was an incredibly efficient system for delivering mail at high speed, requiring only the installation of pipes and development of air pressure stations. However, if the system had had to dig up the streets of Paris to lay the pipe, the cost of the system would have so far outweighed the increased efficiency in work flow as to have made the project merely a pipe dream.

In essence, this is a matter of diminishing returns. With improving work flow, as with any other element of the layout development process, there is a time when further improvements are more than outweighed by the increased cost of producing them. It is only where cost and profit are not the major issues, as in wartime government spending on military weapons systems or strategic defense development, that efficiency takes precedence over economy.

6. Computerized methods. Thanks to the advent of the computer age and the arrival of effective and user-friendly computer-assisted design (CAD) and computer-aided manufacturing (CAM) programs, the facilities planning process has been greatly simplified. With computers to do the number crunching, alternative possibilities for layout can easily be constructed, evaluated, and discarded until the most efficient overall structure is found. A number of computer methods are available for doing this, many based on the models already presented; four of them are mentioned here.

a. Successive Maximum Elemental Time Technique. This is a computer-based model that operates on the PERT approach, determining the maximum amount of time each element in the production system will take, and from that, developing an overall model of allotted and required time to complete the manufacture. For inputs, time information is fed in, and the resulting output is an overall manufacturing model based on the maximum efficiency of time spent in the process.

b. Computerized Relative Allocation of Facilities Technique (CRAFT). The CRAFT model uses a set of design parameters by which it can define acceptable and unacceptable juxtapositions of elements in the production process. Because it follows rules that determine acceptability, it is termed a heuristic model that arranges and rearranges system elements in a search for the optimum layout. That layout is determined by a minimization of materials-handling cost in terms of distance, time, equipment, and so forth. CRAFT takes into account the limitations of the facility itself, such as overall dimensions, construction characteristics, and shape. The program goes through a series of rearrangements, trading off one cost against another until the materials-handling cost is at a minimum.

c. Computerized Relationship Layout Planning (CORELAP). CORELAP operates similarly to CRAFT in that it is heuristic. It determines the most effective overall layout based on the relationships among the equipment and steps in the production process. The essential parameters in this approach are the relationships that exist among production processes, as well as time, distance, and cost factors. The layout is created in terms of the relationships that exist, and serves to determine the minimum amount of space and time needed to produce the product.

d. Automated Layout Design Programs (ALDEP). An ALDEP system, which is a version of a CAD program, integrates the full set of layout design characteristics, using some or all of these characteristics as parameters to achieve the best facility design. Possible parameters include product characteristics, product flow patterns, personnel requirements for both productive personnel and secondary staff, material-handling requirements, inventory control and storage, space usage, capital needs, and cost of production. By combining these elements, ALDEP compares various combinations with minimally acceptable or ideal states and improves overall design until further refinement is not possible. The choice of factors to use as parameters is at times a matter of program design and at times a matter of what is considered to be of critical importance by the firm.

7. Layout in nonmanufacturing enterprises. Such circumstances can present unique problems. In service organizations, there is no product, per se, being produced, but rather services rendered for the customers, who may be either present or absent, depending on the nature of the service. Even if the same service is being supplied to each customer the circumstances surrounding the delivery of services may be different, with one "run" for one customer followed by another run for some other customer. Consider three cases, a maid service, a laundry, and an accounting service. In the first case, the service is carried to the customer, since the service is normally to be rendered at the residence or place of business of that customer. In this case, layout of the facility refers not to the customer's property, but to the mobile structure by which the service is rendered. It could involve the organization of the trucks that take personnel and equipment to the locations, the design of special equipment to do the job, or the determination of the order in which the activities are to be carried out. Like process steps, there is probably a routine to see to it that all activities necessary to cleaning the building are carried out.

In the case of a laundry there is a degree of processing taking place, since the customer generally brings goods to the business location to be processed and then returns to recover those goods, though pickup and delivery are often a part of the service rendered. Here, a great deal of physical layout is required for the processing part of the service, but no product is actually manufactured. The flow of goods through the system is decidedly different from the flow of raw materials. It has to be done in batches, and the batches must be closely controlled to avoid embarrassing and expensive mistakes. Here, more traditional layout activities are expected.

In the case of the accounting service, a wide range of input and output forms may be involved, with information coming in over telephone lines, through the mail, or in hand-delivered parcels. It may even be picked up from the client's place of business and later returned with the finished product, which in this case consists of financial statements and other data. Yet the processing that takes place is nonphysical, being a combination of electronic and manual information transfer. Also, different services may be offered to different clients, depending on their needs and situations. Again, batch processing is involved. The layout may involve locating computer operators, printers, readers, and designing networks for efficient work flow, all of which shifts the facility design from purely physical to a combination of the physical and the nonphysical. A computer program designed to turn raw data into a set of financial and accounting reports is nothing more than a design of how information will flow through an environment, be processed and transformed into something more useful, and then stored for delivery as a finished product. The more we center our economy on information manipulation and communication, the more this type of facility design will become important.

Whatever the form of the service, the critical element in the facility design will be the needs of the customer. Design is greatly dependent on what the customer wants from a service, and there must be sufficient flexibility in the facility design to allow for the differences in customer needs. One laundry customer may want only washing done, while another may require washing, dry cleaning, some minor alterations, and winter storage. Or an individual may require house cleaning once a week by a maid service, while another requires true maintenance of a facility by an outside janitorial service. The "facility" involved in these types of services must be flexible enough to satisfy a wide range of desires and yet do it economically, operating as they do in a widely disparate batch environment.

V. MAINTENANCE

Maintenance is a control function. Its purpose is to control the cost due to machine failure, repair or replacement, and downtime while necessary repair or replacement is taking place. We have already mentioned that designing controls into the productive process is a necessary element in the facilities design creation, and carrying this further, it is a matter of course to see to it that maintenance is incorporated as well. The facility design should facilitate not only production but maintenance as well. This is done by first ensuring that there is adequate physical space for maintenance to take place without disrupting production unnecessarily, and by developing a program designed to minimize downtime. This program design involves three ideas consisting of repair policies, maintenance policies, and maintenance decisions techniques.

A. Repair Policy

A definite repair policy must be established that states the manner in which the firm intends to deal with maintenance. This usually involves a statement of the firm's philosophy concerning maintenance and a specific plan, complete with criteria for action, schedules of action, and parameters of operation for the maintenance program that allows those involved in the operation of the facility to know what is and is not expected of them.

B. Preventative-Maintenance Policy

This is a specific plan for preventing unnecessary downtime due to equipment that fails unexpectedly. The philosophy here is that if the machinery and equipment are maintained on a regular basis, major repairs can be prevented. Periodic prescribed programs of care are specified in preventative-maintenance policies and the degree to which machines are to be inspected, maintained, or overhauled is clearly defined.

C. Maintenance-Decision Techniques

Equipment eventually wears out and has to be replaced. The whole concept of depreciation is centered in an acceptance of this cost as inevitable. But the purpose of the maintenance program is to minimize the cost of that replacement process and to ensure that it does not take place other than as scheduled. This is a problem of prediction. The firm is dealing with a series of unknowns, since the exact date and kind of failure it is likely to experience with equipment cannot be factually known. To combat the uncertainty of the situation, techniques are designed to estimate as closely as possible what can be expected, and then to plan for as many eventualities as possible in an attempt to control expected but undefined events. Two methods for doing this involve simulation of maintenance problems and waiting-line theory models.

1. Simulation of maintenance problems. By simulating conditions that might arise in maintenance, the designer is able to discover what is most likely to occur, and how best to design for it. Which items are most likely to experience unscheduled downtime? How does that downtime affect the overall production process? Where are the bottlenecks in the design that can hamper the operation when downtime occurs? Which are the critical points in the process that offer no lag time? These types of questions can be explored through the simulation approach to discover how most efficiently to design for error. The more complicated the product process, the greater the need for this type of analysis since, as complexity rises, hidden elements seem to rise geometrically rather than arithmetically. The simulators play "What if" and then work on what to do in a given situation, developing contingency plans in the process. In this way, in any given situation, decisions have already been made and action can be taken immediately.

2. Waiting-line theory. Usually thought of as applying to queuing theory and retail service organizations, waiting-line theory can be applied to the maintenance problem as well. The purpose of this tool is to minimize waiting time by eliminating bottlenecks and specifying the most economical yet efficient number of channels that are needed to keep a flow moving. This applies to information flow, to customer flow as in checkout lines, and to production flow, experienced on both a planned and unplanned basis. The facility designer can utilize queuing theory to work with reducing bottlenecks in the manufacturing process due to routine maintenance and unexpected breakdowns. Starting with statistical data on the number and nature of expected maintenance operations, the maintenance department can be designed to accommodate the needs of the facility with a minimum of wasted time and effort. Too many workstations, for instance, would result in idle workers and equipment that would increase costs of operation unnecessarily. Too few would create bottlenecks and long waiting lines to complete repairs on a regular basis. The goal is to keep waiting time to a minimum while not "overengineering" the department so that it can handle much more than will normally be required of it. The catastrophic event may still occur and repair facilities may fall short of need, but the vast majority of repairs can be handled with ease if the department is able to adequately predict average need and design for the fulfillment of that need.

VI. FLOW SCHEDULING AND CONTROL

A. Line of Balance

Line of balance, or LOB, is a controlling device for production to ensure proper delivery schedules. When the firm makes a commitment to deliver given numbers of units at given times, any disruption of the normal flow of parts or noncompletion of production steps can result in slowdowns and delays that can jeopardize the firm's ability to keep its commitments. By using LOB, the manufacturer is able to know in advance when these possible delays will occur and head them off at the pass, so to speak, so that delivery schedules can be met. In the LOB process there are four steps: first, establishing the planned completion schedule for whatever project is under scrutiny; second, noting the state of cumulative deliveries made to date at the time of analysis; third, determining the line of balance, that is, the proper level of completion of production processes necessary to ensure maintenance of completion or delivery schedules; and finally, comparing the present status of each step in the production process and the line of balance in order to determine any delays that might exist. The best way to explain this process is to use an example.

1. Project plan. Assume that a firm has committed itself to produce and deliver electronic subassemblies to a major automobile manufacturer on the schedule in Table 14.4 (for a graph of this data, see the chart in Figure 14.2a).

Further, assume that the planned production process contains six steps, as shown in Figure 14.3.

TABLE 14.4 The Master Production Schedule

Month	1	2	3	4	5	6	7	8	9	10	11
Units (000s)	2	2	2	3	3	4	4	4	4	2	2
Total (f)	2	4	6	9	12	16	20	24	28	30	32

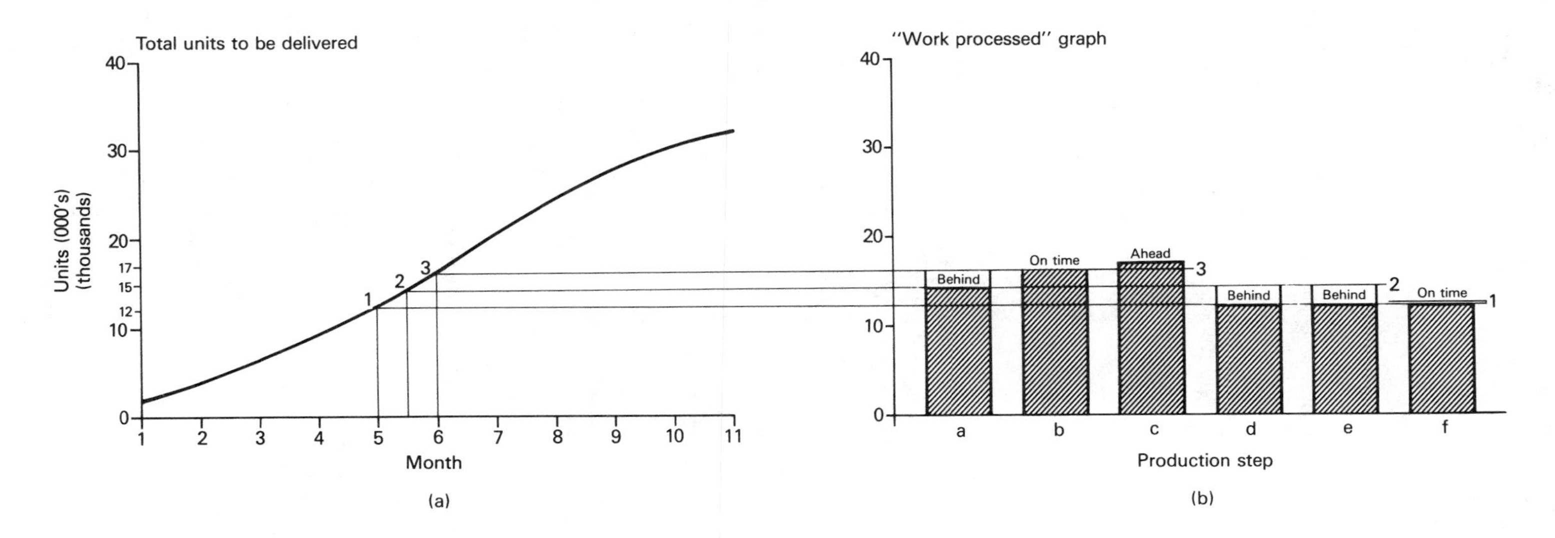

Figure 14.2 The Line of Balance Graphs

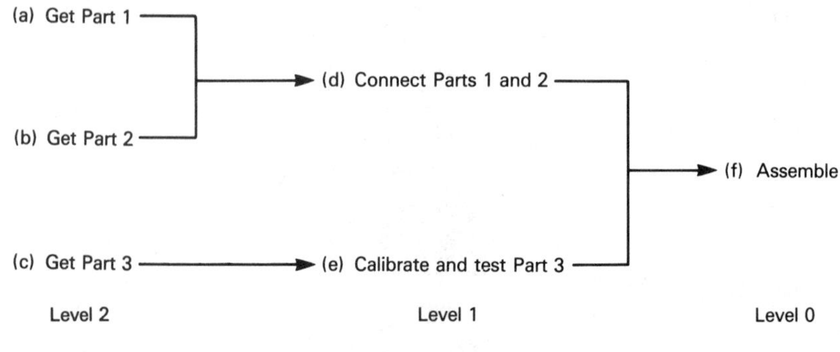

Figure 14.3

In this chart (Figure 14.3), each level takes exactly two weeks, and as indicated, steps (a), (b), and (c) take place simultaneously, as do steps (d) and (e), followed by (f). This would indicate that the entire process requires a total of 6 weeks. With this in mind, let us construct an LOB for the production schedule as of the end of the fifth month.

2. Cumulative deliveries. If we look at the record as of month 5, we find that the total number of units that have gone through each of the processing steps is as shown in Table 14.5 (for a graph of Table 14.5, see Figure 14.2b on page 291).

3. Determination of balance. This process involves determining how much of each step in the productive process would be completed if the project were in balance, that is, if it were totally on schedule. This is done by

TABLE 14.5 The Fifth-Month Work Processed Chart

(a) Part 1 received	15,000
(b) Part 2 received	16,000
(c) Part 3 received	17,000
(d) Parts 1 and 2 connected	12,000
(e) Units assembled	12,000
(f) Units shipped	12,000

referencing the production flow schedule (Figure 14.2A) and the amount of time that has passed. Graphically, we develop a function between the amount of time that passes and the amount of the order that should be completed by this time. This is a simple enough procedure, as is shown in the graphs a and b in Figure 14.2.

4. Status versus line of balance. This is the final step in the process, by which lines are drawn (lines 1, 2, and 3) from the schedule curve (Figure 14.2A) to a simultaneous chart of production steps units completed (Figure 14.2B), showing where each step in the process should be. These lines are drawn at the end of the fifth month (1), two weeks from then (2), and four weeks from then (3). We chose this spacing based on our bill of material chart (Figure 14.3) timing of two-week intervals. By adding bar graphs of the actual completion rates taken from Table 14.5, it is possible to see if there are any areas that are either behind or ahead of schedule.

As seen in Figure 14.2B, present shipments are on time [step (f)], but it will be difficult to keep up the present rate of delivery if something is not done about the production steps (d) and (e), which are behind schedule (their bar graphs should be up to line of balance 2). Step (a) is also behind schedule, so we need to check on the problem with part 1. Steps (b) and (c) are on schedule. By utilizing this process, the manufacturer is able to keep up to date, checking and rechecking present conditions against the original plan, and can maintain shipping schedules as agreed upon with the purchaser.

B. Industrial Dynamics

This is another aspect of flow channeling and control. As the name indicates, this concept deals with the fact that changes are both inevitable and necessary if we are to continually improve our efficiency and level of output. Dynamics can be seen as a matter of two stages: one, the acceptance of a philosophy that says all productive processes are evolving and thus dynamic in nature, and the other, the practical development of a framework that allows for changes to be made in positive and cost efficient ways.

With the first of the elements in the industrial dynamics approach, the philosophy of change as necessary and inevitable, comes a kind of industrial Darwinism, which views productive processes as being constantly developed and refined, with the most efficient aspects surviving while less efficient aspects fall to the wayside. Since there is a basic competition in the marketplace that is dependent on efficiency, this development never ceases but is spurred on by new techniques and new technologies as the changing tastes and needs of the public become known. Particularly in the latter half of the twentieth century, when even the rate of change is increasing, it becomes obvious that this is an appropiate philosophy to follow. Dependence on static production methodology dooms a firm to fall by the wayside, or simply to hope that it fits some unchanging niche in the industrial environment, although such niches are few and far between.

The second element in this dynamic approach, the development of a framework allowing changes to be made easily, is a logical outgrowth of the philosophical orientation. Since change is inevitable and, indeed, to be sought when it leads to efficiency, the entire issue of facilities planning and design becomes a matter of building in flexibility, so that changes can be made when they are appropriate and when the opportunity arises.

To prepare for change, one must see it coming. In designing production or service facilities, such foresight involves looking at internal signals and environmental indicators where changes occur. Internally, such signals come from a monitoring and analysis of several different aspects such as (1) Costs: Are costs rising? Are specific aspects of the productive process out of line with the budget? Are there new methodologies that promise reduced costs? (2) Delivery schedules: Are customers receiving deliveries on time? Are delivery schedules becoming more and more a problem? Are there new ways of scheduling and moving finished goods that can improve efficiency, like technological changes in the plant or computerized tracking systems? (3) Product life and product quality: What has happened in quality contol? Are we maintaining our quality? Are we improving it at a cost-effective rate? How do the customers view the quality of the product? (4) Flow of ideas through the system: Are we receiving ideas from employees and from the staff responsible for idea development? Are we being original? Are we innovating internally? (5) Behavioral dynamics of both individuals and groups: What do the absenteeism figures look like? Are employees generally pleased with their conditions? Are they willing to go the extra mile for the firm? Are we up to snuff on benefits and salaries? Are we experiencing favorable learning curves, or is this factor eroding?

Externally, the factors to look for involve changes in the market, changes in the industry and, particularly, changes in the technology of producing the firm's service or product. In all of this, it is imperative that the design of the original facility be such that it is a relatively simple, cost-effective task to make the changes that are needed. Note that this has particular importance to facilities that are relatively permanent and specialized in function and design, as in the case of assembly-line layouts, where specialized machinery and fixed relationships abound.

The problem becomes one of balancing the permanency inherent in such a facility with the capacity for initiating change as the need arises. To balance the two, the planner must determine what part of the process is permanent and what part is subject to change. For instance, in the case of an automobile production facility the assembly line may be fixed and highly specialized while certain areas of the process, such as the stamping of body parts, are flexible in that dies can be changed as needed to accommodate design changes in the vehicle. Assembly is permanent; the individual dies are not. Planners must allow for change within a permanent production system.

In all of this, a prime human factor must be considered: the natural resistance to change that will be encountered as the system is altered. People are naturally averse to change, and altering their job structure will result in resistance to the process. Such resistance needs to be minimized, though doing so will be difficult. Industrial dynamics is nothing more than a special case of systems dynamics, and, as we have noted many times in this text, all systems are constantly in the process of achieving and maintaining dynamic balance through alterations (changes) in the system in response to changing conditions.

To deal effectively with the natural tendency to resist change, management can do several things. First and probably most effective is to foster the idea of dynamic change within the organization, instilling in the employees the idea that change is a necessary and desirable part of operations and that it can be expected to occur in a periodic though orderly way. By dealing with this issue up front, there is an expectation of change that will help alleviate the fears of workers connected with sudden changes in their circumstances, and will further encourage them to not only accept but participate in the creation of changes in the system, since it is viewed as a natural part of the work environment. Second, the negative effects of homeostatic reaction can be minimized by involving employees in the change process, proceeding at a pace that allows them time to adjust, and informing them of every step in the process either as it occurs or in advance, so that they can become accustomed to the idea. People are not technological devices, though they may fit into a neat slot in the productive process. They function differently from equipment and gears, conveyer belts, and fork lifts. They are variable and very sensitive to changes in the environment. By properly caring for people and recognizing their special needs, the firm will find them to be flexible, capable, and surprisingly adaptable elements in a system already committed to periodic change.

QUESTIONS

1. What is meant by *operational process*?
2. Name three types of numerical controls. How is each useful?
3. What is a line-balancing problem and how is it solved?
4. Name three types of operations systems.
5. Name four processes in operations systems.
6. What are the differences among a casting process, an extrusion process, and a forming process?
7. Name three methods of joining and assembling.

8. Name five machining processes and describe each.
9. Offer examples of each of the three types of layouts discussed in the text.
10. How is material flow analysis utilized in solving layout problems?
11. Is preventative or corrective maintenance the most efficient approach? Why?
12. What is CORELAP? How is it used in process design?
13. Travel distance analysis seems to be a primary concern in both manufacturing and service facilities. Why is this? How are productivity and efficiency improved through such analysis?
14. Discuss the concept of *line of balance*.
15. What is industrial dynamics?
16. What is the difference between special-purpose machines and general-purpose machines?
17. Explain how contouring control and positioning differ.
18. Where are industrial robots most valuable? Where are they least likely to create advantage?
19. How does the analysis of travel distance contrast with improving flow of materials in process designs?
20. Explain the difference between CRAFT and ALDEP.

PART FIVE
Workstation Analysis

15
Work Flow Analysis

Every time we attend some sporting event, we are in the process of measuring the sporting abilities of the people involved in the event. Whether it be a single person competing against the clock or some established record, a pair of tennis players competing against each other, or teams performing in competition, we are still measuring their abilities. We do this by keeping score. Essentially, there is no difference between this process and what goes on in industry when we analyze the content of work. We are measuring the abilities of individuals and teams. The only difference is in the criteria we use for that measurement.

Work analysis is not a pleasant process for the workers who are being measured. People are naturally self-conscious about being scrutinized, particularly when their livelihood may be threatened by the outcome, though this is seldom the case in measurement activities. Yet the facility designer and those who will manage a facility once it is built must see to it that the proper amount of work is done in the proper amount of time. It is simply a matter of determining what is meant by the "proper amount" to achieve this purpose. There is a fine line between too much allotted time for a job and too little. A narrow range of acceptable limits exists, and judging that narrow range becomes a problem because, as the saying goes, work expands to fill the time available for it. People tend to use whatever time is allotted to get things done. If the time allotted is too great, the efficiency of the work process goes down, and if not enough time is allotted, then the quality of the work performed decreases.

Facility designers must determine standards against which facility performance can be measured. This is not simply a matter of averaging, for a large number of variables may affect the outcome of such an analysis. The law of reciprocal compensation says that nature makes up for deficiencies in one area by providing an overabundance of some other ability. This is true of people, as it is of any other living thing. Some people have incredible eye–hand coordination, while others have highly developed musculature. One worker may be quick at calculations while another exhibits an uncanny ability with machines. Yet standards have to be set for all of these individuals and thousands of others. To design a facility properly, we must not only understand how to physically construct the plant but also how to develop and measure standards of action within that structure that are reasonable and representative of the true nature of things, and that above all allow us to maximize efficiency without sacrificing quality. That means analyzing the flow of work through the facility in terms of mechanical and human performance.

I. METHODS OF WORK MEASUREMENT

A. Time Study

Time study is that part of operations standards development which investigates and analyzes operations in an attempt to determine standard time, which is the level of performance that the average worker, operating at normal speed and efficiency, is expected to maintain. That is, it is the purpose of the time study to determine and establish standard times for performing the various operations of the production process.

Time study can be applied to any job activity. It is done by observing and recording the job being performed. By establishing times for each step in an operation, it is possible to determine what can be expected of employees during actual production, thus determining the number of workers and time necessary to complete a job. It should be noted that time study is not the same as motion study, though it is inevitable that some motion study will take place during the time study process. The purpose here is not to analyze the physical motions of the operator, but rather to determine how much time, on the average, already established motion patterns will take. The simplest time studies involve the determination of no more than an overall time for the accomplishment of some job. Individual movements and substeps are not taken into account. In more elaborate time studies, such as those that may lead to a modification of the motion components of some job, each operation is divided into logical suboperations and these are studied in detail as to elapsed time, lag times, if any, and ways of improving this performance. It is in this type of time study that some motion study may be found.

There are three primary methods of conducting time studies: (1) the observed time study, in which the operator is actually observed and timed, either in person or by use of motion pictures or videotape, (2) the synthetic time study, in which the work is standardized through an analysis of machinery and estimated times based on experience with similar operations but in which no actual observation takes place, and (3) work sampling, in which a large number of samples of work operations are taken at random times and for brief periods to determine statistically the probable set of conditions surrounding the job, such as downtime, average machine speeds, and overall job timing.

1. Equipment used. Though relatively standardized, the equipment used in the time study process is neither particularly expensive nor complicated. For general time studies, no more than a stopwatch and a clipboard with a recording sheet attached is required. For more complicated time studies, two or three stopwatches may be used to record several types of elapsed time at one time, or highly accurate elapsed time chronometers of electronic rather than mechanical design may be used. For intermediate purposes, stopwatches with several sweep hands of different colors may be used in lieu of multiple timepieces.

2. Stopwatch time study procedure. This involves five basic steps: (1) dividing operations into acceptable and meaningful elements, (2) determining how the readings are to be taken and data recorded, (3) determining the number of work cycles that will be observed in order to get a statistically meaningful population of data from which to work, (4) rating the performance of the operator, and (5) computing and analyzing the study data gathered. Following is a brief discussion of each of these.

a. Dividing Operations into Elements. This is an important step that must be completed before it can be determined how much time each element of the operation is to take. Choosing the elements is the important factor here. There are a large number of ways in which to divide a work process into constituent elements. This is a systemic technique, designed to divide complicated processes into a number of simple ones. As with any systemic technique, the elements one chooses and the number of elements one chooses can seriously affect the type and quality of data received from the analysis. To be useful, the elements must be distinct from each other in some logical way; they must be small enough to be distinguishable, yet large enough to be meaningful; they must be significant in that the operation they represent is an important part of the process. If we divide up a process into too many elements, there is a possibility that our time study will turn into a motion study with time factors included. It must be remembered that these two techniques have totally different purposes and should therefore not be used interchangeably without good cause.

b. Reading and Recording. This procedure involves actually observing actions and taking measurements. The manner in which this is done is predetermined and standardized to ensure efficiency and consistency of the gathered data. Usually, the individual doing the measuring holds a clipboard and time measurement device in one hand and records data on the form with the other. The sequence in which this takes place is designed to minimize error and lost time while ensuring that the observer has sufficient opportunity to gather the required information. Each step of a process is measured and recorded on the form in turn, which has generally been blocked and sectioned prior to the observation session. Time and motion recording sheets are usually laid out in a matrix grid, indicating the operations to be measured and the number of observations made. The sheet includes information such as the location and designation of the operator, the overall job being observed, page numbers, dates, observer's identification, and other information for isolating the event in some logical manner. Table 15.1 offers a sample time study recording sheet.

c. Determining the Number of Cycles to Be Timed. This step is important because of cost and accuracy considerations. It is expensive to carry out time studies. Materials and observers must be provided, and with a large number of operations to time, it can become a long, drawn-out process. Reduction of even a few unnecessary repetitions of the observations can result in large savings over a number of operations. At the same time, it is necessary to be certain that there is a statistically significant population to work with if the data are to be accurate. As the saying goes, once is an incident, twice is a coincidence, and three times is a pattern. Patterns do not develop until there are a number of observations to measure. Choosing the correct number is usually determined statistically by measuring the degree of confidence (accuracy) that the individual firm deems appropriate.

d. Rating the Operator's Performance. This procedure involves working with the gathered data along with the judgments of the observer. It is necessary to know how typical the worker observed is considered to be if the information gathered is to have validity as a determining factor in expected plant behavior. If the employee observed is considered (or should be considered) slow, then the data reflect a time toward the maximum acceptable time part of the spectrum, whereas a highly efficient worker's scores would more properly reflect the minimum expected time for a job. What is sought is a standard, and without an understanding of the quality of the worker observed, that standard may end up being too high or too low. As to the actual rating of the worker, this is usually a matter of judgment on the part

TABLE 15.1 Simple Stopwatch Time Study Record Sheet

Sheet __ of Sheets		Get Work	Put in Vise	Turn Bolt	Drill Hole	Take from Vise	Put in Box	Outside Activity	
Number		1	2	3	4	5	6	T	Descrip.
Notes	No.	seconds	seconds	seconds	seconds	seconds	seconds	seconds	
	1	9	5	6	10	4	8	93	Get tool
	2	8	4	6	12	4	8	387	New bit
	3	10	5	6	11	3	8	312	Get drink
	4	7	6	6	10	4	9		
	5	8	4	7	10	5	8		
	6	9	5	7	11	5	10		
Slipped	7	20	4	5	9	3	8		
	8	7	4	5	11	4	7		
	9	9	5	6	10	3	8		
	10	8	7	5	12	5	9		
Total Time		95	49	59	106	40	83		
Number of Observations		10	10	10	10	10	10		
Average Time		9.5	4.9	5.9	10.6	4.0	8.3		
Minimum Time		7	4	5	9	3	7		
Maximum Time		20	7	7	12	5	10		

of the observer and of the worker's supervisor. Only exhaustive comparative studies can determine the true relative performance of the worker observed.

e. Computing and Analyzing Study Data. This step is carried out by statistical methods in which the statistical mean (average) is determined, usually along with the median time (the time halfway between the longest and shortest times recorded) and the modal time (the most frequently recorded time). In developing these measures, the longest and shortest times, along with any unusual times that are considered atypical for some reason, as noted on the time recording sheet, are thrown out and not used in the analysis. It is then a matter of determining which of the three measurements, the mean, the median, or the mode, is to be used as what is known as the *base time*. It is from this base time for the operator that normal times are derived.

To derive normal time from base time, the performance rating is used. As we noted earlier, this is a matter of deciding the relative efficiency of the worker observed as compared with the "average" worker. That is, the performance rating determines if the worker observed is slower, faster, or about on par with what is considered to be an average worker. These performance ratings are given as decimal ratings with average being equal to one, above average equal to less than one, and below average equal to some number greater than one. By multiplying the performance rating times the observed base time, a normal time for the operation is established. Note that superior workers are given ratings less than one to bring the time back into line by downgrading the final figure, while slow workers are given ratings greater than one to inflate their figures toward the average. Like a good game of golf, a low score implies skill.

Only one step remains in determining the standard time, and that is to convert the normal time after consideration of allowances for unusual or abnormal time-consuming activities. These allowances take into account the fact that there are job interruptions and maintenance activities that periodically serve to delay normal operations. Among the allowances usually cited are fatigue, rest periods, personal need interruptions, periodic process steps such as changing tools or receiving and removing work in process, and other factors that would be encountered during a normal workday flow. These allowances are added to the total normal time for the job to determine standard time for the job. The formula for standard time determination then becomes:

$$\text{Standard time} = (\text{Base time} * \text{Performance rating}) + \text{Allowances}$$

In an alternative method of determining standard time, the normal time is divided by [1 − Allowance fraction]. In this case, the allowance is not in time but is expressed as a decimal representation to account for the total time for extraneous activities. The fraction is generally a small number, and the result of the subtraction raises the value of the normal time to achieve standard time. In either case, the allowance is considered in the last step to convert normal time to standard time for completing the job in question.

B. Elemental Standard Data Systems

These are synthetic time study methods that determine standard or normal time from previously gathered data. The standard data comes from two sources: information on the performance of equipment, and information about the predetermined time allowable for basic human-motion operations. In the first case, data about machine and equipment performance the standard of normal time for carrying out some mechanical operation, can be gleaned from past experience with those machines performing those activities, from specifications developed by industry models, or from the manufacturers of the equipment itself. The speed at which machines should be operated, the time required to perform work on specified types and gauges of material, and the overall time required for specific jobs can all be found in literature from manufacturers or in trade journals and studies commissioned by industry organizations. By utilizing this data, a predetermined normal or standard time can be found for many operations.

In the case of physical worker actions there are also standard estimates available for the performance of specific activities by operators, particularly if the physical actions are common practices encountered in most work situations. These measured motions include grasp, release, move, reach, disengage, position, reach, and turn. Several publications supply standardized estimates for times necessary to perform these movements, and they have a high degree of accuracy. It is possible, by using these data in conjunction with the specifications and recommendations for machine operation, to determine the standard or normal time of operation for a process without ever actually observing the process taking place, hence the name *synthetic*. If an individual plant situation is being investigated, synthetic time studies such as standard data determination should be done only if individual allowances for that particular location can be logically measured and included. That is, because conditions will vary from one plant to another with design, peculiarities of the manufacturing process, and the capacities of workers, it is advisable to start with a standard time determination and then alter the expected times in accordance with expected and reasonably measurable variations due to the eccentricities of the plant itself. Otherwise, in the general case, the measurement should be a normal time, which, of course, excludes allowances.

The advantages of using synthetic studies are (1) the tremendous reduction in the amount of time and money necessary to perform the studies, (2) the ability to set standards prior to workers actually performing the work, an important factor if the performance standards are for new facilities, (3) the elimination of dependency on value judgments on the part of the observer, and (4) the elimination of the necessity for correcting for individual differences between observed workers. On the negative side, it must be remembered that no synthetic process will take the place of actual observation for gathering information. The "desk-top observer" is more likely than the on-the-spot observer to miss some facet of the job, or not to consider extraneous factors, particularly those leading to allowances.

C. Motion Time Data Systems

These are similar to standard data systems in that they rely on previously created data to reach conclusions. The difference is that with motion time data systems, the most efficient way of completing a process or movement, as well as the expected time for doing so are indicated in the data. Standards are set in terms of how one should complete movements and how one should arrange activities to minimize time spent. These data are then offered with expected performance times to create efficient processes and, at the same time, set standards of performance for workers which assume that performance of the activities will be done in the most efficient manner. It is as if a combination of motion study and time study were done using standard data rather than observation and rearrangement of the physical workstation. Other than this one alteration, the activities and content are the same as standard data approaches. We are still dealing with a synthetic approach here, and the same advantages and shortcomings apply.

D. Work Sampling

The work sampling technique is used in situations where neither work observation nor synthetic time study are possible or appropriate. The work sampling technique is generally used to determine specific classes of information, including the following:

1. Data on the utilization of machinery, that is, the amount of time during a specified work period that machinery is actually in use and not idle or down.
2. How often work delays and work stoppages for necessary or accidental causes occur.
3. Measurement of general allowances for use in computing standard time.
4. Development of time standards for continuous rather than discontinuous operations, a situation in which stopwatch observations are meaningless (such as in the production of steel or plastics, where there is no break in the process being performed).
5. Time studies of individuals and jobs that contain a number of miscellaneous activities, any of which may be predominant at a given point in

time. This is useful for analysis of jobs such as counter clerk or administrative assistant, whose tasks vary notably.

6. Development of production standards for periodic yet occasional jobs, such as maintenance, emergency repair, and special-case production runs.

The technique involves sampling activities over a long period of time and at random intervals, to determine the activities taking place, and the percentage of time devoted to each. Note that the two important differences between this technique and earlier ones is that the sampling takes place over an extended period of time, often as long as four to six months, and that it occurs randomly, rather than on a scheduled basis. Here the idea is not to get a large number of readings of the same activity, but rather to get a reading of various activities within a job over a long period of time. This process allows the determination of the percentage of time spent, on the average, at any one activity by the worker. If the worker has many functions to perform, this is helpful. Additionally, if the need exists to measure activities that do not occur on a scheduled basis, such as machine breakage, a random sampling technique is a more accurate form of measurement.

As an example, consider three cases in which work sampling might be used. First, consider the case of a piece of equipment that is sampled (that is, observed randomly) over a six-month period. During that six months, it is noted that out of 100 hours of sampled observation, the machine was out of commission for minor repair for a total of seven hours. It can be inferred from the sample that the machine is in repair for seven percent of the time it is on the job. Second, assume that a counter clerk is observed for 250 hours over a three-month period. During those 250 hours, the clerk is observed writing customer orders for a total of 50 hours. It can be inferred that the clerk spends 20 percent of the total time on the job in writing customer orders. In a third case, a glass manufacturing process is sampled for 500 hours over a six-month period, and it is found that, in addition to whatever else is going on in this continuous process, molten glass is being poured from furnaces into molds for eight hours of the 500 hours. The sample indicates that molds are being loaded 1.6 percent of the time.

Work sampling issues involve the nature of the process being tested, the type of data desired, and the technical aspects of the sample itself. These are summarized as tallying the observations, determining the sample size, using sampling to determine allowances, and setting standards by work sampling.

1. Tallying the observation. This is a statistical process by which averages are derived from the sample. In its simplest form, observations are made and recorded concerning the work being carried out, and the percentage of time spent on any one activity is found by dividing the number of samples (or duration of sample time) during which a given activity is being performed by the total number of samples or total sample time. This has already been illustrated in the examples of work sample use offered previously. The key elements in the tally are to measure the number of samples, the time expended during each sample, and the number and duration of each event encountered during the total sampling process. At times, the number of times an event occurs is the important factor, and at other times, the total time elapsed is important. Which information is most useful is predetermined or suggested by results.

2. Determining the sample size. The process of determining sample size tends to be as critical in this case as it is for any observation method, and for the same reasons. Sample sizes must be large enough to be statistically significant and at the same time, small enough to be reasonable and cost-efficient. It should be obvious that with sampling determinations time is not of the essence, since samples are to be taken over a long period of time. Yet the cost factor of the number of samples taken is still an issue, since each time a sample is taken, it costs time and effort. The result is either additional employee expense, or the need to detract from the otherwise efficient use of employee time taken off to perform the sampling observations. Of course, the larger the sample, the more statistically accurate the results are likely to be, and if accuracy is more important than economic efficiency, a larger number of samples will probably be taken.

In connection with this issue of how many samples to take is the issue of sampling reliability. Reliability involves not only the number of samples, but the way in which the samples are carried out as well. As difficult as it is to do, a sampling technique can be reliable and accurate only if there is absolute randomness in the timing of observations and if observations are not slanted due to conditions, interval, or time of day. It would be very convenient, for instance, for observers to choose a convenient time of day to take samples so that it fit into their schedule. However, this would skew the results by biasing measurements to a specific time of day. If it is early in the morning, the data could be skewed because workers are fresh; if around noon, because they are more interested in lunch than in their jobs, or specific activities may happen more often in the afternoon than in the morning. Any of these circumstances would reduce the reliability of the data. The actual determination of the number of samples to take and the randomness of their timing is a matter of statistical and probabilistic analysis. As to the techniques required, it is a matter of confidence and of random-number generators making those determinations.

3. Using work sampling to determine allowances. This is a natural use for this technique, because events can be defined in any way we wish before they are measured. This means that it is possible to sample for and measure the number and duration of nonevents as well as events. Since allowances are designed to take into account odd occurrences or nonscheduled random events, the sample technique is perfect. It becomes a matter of simply defining one set of occurrences as events and everything else as non-events, and then sampling activities. By observing on a random basis, the nonevents and unexpected events that need to be included in the allowancing factor can be captured, and the frequency and duration with which they occur can be predicted. We can find the percentage of time the events occur and we can find the percentage of total time that they last, as well as determine the average length of time a single event of this nature is expected to take.

For instance, if 500 hours of observation sampled at one hour per sample show that an employee had to spend 20 hours adjusting the equipment, then we can infer that adjustment represents approximately four percent of the total observed activities time involving machinery adjustment. And if the number of adjustment periods observed equaled 40, we can infer that the average amount of time necessary to carry out a single adjustment is one-half hour. This determination is illustrated as follows:

$$\text{Adjustment time} = \frac{\text{Total adjustment time}}{\text{Total sample time}} = \frac{20\text{ hr}}{500\text{ hr}} = 0.04, \text{ or } 4\%$$

$$\text{Average adjustment time} = \frac{\text{Total adjustment time}}{\text{Number of adjustments}} = \frac{20\text{ hr}}{40} = 0.5\text{ hr}$$

4. Setting standards by work sampling. This is a very shaky business, chiefly because of the questionable accuracy and reliability of the technique. To set standards using the work-sampling technique, the first step is to determine the percentage of total time a worker is engaged in a specific activity (or the proportion of time, if a ratio is preferred), and then use this figure to determine normal time according to the following formula.

$$\text{Normal time} = \frac{\begin{pmatrix}\text{Total sample}\\ \text{time}\end{pmatrix} * \begin{pmatrix}\text{Percentage of time spent}\\ \text{performing a job}\end{pmatrix} * \begin{pmatrix}\text{Performance}\\ \text{rating factor}\end{pmatrix}}{\text{Number of units completed}}$$

Once allowances are added as a factor, the standard time can be determined. Admittedly, there is more room for error innately embedded in this approach, because the population of events is a sampled population, but with sufficient sampling size to yield acceptable confidence intervals, this problem can be minimized.

5. Work sampling versus time study. Work sampling has some decided advantages over time study, and some disadvantages as well. Neither of the two approaches is inherently superior to the other. Conditions largely dictate which is preferable, as we show here.

On the positive side: (1) As was earlier noted, sampling can be used where true time studies are impracticable. (2) Sampling allows for longer studies to be done in an economical manner. (3) The impact on the worker being tested is eased through the random nature of the sampling process, in that the worker doesn't have time to become apprehensive about being tested. (4) The same objectives are obtainable with a smaller total expenditure of time on either observation or evaluation. (5) Sampling can be used for continuous operations that do not lend themselves to stopwatch observation. (6) Sampling is effective when the observed activity is being carried out by groups of nonubiquitous workers, that is, in conditions where not all workers in the process are equally active and efficient at all times. (7) Sampling can be used in conjunction with the more traditional observation methodology to determine allowances, something that is difficult to determine by using direct time studies or standard data methodology.

On the negative side, the shortcomings to the work-sample approach are the following: (1) If an operation is repetitive in nature and is standardized, a direct measurement process is more economical and certainly takes less time; in other words, there is no need for randomness in this case. (2) Sampling in no way eliminates the need for control planning on the part of operations managers because, unlike the observational method, which can be extended to motion study to take corrective action immediately, sampling does nothing but provide raw data for statistical analysis. (3) Workers are reluctant to accept sampling techniques as a basis for incentive wage and wage rate determinations, because they view the direct observation method to be the only truly reliable measure of productivity. (4) Sampling does not analyze the various individual elements in an operation and therefore contributes nothing to motion analysis or methods research that might improve overall performance by changing process content or event sequencing.

E. Physiological Work Measurement

This is a field of work determination that has become particularly valuable in recent years, in no small part due to the boost the technology of this field has received from sports medicine and space exploration. Due to the knowledge of the human body that has developed in recent decades as a result of efforts in these two fields, a wealth of information has been produced that has been found to be useful in the construction of efficient work systems. With the physiological work measurement approach, the effects of work content and conditions on the worker can be measured, along with information used to predict efficiency and health hazards on the job.

The basic technique is the same as it is in sports and space medicine, mainly observation and measurement of body function and body stress under a given set of conditions that reveal what the reasonable work limits of an individual will be in that environment. In space medicine, the interest is in discovering what can reasonably be expected of astronauts in orbit. In sports medicine, it is a matter of maximizing performance and minimizing the risk of injury. Fundamentally, the purpose of this type of approach in business is the same, that is, measurement of performance to maximize efficiency, to determine what is and is not reasonably possible under the given conditions, and to avoid injury and provide for the general health and safety of the person performing the tasks. Four primary measurements are carried out in the physiological approach, designed to measure oxygen consumption, heart rate, the effects of environmental stress, and the degree of fatigue, including the determination of needs for reasonable rest periods during a work session. With these measurements, management is able to structure a job to achieve maximum worker performance with minimum worker discomfort and/or risk.

1. Oxygen consumption. The amount of oxygen consumed indicates the amount of physical work that is being performed by the worker during given activities. It is useful for two reasons. First, it measures how much the worker can reasonably be expected to do in a work period. If oxygen usage is high on the average, then a job is physically taxing and needs to be adjusted, or forms of compensation for the taxing nature of the job must be included, such as frequent periods when nontaxing work is performed, or more frequent breaks. If the oxygen consumption is too low, it may be that the worker will require heightened physical activity elsewhere to make up for the lack of exercise inherent in the job in an attempt to avoid deconditioning, the phenomenon experienced by patients who find themselves uncharacteristicly weak after long confinement to bed (or for that matter, that sluggish feeling one gets from oversleeping). The muscles actually start to deteriorate, or decondition, if not used regularly.

Second, the oxygen consumption rate of workers in general can yield information about the overall physical condition of workers in the facility, thus pointing the way to any corrections that might be necessary to increase their alertness and output while reducing the chance of accident on the job or absenteeism through health problems. In addition to measuring the amount of oxygen consumption, the degree to which the consumption rate changes during the performance of duties is also significant; a reasonable amount of change is considered desirable, whereas excessively high rates or low rates of change are considered counterproductive to efficiency and general health.

2. Heart rate. Measurement of heart rate is closely tied to oxygen consumption, but in this case the measured phenomenon is the rate at which blood is circulated through the body. Frequent changes in heart rates can be taxing on the body, while steady heart rate patterns indicate no significant change in the amount of exertion necessary to do a job from one point in that job to another. As with oxygen intake, either too low a heart rate or too high a heart rate is considered counterproductive. There is a reasonable range within which the worker is considered to be operating in a healthy environment. The determination of those limits is beyond the scope of this book, and should be dealt with through a physician well versed in the field.

3. Effects of environmental stress. Such effects on the worker are also of prime importance, since excessive stress can lead to fatigue, error, and deteriorating health. It should be noted that mild periodic stress is practically unavoidable on the job, and is considered a normal occurrence. This is not the concern that is being referred to here. Stress as dealt with here refers to extreme or continuous stress brought on by the work environment itself, which might occur from excessive noise, vibration, heat, moisture, and changing light conditions, or for that matter, any other extreme in physical working conditions. Additionally, stress can occur as a result of too high an expected rate of output, particularly on assembly-line stations, where work proceeds at a predetermined pace and the worker is expected to keep up with the predetermined flow. All of these conditions can lead to excess stress.

When encountered, excess stress results in alterations in physical body functions, normally leading to rapid heartbeat, muscle tension, increased blood pressure, and possible physical damage to any part of the body bombarded with excessively high stimuli. The worker can become overly fatigued (and therefore careless), may experience frequent need for rest, or may exhibit a lower production rate because the stress interferes with normal activities. By measuring and determining whether the amount of stress built into a job is reasonable, steps can be taken to maintain efficiency while protecting the worker.

4. Fatigue allowance and rest periods. These must be determined in developing job specifications. As we have indicated, the effects of fatigue can be detrimental for both the worker and the firm and should be avoided. Sufficient rest periods must be included in job content to avoid excessive fatigue in the worker, which could result in "downtime" for the work unit (the worker) while it repairs itself (or is repaired by the physician). By measuring fatigue and the effect of rest periods on efficiency, the firm can determine the proper number and length of rest periods to avoid excessive fatigue.

II. APPLICATION TO FACILITY DESIGN

From the point of view of the facility designer, all of these measurement issues are important, as they offer information that will determine how the facility is to be constructed. If a facility is to have a given capacity (a given level of maximum output), then all of the bits and pieces that go into its design must be capable of supplying at least that output. That also means that the time required to perform the production functions must be such that the job gets done within the specified length of time. How many production lines are required to do the job? Which methods of operation will run quickly enough to reach capacity requirements? How much production can logically be expected from a given station and a given worker at that station? All of this must be predetermined prior to construction of the plant, and the plant design must reflect these limitations. Facility designers need to know within what physical and temporal limitations they have to work, and the various methods of work measurement supply the data for finding those limitations. Whether the plant is to be a copy of some already existing facility where physical measurements can be made, or whether it is a completely new concept where standard measurements are required, the information gained from work measurement studies are primary in the design and implementation of the facility layout. To do otherwise would be like designing a car with no idea of what horsepower the engine develops, how much translation of power the transmission will produce, how to calibrate the odometer or speedometer, what voltage the electrical system will produce, or how well the brakes will work, if they will work at all. Only by knowing the capacities of the constituent parts can we determine what the structure as a whole will do.

QUESTIONS

1. What equipment is generally used in time studies?
2. How do motion-time data systems differ from straight elapsed-time studies?
3. Give an example of a case in which elemental standard data systems would be a desirable method of work measurement.
4. When would work sampling be an efficient method of work measurement?
5. What is meant by *work measurement*? Why is it important to facility design?
6. When is sample size a critical factor in work sampling? When is it not critical?
7. Is work sampling more efficient than time study?
8. List the five steps of the stopwatch time-study approach.
9. List several ways in which time standards are valuable to management.
10. Are time studies more important during production design or during ongoing plant operations? How are they useful to each?
11. How do environmental conditions affect the accuracy and execution of work measurement?
12. Two timing techniques are repetitive timing and continuous timing. How are they different? How similar?
13. What is a performance rating? How is it determined?
14. What are the steps of the work-sampling process?
15. How are operations divided into elements?

16. What are some of the difficulties encountered in accurately reading and recording the results of a stopwatch time study?
17. In time-study operations, is the rating of the operator's performance objective or subjective? Explain.
18. Assume that you are to do a time study of a fast-order cook creating a cheeseburger. What operational elements would you choose to observe?

16
Workstation Design

If you think back to the time when you were considerably smaller and younger, you can probably remember when it was necessary to reach up to take a water glass from the dinner table, or when putting your arms on the armrests of a chair demanded the flexibility of a contortionist. Doorknobs were too high, counter tops could not be reached, and chairs left your feet dangling precariously in mid-air. Yet when you matured, the situation changed, and probably the "correctness" of furniture dimensions is a foregone conclusion in your mind.

The reason you don't notice those inconveniences of dimension any longer is that, in the world of adults, most furniture is standardized, although occasionally a sofa will eat you alive as you slowly sink to the floor in a great puff of expelled air, or a table will snap at your knees as you try to slide under its edge. Standardization creates tables and chairs and counters that are just the right height for the activities carried out there, which is the result of a great deal of purposive effort on the part of designers.

The same effort and sizing that is put into normal household furniture is utilized in the construction of workstations, but in a more sophisticated, and in most cases, more extensive manner. This is a logical approach, since not only the comfort of the user, but the efficiency and productivity of the user is in question. We are dealing with the most basic level of facility design when we delve into the creation of individual workstations. Since it is important to ensure that workstation designs result in a minimum of fatigue, error, wasted time, and physical effort, facility planners take special care to be certain that the necessary elements which accomplish that goal are present in the designs.

I. DESIGN AND LAYOUT OF WORKPLACE AND MACHINES

In the discussion of the layout and design of workplaces and machinery, three primary areas will be dealt with: seating considerations, instrument display designs, and controls. From the point of view of the human engineering specialist, these are of prime importance in workstation design.

A. Seating

In designing workstations, the seating arrangement is of primary importance if the operator is to have a degree of comfort and still be able to perform required tasks. Although providing such comfort seems to be a kind and altruistic act on the part of the firm, the reason for it lies in the reduction of fatigue and downtime resulting from a poorly designed seating arrangement. Different jobs require different types of seating, and different treatment of such factors as the spread of arm movement, type of arm and hand coordination required, importance of stability of the seating versus flexibility to turn, pitch, and flex, and amount of support required for safety and comfort. In one operation, a simple circular stool of the appropriate height may be best, while in another, an entire integrated console with reclining contour formed chair is necessary. One would not want a reclining couch to operate a lathe, nor would one want to operate the pilot's station of a space shuttle from some precarious perch atop a bar stool. It is imperative to match the seating with the operation. The primary considerations in designing a seating arrangement are (1) height, (2) contour, (3) position of backrest and seat, (4) motility, (5) ability to maintain proper ratio of seating to workplace height, (6) minor corrections for shoe height and slump, and (7) footrest design.

1. Height. The height of the seating must be appropriate to two factors, the general body proportions of the operator and the manner in which the seating is integrated into the job function. If countertops are wide and deep, then seating must be high enough to allow the employee to reach comfortably to the full extension of the arms, which generally means higher seat height or lower bench height. For typing and other keyboard activities, the seating height should allow the employee to comfortably reach the keys without the necessity of raising the arms at an angle, which causes fatigue and reduces the efficiency of the typing stance. This means lower keyboard surfaces or higher seating. If it is necessary to look over the console or work surface, as in the case of a supervisor who needs a clear view of the work floor or an operator who must see the results of commands issued to the console, elevated seating can raise the line of sight and bring the operator to a height that towers over the work in progress. This is often the case of industrial functions that are remotely monitored or controlled. At a minimum, the height must be sufficient to keep the knees from rising to interfere with the hands and arms, and low enough to keep feet from dangling or otherwise reducing the maneuverability and support of the remainder of the body. As is indicated in Table 16.1, there is a range of proper seating dimension for the human anatomy which can be accommodated either by taking average dimensions or, if seating is particularly critical to the design, by allowing for adjustment within the mechanics of the seating design. The dimensions offered are for standard desk and work surface arrangements used in furniture design. It should be noted, however, that like all other aspects of the human experience, there are changes in the way people sit and work through time, and as this takes place, there is a change in what is considered the proper dimensions for furniture and body support. The range given involves the standard dimensions for the furniture listed, and the second number represents a more approximate number of general comfort, based on studies of individual preferences over a ten-year period. Both should be considered in deciding on the proper dimensions for a particular workstation.

2. Contour. Like height, the contour of the seating should reflect two factors, comfort and function. Workstations that are to be occupied in a seated position for extended periods of time usually require more attention to contouring to match activities, thus minimizing fatigue and physical discomfort. Referring to the earlier example of the flight couch in a space

TABLE 16.1 Anthropometric Data for Design of Workplace to Accommodate Both Sitting and Standing Operators

	Height in Inches by Percentile		
Dimensions	5th	50th	95th
Standing			
Eye height	60.8	64.7	68.6
Shoulder height	51.0	54.8	58.8
Elbow height	40.6	43.5	46.4
Sitting[a]			
Eye height	28.4	30.3	32.1
Shoulder height	21.6	23.4	25.0
Elbow height	8.0	9.4	11.0

[a]Measured from seated surface.

Source: Reproduced with permission from *Production and Operations Management* by Arthur C. Lauter, © 1984 by South-Western Publishing Company, Cincinnati, Ohio.

shuttle, the long hours under conditions varying from extreme g-force to weightlessness require a custom contouring to match the body of the person in the flight position. Though less critical, the factor of contour can be seen in automobile seating, particularly touring car seating, and in the design of airline seating, where the passenger may remain in a certain position for any number of hours, depending on the length of the flight.

In general, the primary joints of the body torso, neck, and legs must be supported by such long-term seating to eliminate the possibility of cramps and muscular fatigue. Proper angle of the back and seat junction and at least a minimum of adjustability (to be discussed later) are imperative. In special cases, partial or total support for the lower legs may be necessary. An extreme case of this approach is the seating in the cockpit of a military fighter aircraft, which is as much a facility as any plant or factory, where contouring and sizing are so critical that pilots must fall within strict dimensional standards to ensure that they do not incur injuries from g-forces and long flights. In this case, it is easier to build a standard contour and find individuals to fit it than to function the other way around.

3. Position of backrest. This is an extension of the contouring problem. Backrest position is indicative of the type of strain to which the body is to be put. If overhead controls or frequent glances overhead are a normal part of the job, the backrest must support the back in a comfortable position and extend to support the head and neck in order to avoid strain. A backrest that is angled too far toward the horizontal offers a minimum of support except for those in a nearly reclining position, and may actually increase strain by forcing the employee to lean forward unnaturally to reach instruments and equipment. On the other hand, a backrest too close to the vertical may create lower lumbar pain and unnecessary strain to the central back by restricting the worker's ability to slump naturally or curve and arch the back, or by putting unnecessary pressure on the hips and lower spine. While a keyboard operator needs nearly vertical support in order to type for extended periods of time, the same type of seating for a control engineer in a remotely controlled lumber operation might reduce efficiency and increase fatigue. Again, in jobs that require a variety of positions, adjustability is frequently the answer to backrest positioning.

4. Position of seat. This is often an illusive element in seating design, because it can create problems in subtle ways. The positioning of the seat contributes substantially to the flexibility and maneuverability of the worker, as well as affecting overall posture and positioning of legs, hips, and back. With a small, short-shanked seat, much of the upper leg as well as all of the lower leg will be largely unsupported, as in a tall stool. In these cases, the legs are usually hooked backward under some brace or around the chair legs to maintain stability and support. This may be appropriate for jobs involving a substantial amount of standing and sitting in combination, where extricating oneself from the seating quickly and easily is more important than long-range comfort. In the case of job content requiring long periods of sitting, the seat needs to offer support and at the same time allow the worker to easily reach the working surface and maneuver. Seats sloped forward naturally "dump" the worker in the direction of the work surface, while seats angled backward force the worker to maintain contact with the backrest. Combining this angle with the size of the seating surface itself, the seat position will determine the worker's stance while on the seat. Properly designed, it relieves the employee from exerting energy for maintenance of an efficient working position; if improperly designed, it causes fatigue and possibly physical damage by constantly having muscles under tension to maintain that working position. This muscle tension is often largely unconscious. Consider the feeling of exhaustion experienced after a day of boating. Even if one is relatively inactive during the outing, there is still fatigue, due to the same type of muscle tension, from the constant shifting and balancing one does to maintain balance in a moving, rocking boat. Now imagine that effect over the length of a workday, day after day. This is the effect that an improper seating angle can create.

5. Adjustability. Already touched upon, adjustability is important because no matter how standard a seating design is, the people who occupy that seating are going to vary slightly and thus need some element of adjustability to correct for those variations. In addition, many jobs require variations in activities, even in tightly defined workstations, and it is useful to be able to adjust equipment to meet immediate needs. Finally, there is the fact that sitting too long in any one position will create discomfort and fatigue, and what is comfortable and proper at the beginning of a workday may not be by the end of the day. Hence the need for adjustments in seating. Most needs for adjustment can be handled by the shifting of height, angle, and length of components, though variations in tension and the ability to periodically lock and unlock elements of the seating arrangement are also valuable.

6. Ratio to workplace height. This is one element of seating design that is imperative for a successful workstation design. It must be remembered that the reason for designing the seating in the first place is to accommodate a worker who is performing some set of functions. Those functions have requirements of space, dimension, and maneuverability, much as the seating itself does. Because of this, the seating must be designed in conjunction with the other elements of the workstation. This was alluded to earlier when it was noted that the hand and arm position on work surfaces and consoles are largely a matter of either altering the seating height or adjusting the work surface. In reality, a combination of both is needed to find a proper, cooperative fit. There are times when the work area height is a nonnegotiable factor, being part of a larger system that cannot easily be altered. If this is the case, then the seating must adjust to the more permanently fixed work area dimensions. Some seating may require nearly prone positions close to the surface of the work area. An odd example of this would be sports car seating, particularly in low-profile, high-performance vehicles, where the driver is nearly prone, supported by contoured seating and neck and back supports.

At other times, seating may of necessity be raised in such a way that the operator needs to literally climb into the seat to get at the controls. Consider the workstation designs on heavy industrial machinery such as cranes and materials-handling conveyors. Often the operator is above the general work area, looking down. Other examples would be the case of camera operators on boom platforms, and military tank gunners. Although these latter two examples are probably not the sort of situation that one would expect to run into in an industrial setting, tanks, sound stages, and the like are facilities like any other environment designed for production of a given end result, and they follow the same rules of facility design as a paper mill, an automobile factory, a marina, or a laundry.

7. Corrections. The ability to make certain corrections must be inherent in workstation designs, and though they represent minor alterations, they are necessary if the efficiency of the station is to be maintained. These include corrections for shoe height, which is about 1.1 in.; for a normal slump while standing, about −1.2 in.; and for normal seated slump, which amounts to −2.0 in. in human engineering design. The reasons for the slight variations are evident to anyone who has a tendency to perform certain functions without shoes on, or who works at a counter that does not allow for slump variations. A matter of one or two inches in dimension can have a great deal of effect on one's health and strength over time. The importance here is in cases where the type of work done is likely to dictate a certain type of footwear or create body slumps.

In the case of shoe variations, consider the difference in dimension between the type of dress shoe worn by executives or office workers as opposed to those worn by someone in a production job, a carpenter or machine operator for instance, whose footwear is designed to protect against blows by foreign objects, electrical shock, or moisture. The thickness of the soles and height of the heels will vary, and even that slight variation can reduce productivity. The nature of footwear expected to be used by workers needs to be considered.

Likewise, the expected standing and seated slump of employees engaged in their craft must be considered. Anyone who has seen an architect or draftsman at work is aware of the slump imposed by bending over a drafting table while perched high on a stool. Note that there is no need for back support in these cases, since the worker is bent forward, and the height of the stool is usually slightly higher than standard stool height to accommodate the tendency to lean forward. In a standing position, the height of the work surface and relational height of the seat also need to take into account the tendency to slump while working in a bent-over position. This need not be explained to anyone who has been standing at a counter or work table and has moved back to sit on a stool, only to have it skitter away because it is too high to comfortably reach, or who receives a jolt as he or she drops into a seat that is lower than expected. Even a difference of one or two inches can create jolts and bruises that are painful and can cause dysfunction over time.

8. Adjustable footrest. Although this may seem to be appropriate for a barber's chair or the back seat of a luxurious limousine, it is in fact a widely applicable technique in station design. Particularly for individuals who are required to sit at a height greater than normal seating, the prospect of dangling legs is both mildly irritating and potentially dangerous. For such cases, seating is often outfitted with circular footrests that surround stool-type seating, or which attach to either the seat or the workstation to provide both support and stability. A surprising amount of leg muscle contraction is involved in seated movement, and if there is not bracing for the feet and legs, this action is transferred to the muscles of the stomach and abdomen, which are unused to carrying out balancing and support functions from the seated position. Fatigue and cramps can result. With the addition of a footrest for bracing and support, much of this is eliminated. In addition, footrests increase the circulation in the extremities to reduce fatigue and the possibility of legs falling asleep or aching from poor circulation.

B. Instrument Displays

The second set of dimensions to consider in workstation design are those involved in instrument display. Like seating, the goal here is to produce a workstation that is both functional and comfortable for the worker in order to create conditions that will promote low fatigue and high productivity. Instrument display considerations, however, involve the head position and eye position of the individual, rather than overall body position. Instruments are designed to provide information and guide actions that will achieve the goals of the workstation. Designing displays is fundamentally a matter of eliminating characteristics that may hinder the achievement of those ends and fostering characteristics that are supportive of them. Such factors as readability, speed of delivery and comprehension, and logical interrelationships among the various instruments are primary concerns, as well as the productivity of the worker working with those instruments.

1. Location. Instruments should be located where they can be easily read and comprehended. This translates into positioning them near enough to the employee to be legible without being so close that they interfere with other activities. In addition, they must be as close to normal eye level as possible and generally positioned in an arc commensurate with the span of the worker's vision. The closer the instruments, the narrower the array must be, since the normal field of vision of the worker will represent a smaller physical width at close range. There is a tradeoff here that must be observed. Problems to be avoided involve placing instruments to the left or right of the observer in such a way that frequent shifts of the head are required to read them, since this could result in neck strain and fatigue, and placing them too low or too high for much the same reason.

In addition, the arrangement of an array of instruments before the operator is an important consideration. Instruments should be displayed with those most often or most critically observed in the center of the array, and those less often dealt with along the periphery. Beyond this, they should be displayed in a logical fashion, all those dealing with the same activities grouped together wherever possible, depending on the way their commonality is defined. For example, commonality might be illustrated by one group of instruments that deal strictly with power supply and delivery, another group that deals with the feed of raw materials, and a third group consisting of an array of temperature-sensing devices. By knowing the type of information sought, the operator knows what part of the instrument display to observe. On the other hand, an array of different informational instruments all connected to one machine or device might be placed together, and the same arrangement repeated for every other machine or device. An example of this type of approach can be found in aircraft that offer information on engine performance clustered for each engine, the position of the various arrays commensurate with the positions of the engines on the plane, starting from far right and moving across the aircraft to far left. On such facilities as ships or power boats, instruments such as tachometers, throttles, and pressure gauges are similarly arranged to the left for port engine(s) and to the right for starboard engine(s). Whatever locational approach is used, it should be logical and deliberate to facilitate the memorization and mastering of instrument positions. For an alternative example, think of the arrangement of letters on a keyboard, in the case of typewriters known as the "QWERTY" arrangement, supposedly created in accordance with which letters are most often used. In this particular arrangement, those letters most often used are placed in the center of the board and those least often used are placed on the outside, and in rows one and three rather than the primary row, number two.

2. Shape of mounting surface. This is a factor that can add considerably to the readability and utility of instruments. The shape of the mounting surface should be designed with the array of the instruments in mind, and if need be, wrap around the vision of the operator in an effort to equalize the accessibility of instrument readouts. What is not wanted is a wide expanse of instruments that requires considerable variations in effort and focus to read as the distance from the operator increases. This is what would happen with a wide array of instruments close to the worker. In cases where numerous instruments are displayed, a wraparound arrangement would be preferable, with multiple banks of instruments rather than one or two long lines. If the instruments are farther away, as in a flat-mounted control panel such as those found at nuclear power plant control facilities, then curvature is not as important as banking or stacking instruments in a logical manner. The shape of the actual mounting surface can vary from square to round with other shapes in between, such as U-shapes, kidney shapes, or odd geometric shapes, as long as they fit the pattern of the display. Of primary importance here is economy of design (no wasted space) and adherence to the instrument layout, so that the shapes of the surface and the layout itself coincide.

3. Contour of mounting surface. This is a similar consideration to the shape of the surface, though here we are referring to the topography of the surface rather than to its outer shape. Contours can add to readability by placing instruments at the same focus for the observer, particularly true of wraparound designs, and by reducing glare from lights over the expanse of the reading surface. Contours can also bring some instruments closer to the observer while receding others so that the proper dimensions for reading are maintained, no matter what the size of the instrument itself. This is like terracing or stacking, but with smoother contours. A combination effect could be achieved by contouring the instrumentation surface so that it angles away from the observer, with the top farther away than the bottom and the instruments tilted in banks for easy readability.

4. Direction of alignment. People have a tendency to deal with information in a linear fashion, usually from left to right or from top to bottom. The alignment of instrumentation in a display can either encourage or discourage this tendency in the worker, depending on what is appropriate to the individual situation. (At times, it may actually be advantageous to do the apparently illogical in order to encourage active alertness as opposed to passive scanning of instrumentation.) In general, unless other information indicates an advantage to the contrary, instruments should be grouped either left to right, top to bottom, or, if circular, in a clockwise fashion so that each reading is connected to the others in a logical sequence. For large arrays, alignment should be from the center radiating outward with the most important or most often used instruments directly in front of the observer's line of sight, as indicated earlier. Groupings should likewise begin in the center and radiate outward, first to the immediate left and right and then above and below, although individual circumstances produce logical sequencing patterns that could vary from this pattern. In some cases, clusters of instruments are logically patterned so that the central instrument is a primary indicator and the others are secondary and important only in relation to the information of the primary indicator. In this case, the clusters are placed center, left, right, top, and bottom in a pattern that centers the primary reading in the midst of the cluster with the secondary instruments surrounding it in a circular or polygonal pattern that follows logical sequencing. The designer needs to remember the goal of the instrumentation, that is, to deliver information quickly and easily in a meaningful manner, with a minimum of error and a maximum of productive efficiency. This prime directive will go a long way toward defining what will and will not be acceptable in surface and instrumentation layout.

5. Information displays. There are different types of information that are made available on instrument displays, and to an extent, the type of information desired will dictate the shape, layout, and arrangement of instrumentation. Basically, the number of types of information made available can be grouped into (a) quantitative readings, (b) check readings, (c) settings, and (d) tracking. Each of these offers different problems and different opportunities.

a. Quantitative Reading. This is a major type of instrument information. Quantitative readings offer indications of dimensions in terms of measurable phenomena, and just about any phenomenon that can be measured physically can be displayed in this manner. Such elements as height, width, weight, volume, elapsed time, ratios of component parts, rates of flow, movement or change, pressure, and temperature can be measured quantitatively, just to name a few. Quantitative readings tell the operator what the physical characteristics of a process are so that corrective action can be taken if needed as the process continues. They may also reveal counts that tell the operator when to start or stop one or another of the operations involved in the particular job.

b. Check Reading. Check readings usually deal with the limits of productive parameters; that is, they tell the operator when certain limits are either reached or approached. Check readings are designed not to give specific information but to inform about the acceptability of present performance in the production system. This is the case, for example, of a telltale that announces critical conditions of pressure or temperature by changing a readout light from green, indicating everything is okay, to red, indicating that it is not. Since this type of instrument is not constantly offering new information, but is rather only offering important information periodically, it may be observed only passively. This often means its nature of change is more extreme than in the case of an informational instrument, sometimes relying on blinking lights, radical changes in color, or audible signals to draw attention to the change in condition. Often, quantitative instrumentation and check reading instrumentation can be linked, as in the case of a pressure gauge that reads pressure in pounds per square inch but also sounds an alarm if the pressure becomes critical.

Another form of check reading is one that does not offer information continually, but only when consulted. This is the case of readouts that respond to a command, such as the flick of a switch or push of a button, by indicating "okay" or "not okay" conditions to the observer. In most cases, check readings are ignored as long as they are registering within certain critical limits.

c. Setting. This type of readout indicates the condition of machines and processes with more than one possible state. Settings indicate to the operator how the workstation's processes are presently programmed to perform, including such information as whether certain valves are opened or closed, or whether machines are on or off, as well as the dimensions to which cutting blades are designed to trim work, in process or not, just to name a few. Settings inform the observer of the set of parameters under which the workstation is presently operating, so that any necessary or desirable changes may be instigated. The idea here is that it is difficult to know how to get where you are going if you do not first know where you are. Settings are controls, and setting instrumentation indicates what controls are currently being enforced. This is particularly important in processes that may be changed to create different results, such as machine shops or, for that matter, power plants, which need to periodically change delivery rates with changes in demand.

d. Tracking. Just as settings let you know the limits of the system, tracking instrumentation lets you know where you are in the process. That is, they trace the movement of work through the system so that it is known at all times where in the process those goods are. Tracking instrumentation can be designed to trace not only the work in progress, but also any other element of the productive process. They can be set up to follow both backward and forward and to maintain conditional relationships among various aspects of a process that are going on simultaneously. In concept, this is similar to a Gantt chart, or a traffic control system that keeps tabs on all the steps (or bits and pieces of work) simultaneously. This has particular value in production or project management and where momentary changes in scheduling of work may be important.

6. Dial indicators. These instruments are designed to present information of the types we have discussed, and to do it in a logical and easily readable form. To this end, a wide range of designs have been developed over the years, each with its own set of advantages and disadvantages, and each representing the "best fit" for specific types of jobs. The main types of dial indicators in use today include (a) moving pointers with fixed scales, (b) fixed pointers with moving scales, and (c) digital counters.

a. Moving Pointers. Dials with moving pointers have the advantage of adding analog information to quantitative information. As the pointer shifts farther and farther to the right (or in a clockwise direction), the length of the line or arc it inscribes is analogous to the magnitude of what it is measuring. This facilitates the reading of the information in terms of ratio as well as quantitative magnitude. A dial is easily read by simply looking to the place that the arrow is pointing to take a reading. Speedometers, pressure gauges, and flow rate gauges are all examples. Moving pointers are generally used in conjunction with fixed-scale devices, in which the scale is stationary and has a specific analogous ratio of change to distance or angle. Scales are therefore numbered progressively upward in most cases.

b. Fixed Pointers. Fixed-pointer devices are designed so that the indicating pointer remains stationary while the scale of measurement moves, usually either turning in a circular manner, or moving past a vertical pointer on a drum-like wheel. The number of indicated value that is aligned with the pointer is the measurement that the instrument is taking. The advantage to this type of instrumentation is that the observer is always looking in exactly the same location to find the reading, rather than having to search the dial to find the position of the arrow. The main disadvantage is that often it is not immediately obvious what the full span of the scale is, that is, what the high and low values of the scale are, and thus it only indicates magnitude rather than ratio and is not analogous in nature.

c. Digital Counters. These types of displays have experienced recent popularity, in no small part due to the rise in use of LCD (liquid crystal diode) and LED (light emitting diode) displays used in electronic measuring devices. Digital displays offer numeric information only, although higher degrees of accuracy are often achievable with these types of displays without high increased cost. With digital counters, a large number of displays can be incorporated into a single layout, and the same display can offer different types of information on command, if programmed to make the necessary shift. Where size is important or the accuracy of data is not of major concern, this approach is valuable. The chief advantages of digital displays are accuracy, convenience, and low cost. Compared to mechanical display technology, they are much more cost-efficient and faster as well. The chief disadvantage is the loss of the analog model that often provides information on the subconscious level to the observer, which, even though subconscious, is still a factor in making value-judgment decisions.

C. Controls

Controls, as the name implies, involve methods of manipulating the activities of the workstation to perform the various steps in the production process. They are initiators and adjusters that either begin or end activities, or change the conditions under which activities are taking place. When one is designing physical controls for a workstation, certain standard principles need to be observed. Among these are concerns of (1) the relationship between controls and displays, (2) sequencing, (3) the use of concentric knobs, (4) the spacing of controls, and (5) the movement, types, and shapes of controls.

1. Conjunction with displays. There must be a coordination between the displays of information arrayed in readouts and the use of controls. The necessity for acting upon information received from displays dictates that controls be available and convenient to the operator. The controls should therefore be used in combination with displays in an appropriate manner. This means clustering controls with the appropriate readouts, laying out controls where they are easily reached and near the hands of the operator (as in arm consoles, or along the lower edge of the display board surface), and utilizing control designs that facilitate flexibility and accuracy. Whatever arrangement is finally decided upon, the principle of logical placement and design coinciding with the functions being carried out is imperative.

2. Sequencing. Controls should be arranged and used in sequence where appropriate. If certain activities normally follow a logical sequence of events, then the controls designed to carry out those activities should be located accordingly. This could be a simple matter of banking toggle switches in a row if they initiate a series of events that are connected. An example of sequencing might be found in the arrangement of controls in the cockpit of an airliner.

3. Concentric knobs. These offer an opportunity to nest controls around a central hub to facilitate easy access and use. Generally, the functions of concentric knobs are connected, as in the use of a central volume knob with a knob for adjusting tone placed concentrically around the outside.

4. Spacing. The issue of spacing is commensurate with the type of control being used, since different control types require different degrees of space and different initiating actions. Pushbuttons, toggles, pedals, and knobs all require different actions, as do levers and cranks. The limiting factor here is the amount of spacing required for successfully carrying out the operation.

a. Pushbuttons. These can be placed rather close together, since the initiating action involves simply pressing inward and requires little or no additional surface on the mounting board beyond that necessary to accommodate the physical characteristics of the button itself. In addition, pushbutton controls can be clustered together and are excellent for sequencing, since three or four fingers can be used to manipulate them. In the event that sequence is vital, however, they may be spaced more widely to avoid inadvertent initiation of action out of sequence. The spacing of buttons on telephones, computer keyboards, and cash registers are examples of the small amount of space needed to accommodate this type of control. The size of the average finger pad is the primary limiting factor in designing the size of a push button.

b. Toggle Switches. These require slightly more space, particularly in the direction of motion. Either up–down or right–left movements are generally required for toggles, which consist of a small lever extending out toward the operator that is "thrown" into one of several positions to initiate action. Some toggles are simple push–pull devices with an off-and-on switch. Technically this is what is present in normal wall switches. Others may offer multiple conditions, as in a double-pole, double-throw toggle switch, where a central position initiates no response, while a movement in one direction or the other from that central position can create one of two alternative actions. Such a switch is seen in a control designed to move some surface along a track in either direction. Normally in a neutral position that initiates no movement, a push to the right creates a movement to the right, and a push to the left creates a movement to the left. With such switches, enough room is needed to allow the throwing of the switch to its various positions, and clustering in rows must allow for extra room in the directions that the toggles are thrown. For instance, with vertical throws, where the toggle is moved either up or down, there must be more room vertically than horizontally, since no sideways movement is possible, and the array can be put into rows at close proximity, but the columns in such an array must be farther apart to accommodate the movement of the toggle. One method of getting around this difficulty is the use of a variation on the toggle theme, namely, rocker switches. Though more space-consuming in nature, rockers rock around a pivot point and simply expose one surface of the rocker higher than the other as it is depressed. This is a combination of the pushbutton and the toggle that allows for more uniform spacing of controls.

c. Cranks and Levers. These controls pose different problems. They can be viewed as the most space-consuming of the controls, since they require a relatively large amount of positional change as they pass through either an arc, in the case of a lever, or a circle, in the case of a crank. Large sweeps of attached arms create controlled change and thus are usually used where it is important to have mechanical advantage. Such devices as winches, automobile jacks, and electrical master panel on–off switches utilize this type of design. Spacing, due to the importance of being able to make relatively large changes in the positions of the controls, must be expansive. Such controls are normally isolated from other controls in some way to avoid interference with operation or accidental activation, and they are seldom placed in clusters or arrays unless the lever movement is toward the operator, as in electrical contact controls.

d. Knobs. Unlike buttons, which are essentially the same basic shape, a knob is moved in a circular fashion, thus requiring enough room for the operator to grasp and turn or push along a single surface. In addition, many knobs have scales and pointers attached, which further increases the need for space around the control itself. Usable in clusters and arrays, the knob is a general control that can be applied to a wide variety of control problems, from on–off switching to infinite adjustment, as in volume controls.

e. Pedals. Pedals are almost always placed near the floor and in easy reach of the operator's feet. Their purpose is to initiate control functions without occupying the operator's hands. The spacing is critical in cases where confusion over a series of pedals is possible, or where it is necessary to maintain pressure on the pedal while other activities are being carried out. Usable in arrays and in some clusters, pedals require enough room for the operator's foot and spacing adequate to avoid confusion. Beyond that, they behave in a similar manner to pushbuttons. One specific case should be noted, however, and that is in the case of momentary pedals, where action is initiated or stopped only for as long as pressure is applied to the pedal, with a release of pressure creating a return to some predetermined state. This is true of, for instance, clutches that disengage drive trains for only the length of time the pedal is depressed, or so-called "dead man" switches, which allow machinery to operate only as long as the operator is able to maintain pressure on the pedal.

5. Control movement. Like control design, movement follows certain precedences. In engineering, standards are useful in seeing to it that everyone understands what is taking place and, in the absence of other instructions, is able to initiate indicated action as necessary. For that reason, the movement of controls is generally expected to follow one of five basic patterns, unless special circumstances dictate otherwise. These patterns allow controls to move upward, to the right, forward in the case of three-dimensional controls, clockwise if rotational motion is involved, or to be pushed, as in the case of pushbuttons. The reason for these conventions should be obvious to anyone who has operated machinery. Up is analogous to an increase in activity. Movement to the right follows our understanding of the sequencing of successive events, as when we read a page a word at a time from left to right in order to understand the meaning of the words. Forward is indicative of progress and, similarly, initiation of forward action. Clockwise follows our general learned tendency to read circular dials in that direction, a propensity that we gain from the way in which we interact with clocks and, for that matter, turn screws. Push is an inward movement that again is indicative of progress, but more importantly is more comfortable and requires less physical effort in most instances than does pulling. All of these conventions are so ingrained in society that, lacking other information, the average person will automatically use one of them in order to deal with machine controls.

6. Control types. Control types are important, and it should again be emphasized that they behave either linearly (that is, in a straight line), or in a rotary (that is, in a circular, probably clockwise) motion. It is often the case that the choice of motion is designed to create an unconscious analogue in the mind of the operator that helps in mentally making control decisions. In addition, the linear and rotary motions are the most natural for the human anatomy, and therefore least fatiguing to the operator.

7. Control knob shapes. Just as the shape and contour of the controls surface is important to effective station design, so too is the design of the control knobs themselves. As we have noted, there are a number of alternatives in how controls move; in most instances the choice of the movement is dependent on what is most commensurate with the nature of the control action and, as much as possible, on what is analogous to the type of action to be taken. The shape of the control knobs will further facilitate this purpose if it aids in carrying out the desired motion. That is, the shape of the control knob can help the operator perform the desired function in helping to avoid confusion by supporting the performance of the correct action. This is particularly valuable in emergency conditions or when the operator is engaged in a number of simultaneous operations and may need to depend on feel to ensure that the proper control has been chosen when other duties prohibit a visual inspection of the controls array. This is true, for instance, in the case of a pilot or astronaut during launch or reentry.

Normally, knob shapes are divided into three classes, labeled simply A, B, and C. Each of these classes defines the general shape of the control and the type of movement involved.

a. Class A. Class A knobs are for controls that are to be turned or spun under conditions where position is not important. Examples of this are volume controls and power controls on variable-speed drills. Changes are relative, and as long as the operator can turn the knob, other cues, such as the loudness level, pitch, or vibration of a turning drill, serve to supply information as to the position of the control. Class A controls are usually circular or conical, though they may at times be spherical, and the uniformity of the surface of the knob lends itself to rapid or gradual turning through a continuous arc. Where grasping may be difficult or resistance high, there are often ridges along the surface that increase contact with the fingers and thus facilitate the turning of the control.

b. Class B. Class B knobs are for controls that are moved less than one full turn and for conditions in which position is unimportant. This type of control surface requires less uniformity and is generally square or triangular, making the control easily identified and grasped, and designed to respond to movement that is neither gradual nor continuous. The turn is less than a full circle, often represented by from one to four "clicks." The smooth surfaces characteristic of class A knobs are not desirable, so class B knobs are made so that a firmer grasp of the hand is possible by virtue of a more extensive flat surface offered by polygonal shapes. With these controls, there is still a continuous shift in response value, but since the rotation is less than a full circle, the ability to spin the knob is unimportant, and other feedback offers confirmation of the degree of change in process. A light dimmer that moves only a short distance from high to low would fit this category. The change in light intensity supplies the feedback, while the polygonal knob offers solid contact between the knob and the operator's hand.

c. Class C. Class C knobs are for controls requiring less than one turn, but under conditions where position is important. In this type of control, shape is important, so the knob is usually elongated or of some odd nature. This allows the operator to quickly and easily identify not only the control grasped, but also the position that that control is in, so that it can be determined quickly which way to turn the control in order to achieve a change in process. Lever knobs, oddly shaped knobs, and eccentric cam-type forms work well for this type of knob. An example of this type of control would be an air conditioning control that switches the system from recirculate to fresh air intake. This is particularly important in automobile air conditioners, where the driver's attention cannot be visually diverted from the road and the feel of the knob in the left or right (or higher or lower) position would inform the operator of the present position of the control, and thus which way to turn it to initiate change. More critical examples are knobs used in arming devices for weapons systems (usually controlled-access toggles or levers), or the fuel-tank switching control aboard a small aircraft. Turning the control to the wrong position in either of those cases could be literally disastrous. Figure 16.1 offers illustrations of these three classes.

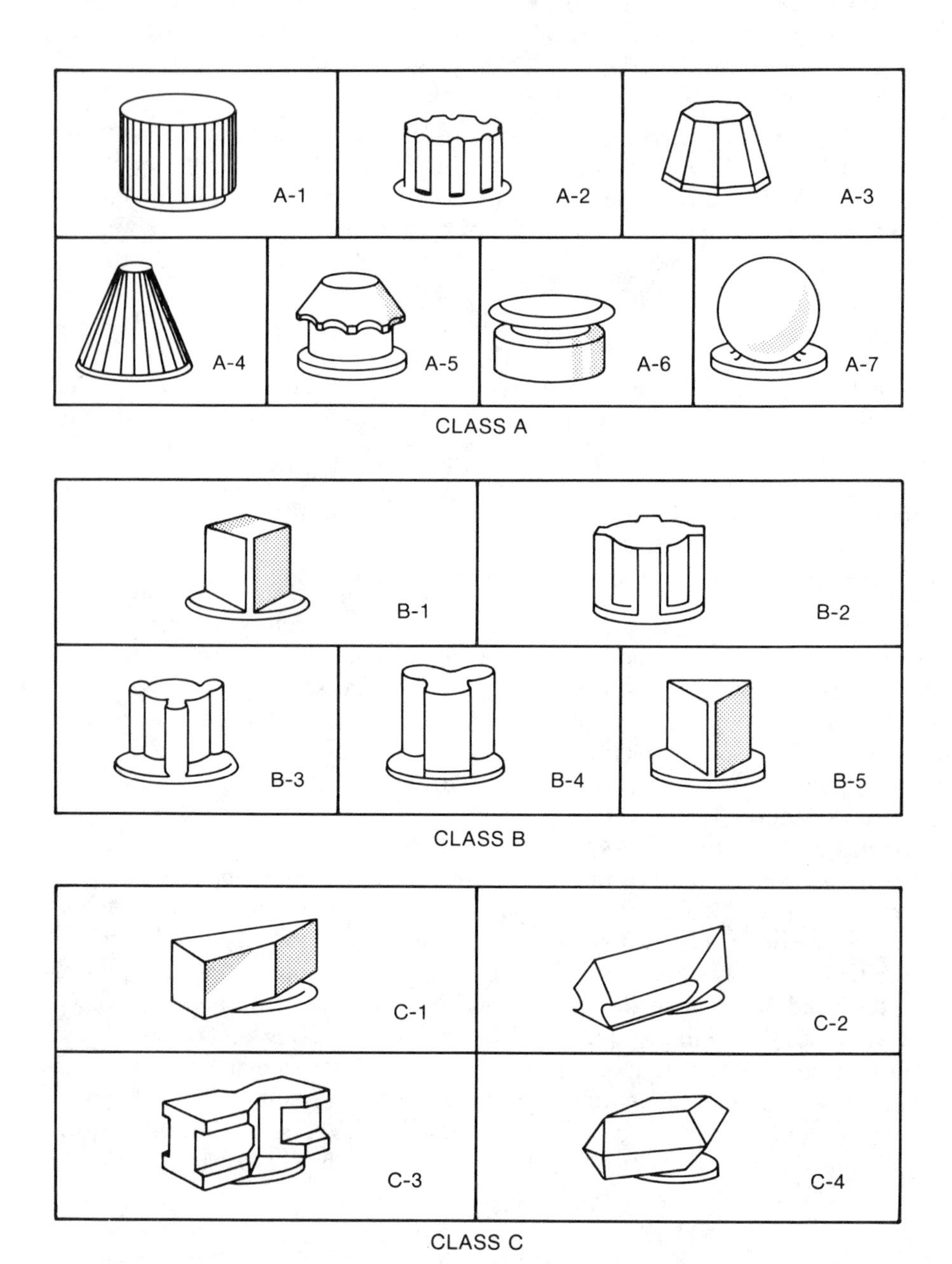

Figure 16.1 Control knob classifications. Source: Reprinted from *Human Engineering Guide to Equipment Design* by Morgan, Chapanis, Cook, & Lund. Copyright © 1963 by McGraw-Hill, Inc., p. 273. Used by permission of McGraw-Hill Book Company.

II. APPLICATION TO FACILITY DESIGN

A primary concern in effective facility design is the proper engineering of the workspace, which naturally includes the proper dimensioning of workspaces and workspace layout. Two distinct types of workspaces require careful structuring, those in a service facility and those in a manufacturing environment. Although the principles of design are essentially the same, the details can vary considerably with the type of operation being constructed.

A. Service Facilities

The peculiar requirements of design in the workstations of service facilities stem from the fact that the customer as well as the worker are often present at the time of service production, and therefore both must be taken into account in the station design. Consider the service counter at an appliance repair shop, or the sales counter of a large department store. In each of these cases, the comfortable size and height of the workspace must reflect the work to be done. Merchandise must be displayed, handled, and sometimes probed in a manner that is comfortable and instructional for both the customer and the service person. Lighting must be designed to either show off the item to advantage, if it is being sold, or to allow for accurate determination of required service in the case of repair. Counter height must take into account both the necessary ease of handling and the avoidance of discomfort from bending over or reaching above normal grasping height. The long hours standing behind a counter require cushioned floor coverings or adequate high surface seating for workers to avoid unnecessary fatigue. In other cases seating may be central, as in the case of an optometrist or dentist, both of whom provide medical services to seated customers. Again, comfort and correct ergonomic design are very important. The peculiar shape and adjustability of a dentist's chair is by no means accidental. Even the mirrors in an optometrist's shop must be properly positioned to reduce strain for the customer and present frames in a favorable manner. Proper design is essential.

B. Manufacturing Facilities

In the case of a manufacturing environment, more traditional types of problems exist for the workstation designer. By the time the facilities planner has reached the stage of actual workspace arrangement, the activities to be performed are highly defined and the steps that must be taken to create the product are well-known. In many industries, standards of design are often extant, gleaned either from past experiences of similar operations or from the manufacturers of the machinery and equipment used in the manufacturing process. This makes the planner's job somewhat easier by eliminating a number of possible combinations of workstation elements that have already been found to be faulty for whatever reason. And yet, in accepting the predetermined approaches of others, the planner foregoes the opportunity to creatively find a better way to do the job. Preferably, what has gone before and been found effective should merely represent a starting point for facility planners in their attempts to find the most efficient methods of operation. By careful analysis of already useful methodology, it may be possible to discover new, innovative patterns of action. And if equipment is to be engineered from the floor up, beginning with similar designs can perform the same function of heading the planner in the right direction. In other words, no amount of standardization can eliminate the necessity for planning and design.

C. Combination Service and Manufacture

This represents a class of facility that is growing in importance, particularly as the service industry itself grows. In truth, most service facilities represent a combination of service and manufacture, with both customer interaction and actual development of the product taking place in the same location. Consider fast-food restaurants. Although principally service oriented, they manufacture a product to be supplied to the customer. Consider also the dry cleaner, or an automotive customizing sales and installation firm where products are created and installed, representing the need for a form of manufacturing facility attached to the sales area. And what of lawn services or mobile tune-up businesses? They are providing services, but the facilities, albeit in the back of a van or truck, are of a productive nature, unlike a showroom. Here we have to design for both customer service and for production simultaneously, which can be done either as two separate operations within the same facility, or as a combination facility performing both functions together. In both cases, the design requires proper layout in order to perform all of the firm's operations efficiently and economically.

In truth, each firm is unique in its needs and in the way it views itself, giving rise to a wealth of opportunity for fresh, creative facility designs emphasizing those characteristics that the firm chooses to exemplify.

QUESTIONS

1. What are the advantages of linear types of controls? Where would you expect to see them used?
2. Name some applications of rotary controls and explain when they would be preferable over some other type of control.
3. Delineate class A, class B, and class C control knob shapes. What are the advantages of each?
4. Why is spacing so important in the placing of controls on a panel surface?
5. Name five types of dial indicators. What are the advantages of each?

6. When are seating corrections for shoe height and variation in arm length most critical?
7. In many cases, pilots for military aircraft are chosen on criteria including minimum and maximum height and ratio of leg length to torso length. Why do you suppose this is?
8. How can the contour of a mounting surface promote or frustrate the proper use of controls?
9. Why are controls and readouts often clustered together within an overall mounting surface layout?
10. Define and describe the slump adjustment for sitting.
11. Define and describe the slump adjustment for standing.
12. Name the major criteria that must be met for proper work seating design.
13. How would the workstation designs differ from a manufacturing facility to a service facility? Give examples of each.
14. Why is the relationship between lighting position and eye position so critical in workstation design?
15. From which direction should light fall on a workstation? Why?

PART SIX
Facility Alterations

17
Facility Redesign

Throughout this text, the main thrust of the discussion has been directed toward the creation of new or expanded facilities. Every element in the process from development of the financial plan through location determination and procurement of property to physical design has been discussed. Yet this is certainly not the only option or the only road down which a firm may travel in creating more production capacity. Indeed, more often than not, the possibility of redesigning existing facilities is the first option that firms investigate due to the savings inherent in beginning with an already existing structure. That is the subject of this chapter.

The redesigning of existing facilities presents special problems while solving others. With a redesign approach, most of the preliminary work is already done, since there is no need for analysis of potential sites, or for procurement, or for the building of major structures. As for the actual physical redesign, it will follow the lines of attack that we have discussed earlier in the text. The main issue here is one of homeostasis, or resistance to change. The issue is not a matter of technical expertise or available technology. It is a matter of attitude and psychology, and of how easily the actual redevelopment may be accepted by those involved.

We all have habits that we follow, and this is a logical consequence of experience, which teaches us what works. If we did not habituate the behavior patterns in our lives that support our happiness and development, then each decision we make would entail "reinventing the wheel" just to get something done. Unfortunately, the other side of this is that the habituations tend to remain long after they are useful. Times change. Situations are subject to shifts and new input and the loss of old elements, and as a result, new methodology must be used in order to adjust to these changes. Yet the habituation process has locked us into thinking along only specific lines and has built an automatic response pattern into our behavior that predisposes us to a specific way of doing things. That is, people don't like to change the way they do things. And the redesigning of the facilities in which people have been operating for an extended period of time will certainly require that they change the way in which they do things.

Resistance to change is a natural survival trait, designed at the base level to prevent us from blithely accepting just any new idea without first testing it and thinking about it to be sure that it does not lead to some unforeseen disaster. That is why change is more easily accepted by the young, who have less habituation experience, and by those who have been in the environment for only a short period of time. It is much easier to convince a new employee, who is already involved in the process of trying to establish new habits in a strange environment, to accept redesign than it is to convince a long-time employee who is familiar with the present system, wedded to it in the sense that operating in the traditional way has become an habituation, and extremely reluctant to change now, after such a long time of operating in what has proven to be a successful manner. It is difficult to teach an old dog new tricks. Fortunately, however, it is not impossible.

It seems that one particular option for the redesign of an existing facility stands out as being an exceptionally flexible, and thus useful one. The just-in-time method, or JIT, is becoming very popular not only for its flexibility, but also for its inherent cost savings and profit potential. Its methodology caters to large facilities as well as to small businesses, where one would be most apt to find employees who would fight any redesign process, probably out of a sense of loyalty to the operation. The JIT facility, we believe, is an important upcoming trend and therefore will be further discussed later in the chapter.

In initiating a redesign, what must be remembered is that the designers and planners must deal with the resistance of the employees to the change. Care must be taken in convincing them that the change is for their benefit, that they are a part of it, and that it does not represent some threat to their security or way of life. Employees must also be involved in the process as much as possible. The reluctance of workers to accept the idea of the redesign can be expected to rise as we move further from the center of the redesign function. Peripheral employees will be much more homeostatic than those involved in the planning of the change. The key is to move decisively, openly, and quickly. Explain fully what is to happen, involve as many as you can in the process, answer any and all objections, questions, and so forth, and then make the change quickly. With luck, those who may have been reluctant to accept the change will find themselves thinking that it was all their idea in the first place.

With that in mind, this chapter examines first the factors creating resistance to change, methods of dealing with these factors, and finally, considerations in the planning process specifically inherent in the redesign approach.

I. FACTORS RESISTANT TO FACILITY REDESIGN

The reluctance to accept change in the operational setting comes from a number of different sources. Some are more inherent in upper-level management, others in lower echelons of the company. All must be dealt with in order to effectively create the desired change.

A. Inertia

Just as massively heavy objects have inertia, meaning that their rest mass tends to resist movement and pressure must be applied to put the object in motion, so too do organizations and social systems. This is to be expected. Firms and the various social (group) structures within them have a great deal of history, tradition, and time invested in developing present patterns and ways of doing business. There will necessarily be a great deal of mass to put into motion before any headway can be made toward a facility redesign. Pressure must be brought on the organization to change, thus creating the motive force to overcome the inertia and start the project moving, while communication and explanation of what is to take place will act as a lubricant to reduce the friction that this mass generates in movement. Inertia occurs at all levels of the organization. There may be executives who have put considerable personal investments into developing the present system, both in terms of actual effort and in terms of ego and emotional attachment. If they are to support a new project, they must first be persuaded to support it. Also, those actually carrying out present policy, those producing under current conditions in the current structure, also have great investments in terms of familiarity with the way things have always been done and in the amount of effort put into learning how to operate effectively and efficiently under the present system. They too must be shown the value of a new approach, or a new facility design, before they will be moved toward achieving that goal.

B. Uncertainty

This is a factor closely aligned to the idea of inertia. Not knowing what is to happen next can cause a reluctance to change stemming from fear of the unknown. This is another trait that is natural to humanity. It is as much a survival trait as homeostasis and performs basically the same purpose; that is, it forces us to be cautious with unknown conditions and new experiences. A lack of knowledge can result in a wrong move at the wrong time. It is better to move cautiously until we know what is going on, or so goes the unconscious scenario that people play in their minds when faced with the unknown. Again, note that this is less evident among young people, who are constantly encountering new experiences, than in older, more settled individuals whose lives may not hold the same content of change. Fear of the unknown will stop progress instantaneously, if it is an extreme fear. In its gentler forms, it creates anxiety and reduces efficiency. Yet we all live with uncertainty every day, since no one can really predict what will happen in the next instant in time, much less in the next few days, weeks, or years. So why the resistance? The uncertainty inherent in the redesigning of facilities strikes at the heart of an individual's stability—the environment in which they earn a living. Being uncertain about any one aspect of one's life is inevitable, often helpful, and certainly manageable. Yet when the work environment is changed, there are fundamental differences in conditions that could affect a great many parts of people's lives. Will they have the same job? Will they have a job at all? What type of new learning will be required? Will they be able to master the new process? How are the social interactions going to change? Will authority and responsibility be affected? Will the nature of the work itself be changed? How will the physical environment be different? Will it be more dangerous or more technological and harder to understand? These are the types of questions arising out of the uncertainty of facilities changes. Interestingly enough, the mechanism that supports our survival by allowing us to think in hypothetical scenarios also helps to fuel the uncertainty and resulting anxiety. Think about the way you fantasize when you have a problem or fear something. You run scenario after scenario through your mind, experiencing consequences and reacting to them. This is not an accident. Nature gives us that ability so that we can prepare for the unknown. If the circumstances arrive, we have already thought through how to react in some hypothetical situation that is at least close to present real-world conditions. This characteristic allows us to be responsive, that is, to be able to respond. The problem arises when this type of fantasizing becomes excessive, as it may under the stress connected with the prospect of totally new working conditions, and then the uncertainty creates a fear of change. Note that in nearly all cases, the fearful conditions that we invent in our own minds are much more frightening than what really occurs. The only way to alleviate uncertainty or fear of the unknown is to reduce or eliminate the unknown nature of what is to occur. For this reason, it is imperative to discuss all facets of the change fully, not only with those immediately affected by the change, but with all facility employees.

C. Failure to See the Need for the Proposed Change

This condition will seriously hamper efforts to create change, particularly if the people who are unable to realize the necessity of the proposed change happen to be in positions of power. Anyone who has seen a need for changes in a work area, only to find that no one else seems to realize it, knows how true this is. This is partially an unconscious move to maintain the status quo and avoid the increased energy expenditure (not to mention disruption of habitual patterns) inherent in the needed changes. It is also representative of another phenomenon in the psychology of humans: an individual's creation of paradigms, or models, that explain how things work. People have individual views of the world, based on experiences and the interpretations of those experiences, on what they have been told in formal and informal learning settings throughout their lives, and on the observed opinions of others expressed by their behavior. The result is that people limit themselves in what they believe is possible in accordance with their paradigms, their self-imposed parameters for what can be. For example, they may believe that there is only one successful way of performing some task, since they have experientially determined that other methods do not work, and thus reject any possibilities to the contrary. People have blind spots. They have belief systems beyond which they will not move. If the need for change and the types of change that are required to solve a problem lie beyond the boundaries of those belief systems, they do not see them, cannot see them, and perhaps will not see them. Only by expanding their paradigm to include those possibilities will they be able to see the need for the change. This is as true in facilities planning as in any other aspect of life, and it is sometimes necessary to depend on visionaries, those with large, all-inclusive world views, to convince others that the need exists and that the present situation is unacceptable. The paradigm must be expanded, or at least shifted, to include the view indicating the necessity of change before that necessity can be seen by others.

D. Failure to Understand the Proposal

Such a circumstance is another source of resistance to change. In this case, however, it is not a refusal, conscious or unconscious, to see the need for the change, but rather a failure to understand how the change is to take place. It may be that everyone agrees that a change is needed. It may be evident to all that production is down, that quality needs to be improved, that costs are rising and efficiency is on the wane, and yet some may not understand how the proposed redesign will change any of that. This is where the expertise of a facility designer becomes important. Facility designers are specialists. It is their job to develop a proposal that increases efficiency and solves problems, and to present that proposal in such a way that it illustrates what changes will occur and what the benefits of those changes are to be. The proposal itself must be readable and understandable for the specialist and generalist alike, so that the truth of the matter can be clearly seen and the need for initiating the redesign can be more readily accepted.

E. Fear of Obsolescence

This is an element that is felt by anyone operating in a traditional manner in the face of increasingly new or unique technology. Many people, from managers to workers, will be reluctant to cooperate in a redesign project if they think that they will become obsolete and thus expendable in the process. It is irrational to expect individuals to voluntarily work for the elimination of their own jobs. Part of the redesign project, therefore, must include provisions for the retraining or placement of displaced workers whose jobs are significantly altered or eliminated. This is to the benefit of the company in terms of morale, efficiency, and utilization of tried, veteran workers.

F. Loss of Job Content

People take pride in their jobs and in their ability to perform a useful function. Ever since the beginning of the Industrial Revolution, when the mass-production factor approach began to separate workers from their tools, the issue of pride and identification with the fruits of one's labors has been a major one for workers. People are good at what they do and are effective at their jobs when they enjoy their jobs. When the content changes substantially, the elements of that job that have been found enjoyable may now be missing, which is something employees may fear. An employee may enjoy the personal physical contact with the product that allows for individual expression, and may become less interested, and thus less effective, when expected to perform a task that eliminates that facet from the job content. Retraining that worker to operate automatic machinery to do the same job would eliminate the very element of the job that brings personal satisfaction, thereby decreasing, if not destroying, the desire to do well. For example, a professor may enjoy the personal interaction with class members, or the freedom to do research. Restructuring the education institution to have televised classes would destroy the professor's pleasure and sense of pride, as would an increased emphasis on teaching load and resultant reduction in time for research in the second case. People may fear that any restructure would rob them of the satisfaction they have in their jobs, and therefore may be predisposed toward resisting changes from the present way of doing things.

G. Personality Conflict with Analyst

People are not machines. They have individual personalities, likes, and dislikes, and they have a whole range of attitudes that predispose them toward or against any given individual. If there is a conflict with the analyst who is presenting the redesign proposal, a great deal of resistance may be encountered. It doesn't matter if the conflict is from past experiences, personal beliefs, earlier confrontations, differences in professional approach, or if the analyst is aggressive about defending the proposal in a way that causes others to defend the method they know. The resulting resistance can slow or block acceptance of the redesign idea or any stage of its implementation.

H. Resentment of Outside Help or Interference

This is generally seen as a problem of pride or ego, but it is also a manifestation of the resisting individual's fear. People who are unusually proud of their work may jealously guard their positions or their commercial "kingdoms," resenting any outside interference or attempts to help. Or the use of an expert may be seen as a tacit demonstration of corporate opinion that the employee simply could not handle the job. Such a threat to the ego automatically generates resistance, particularly if the employee is harboring feelings of inadequacy to begin with.

I. Resentment of Criticism

Such a problem is part of this same process, where any criticism, rather than seen as helpful, is viewed as an insult and as an attack on the integrity and capabilities of the worker. Consequently, it is natural and necessary to resist. Numerous examples can be seen, as in cases where suggestions intended to be helpful are seen as one person's telling another how to do a job. Egos are often fragile, barely maintained things, complicated by the social environment of an established workplace that includes a pecking order based on power, experience, and expertise. Any perceived attempt to usurp an individual's position in that hierarchy will be met with a defensive, resistant posture. This is a situation requiring diplomacy and tact as much as technical expertise.

J. Lack of Participation in the Change

Resistance here stems from the feeling of being left out. Incredible opposition can arise simply from thoughtlessly overlooking some key person in the planning process. There is a social dynamic inherent in any organization, since organizations are social systems as well as commercial institutions. That social system includes a group dynamic that defines relationships—in other words, a pecking order. Omitting the input of key personnel disrupts the pecking order and threatens the positions of those bypassed. In such cases, how much cooperation can one expect from those offended? Precious little, to be sure. If there is no participation in the formulation of the change, the level of fear of the unknown rises due to a lack of information and understanding about what is occurring in the change.

Here we are back to creating fear because the employees do not feel that they can be sure of their future. It is, at best, unpleasant to feel that one is at the mercy of the firm or of those planning the redesign, and reluctance to accept or support any such move is almost inevitable. A lack of participation in the formulation of the change indicates to employees that they lack control over their lives, which can present serious situations. Individuals who have little control over the circumstances in their home life may be severely shaken by the realization that guarantees do not abound in the workplace either. Such individuals, whether employed as executives or as workers, will no doubt try to reestablish their confidence by fighting to maintain conditions that previously allowed them to feel that they were in control.

K. Tactless Approach on the Part of the Analyst

Marshal MacLuhan said that the medium is the message. That is, it is not so much what is said as how it is said that is important. A tactless approach by the analyst, or by whoever is presenting the proposed change, is sure to cause problems. Any manner that might indicate a feeling of superiority over those being briefed, or the display of an attitude that the opinions of the participants in the process are not worth considering, can kill a project quickly. People always defend themselves when attacked, whether it be on the street, in social conversation, or in the boardroom. Tact is required to ensure that those who must actually carry out and work with the redesign are treated with respect, and made to feel valuable, as indeed they are if the project is to be successfully completed.

L. Lack of Confidence in the Analyst

Even good projects can fail if analysts are unable to sell themselves as well as their ideas. If those who must make the decisions in the matter view the analyst as untrustworthy or of questionable ability, they are not likely to seriously consider the proposal. This means that an atmosphere of excellence must be created, which can be accomplished by means of a thorough proposal and a confident presentation. The analyst is expected to be an expert. That means anticipating and answering questions generously, going further than the minimum requirements of the project, developing a plan that is both comprehensive and competent, and being certain that there are no hidden factors that have not been uncovered. The analyst's motto should be "Be prepared." So many factors come into play in a facility redesign that it seems impossible to cover them all. Yet an effective analyst seems somehow to do so, or should at least give the solid impression of having done so. Therefore, great care should be taken in the choosing of an analyst, so that such criteria can be met in the process of change.

M. Inopportune Timing

Finally, we have the question of timing. This can often be the crucial factor in the degree of resistance to a redesign plan. Remember that even if you are the person in charge, you are still attempting to sell the others involved on changing the system with which you have all been working. That requires strategy, and strategy requires timing. Failure or success is often predicated on timing, as any military commander or football coach will tell you. A number of factors, such as market position, condition of the economy, financial strength, position in the yearly business cycle, personal loyalties, and the general climate of the workplace will affect the chances of success when a proposal is presented. By waiting for the propitious moment, one can create a maximally favorable climate for the proposed change. For example, when the market for the firm's products is down, there may be a reluctance to redesign. When finances are in excellent condition, the willingness to expend the necessary capital may be higher than it would be otherwise, as is true when the economy itself is heading into a strong upswing. It must be kept in mind that the attitude of the firm's executives will be largely dependent upon the firm's circumstances in the current marketplace. Also, beyond the strategic considerations mentioned, there is the matter of need. Redesign projects must be initiated far enough in advance of need to allow for successful completion of the project prior to that need, yet close enough to allow for a reasonably quick return on investment. Redesigns are like a huge jigsaw puzzle; if all the pieces don't fall into place, there is no complete structure.

II. METHODS TO MINIMIZE RESISTANCE TO REDESIGN

It is abundantly obvious from the previous section that the analyst hired to redesign an existing facility can expect resistance to redesign projects from many sources. Some, if not all, of these sources of resistance must be overcome if there is to be any redesign project, and all objections must be explored and answered. In this section, various methods of minimizing the resistance to redesign are explored. Owners should keep in mind that although it is up to the analyst to research, design, and present the proposal, the business belongs to the owner. Anything the owner can do to aid the consultant in utilizing these methods should be done, and fervently. Being aware of the steps that the analyst will take is the owner's first step toward assisting in the redesign efforts.

A. Convincingly Explain the Need for Change

If those affected by the change fail to see the need for such action, explaining the need for the change is an essential step in the process of eliminating resistance to a proposed plan. It is imperative that those involved be in agreement on the need for change. If not, there will be no motivation on the part of those people to move toward a new, redesigned facility. The analyst must be able to explain in detail why the project is necessary, how the project will improve the condition of the firm, and how inappropriate it would be to continue without the proposed change. If done properly, this process includes not only key executives but also employees throughout the hierarchy of the company, from board of directors to production employees. Once the aid of the human element in the firm is secured, the success of the project is all but assured. Major projects such as redesigned facilities require the cooperation of everyone. Disruptions to routines, physical displacements, changes in job titles, content, and descriptions are to be expected along with the physical changes that will take place. Under such conditions it is imperative that both workers and management support the efforts of the developers if their cooperation is to be expected.

B. Thoroughly Explain the Change

Not only must those involved be convinced that the change is necessary, they must also be thoroughly informed as to the nature of that change and the activities that are to be expected in order to achieve it. This way yields a minimum of surprises when events begin to take shape, and the continued cooperation of employees is more apt to be enjoyed throughout the process. Both the need for change and the logic behind the change's occurring in the way prescribed by the plan must be understood if a smooth transition is to take place.

C. Facilitate Facility Participation

Just as lack of participation creates resistance to change, involvement in the project helps alleviate worker tension. Therefore, a role in the process of change, or at least an opportunity to discuss the change, must be made available to employees. Make it easy for those involved to participate, in some way, in the facility redesign and development. It must be remembered that as participants in the present production process, the workers and the management represent invaluable sources of ideas, information, and subtle methodology. Enlisting their aid in the project should be a priority. The analyst needs not only to allow that participation, but to encourage and facilitate it so that it can be used to the best advantage. The result is a feeling of kinship and of belonging among the employees, not to mention a host of additional data from employees' suggestions that will no doubt contain valuable information for the use of those carrying out the project. This is an area where the abilities of the originator of the facility would be extremely helpful to the analyst, from the standpoint of increased familiarity with the employees and their particular strengths and weaknesses.

D. Be Tactful When You Introduce the Proposal

That is good advice in any case, but in the matter of reducing the degree of resistance encountered, it is of prime consideration. Remembering that how something is said is as important as what is said, try not to cause prejudice against the change by creating anxiety, animosity, or ire among those you are trying to influence. Take individual conditions into account, and remember that a project will not be accepted simply because of your own enthusiasm. Others must be convinced of how it is to their advantage to go along with the project. One does not do this by attacking employees' ideas, by treating them as less than equals, or by attempting to force them into a given position against their will. One uses tact and the display of genuine concern.

E. Watch Your Timing

The analyst should search for the most advantageous time to present any proposals for restructuring present facilities. Remember that preferable conditions are those most conducive to the acceptance of the project, and that does not necessarily mean good times. When a company is experiencing negative performance it may be open to new ideas for turning things around, just as when the firm is doing well, it may be accepting of proposals that would be considered too risky under different circumstances. Further, this search for the right moment is not even dependent on the conservativism or liberalism of the deciding agency. The same committee or board may make a favorable decision whether they tend to be liberal or conservative, depending on individual circumstances. Some 2500 years ago, Sun Tzu, China's foremost strategist, said that success in battle requires knowing your enemy and knowing yourself. In the matter of having a plan for redesign accepted, the enemy can be seen as resistance to change, and knowing it as well as one's own strengths will ensure that the timing of one's presentation is optimal. In the same vein, good timing will ensure that the proposed change meets the needs of the individual facility at that time, with due consideration given to its long-range goals.

F. In Major Changes, Do It in Stages

The greater the change, the greater the need for doing it in stages. This should be immediately self-evident. Remember that people fear change by nature, and that the greater the change, the greater the objection. One way of combatting this type of resistance is to see to it that the big changes are not big. That is, if they are broken down into a series of small changes, they are more easily accepted because they are less threatening, and thus the ability of the employees to adjust to each incremental change is enhanced. In the martial arts there is a technique for acquiring difficult skills that follows the same idea. One such skill has to do with agility in movement, and one of the methods of teaching it is to require the student to walk on the edges of a stiff straw basket filled with rocks. If the rocks were not present, the basket would automatically tip when the student stood upon the edge, no matter how carefully the student balanced. However, with the rocks within, the basket remains steady. The technique involves having the student walk around the outside edge of the basket each day without tipping it over, followed by the removal of one small rock. The next day, the same ritual is carried out and so forth, until the student eventually becomes so perfectly balanced and agile that the basket can be traversed without tipping, though there are no rocks within to stabilize it. This is an excellent example of a difficult task made simpler by virtue of being learned in stages.

In like manner, breaking a major project down into small steps allows for acclimatization of the workforce to the changes, for the discovery and solving of any unforeseen difficulties, and for the avoidance of virtually all resistance to the change.

G. Sell the Features of the Change

John Locke, the British philosopher, and Adam Smith, considered the father of capitalism, both said that people are motivated to do things because they feel that they will be better off in the process. Smith called it self-interest, and declared it to be the invisible hand that always leads a free market to move in exactly the right directions to create efficiency. Likewise, managers and other employees who feel that changes will be to their benefit are decidedly more likely to accept them than are those who see no benefit to those changes. By selling the features of the change, the employees are made to see how the change will benefit them and thus will be more eager to instigate those changes. Caution must be taken, however, not to become overzealous in this process. If changes are oversold, that is, if the promise of benefit is not realized because the feature was in some way misrepresented, or if the benefit simply does not materialize, the change could be disastrous. Resistance is possible after the fact as well as before, which creates a practical, not to mention ethical, obligation on the part of the analyst or designer to be totally honest about both the positive and negative aspects of those features. It is simply a matter of emphasizing the benefits and not giving undue weight to any perceived shortcomings that are expected, such as temporary disruption of work.

H. Let People Think It Is Their Idea to Change

In this case, the support of those who have to live with the change is secured by allowing them the luxury of believing that it was all their idea to begin with. On first inspection this may seem somewhat dishonest, which it may be if not done properly. It may also seem manipulative, and it is, but that is to be expected. The analyst is, after all, attempting to convince people of some point of view, and that in itself is manipulative by nature, in its broadest context. As it happens, if the people in a position to make a decision about the project had all of the facts that the analyst had, they might independently reach the same conclusion, mainly that redesign needs to take place. In other words, if the data were presented to the decision-making body and they were allowed to make a decision of their own, they would probably reach a conclusion similar to that of the analyst. In this case, they can in fact adopt the idea of the change as their own, and that creates a situation in which they are justified in calling the idea their own, with the analyst simply a data-gathering facilitator of the project. Once this has occurred, support for the idea is automatic and resistance is at a minimum.

I. Show a Personal Concern for Those Affected by the Change

Here is another area where the owner of the facility will be able to lend great assistance to the redesign effort. Being familiar with individuals and their circumstances provides a means for knowing basically which employees need special attention at this time. Aside from a personal show of concern, the entrepreneur will also be able to direct the attentions of the consultants to the most pressing situations. The concern shown by the analysts will be genuine if they are at all proficient at doing the job. Everyone affected by the change will experience some discomfort and possibly temporary or permanent displacement. To simply ignore their condition would be unwise on the part of the planners, not only because of the ethics involved but also for the sake of practicality. Part of what makes a redesign plan work is concern for the people involved. After all, they are the ones who have to live with the redesign decisions. They are the ones who are ultimately going to determine whether or not there is a successful transition from the old approach to the new. And they are the ones for whom the change is taking place. That may seem to be a bit of an exaggeration until it is remembered that the individuals who are affected by the redesign include not only employees but also management, stockholders, and the customers who buy the product. The purpose of all this activity is still to increase efficiency and, by doing so, create profits. That means higher-quality products at a comparable or lower price, or an equally high-quality product at a reduced cost. The customer profits by having a better product in the marketplace. The employees benefit by having more security, since they are valuable as producers of these high-quality products. Management wins by achieving higher efficiency and by successfully fulfilling their job obligations, often with monetary rewards in addition to the other rewards incurred. The owners of the firm win by having higher profits from a more efficient operation. The prime directive here should be the creation of benefit for all concerned, because if any one element in the system is not benefited by the process, everyone loses to some extent.

J. Have the Immediate Supervisor Announce Changes

By using immediate supervisors, several problems are alleviated. First of all, there is no disregard of the chain of command, which would be disruptive and short-sighted. Second, the information about the change comes from a source with whom the employees are already familiar, and from whom they are used to taking instruction. This keeps it in the family, so to speak, and adds to the feeling that the change is an internal process, not some new set of rules coming from an unknown, outside source. Finally, by using the normal supervisory channels, the analyst ensures that the information is communicated in a way that is understood, and that any feedback is received immediately and dealt with appropriately.

III. CONSIDERATIONS IN FACILITY REDESIGN PROCESS

What follows is an eleven-step redesign process based on a systemic approach to problem solving. In keeping with systems theory, the production facility is viewed as a group of related elements acting cooperatively to achieve a goal, in this case, the production of goods. As stated in previous chapters, the four chief factors that are involved in the operation of the system are input, process, output, and feedback. This loop is familiar to those who have had experience dealing with either systems theory or communications theory. To further reiterate, the system is then seen to consist of subsystems made up of subsystems and so on, until the final level of elements is completely comprehensible as to content and definition. The steps in the process of analyzing and designing use this approach because it lends itself to dealing with an analysis of current structures and the effects of changes on those structures. The eleven steps in the process are as follows:

A. Determine the Compatibility of the Input/Output System and the Problem Area

The purpose of any redesign project is to improve the design of the elements in the productive process that are problem areas, that is, areas that are inefficient or out of date, or which create bottlenecks for the rest of the operation. This can amount to anything from a single department to the entire facility, or a facility complex covering a large part of a company's operations. Whatever the case, it is necessary to first take a look at the input/output system, which is the part of the system that deals with inputting materials (receiving) and emitting finished goods (shipping), to determine how compatible these elements of the operation are with the actual production process under study. The problem can be viewed as a matter of how raw materials enter the process or how finished goods exit, rather than the process itself. The goal in this step is to know if the problem could be solved by simply creating compatibility among these input-process-output elements without manipulation of the process. For instance, if a production process is highly automated and of a continuous-production nature, large-batch delivery or shipping methodology would be inconsistent with the production operation because materials are not used for the production of goods in a large batch mode, nor are they delivered to the shipping area in this mode. Before any other action can be taken, one must first determine if any corrections of this type are necessary.

One fact illustrating the importance of input/output analysis with redesign stems from the recent trend toward the JIT or just-in-time approach to manufacture, under which there is very little on-premises storage beyond what is needed for loading docks and transportation yards. To the redesign engineer, this offers an opportunity to utilize floor space formerly occupied by storage facilities. In practice, a true redesign of facilities could be undertaken under these conditions without the necessity of additional structures or expansion of the present plant. Because of the tightly controlled scheduling required for a successful JIT installation, such a redesign project does necessitate updating and redesigning transportation systems at the facility along with internal changes. Smaller bay areas, more extensive and more highly automated loading and unloading facilities, and consideration of these factors in work flow and layout analysis all work to create input/output compatibility which, in the case of JIT redesigns, is primarily a matter of synchronization.

B. Determine the Subunits Under Study

This is the first part of the systemic approach in which the processing part of the production system is broken down into its constituent parts, or subsystems, in an organized manner, so that the relationships that exist among the subsystems can be defined and studied. In this stage, any completely standardized or state-of-the-art subunits that are not to be changed can be identified and eliminated from further consideration. Others that represent true problem areas can then be isolated, and a more thorough investigation of their nature and the nature of their inefficiency can be undertaken.

C. Determine the Compatibility of the Problem Costs with the Facility Redesign Model Costs

Problem areas are problems because of their inherent inefficiencies in contributing to the production process. This essentially means that they are too expensive in light of present technology to perform the function economically. In this step, a comparison is undertaken to determine the costs of the problem areas and compare those costs with the costs of any proposed facility redesign models being considered. The idea is to determine where and how money can be saved through the reduction of operating costs, and whether or not changes that are necessary to create the new process structure are sufficiently economical to warrant undertaking of the project. Individual problem areas need to be compared and analyzed, as well as the overall effect of the entire redesign model, in determining feasibility based on cost differentials between present and proposed production facilities.

D. Examine the Impact of Data Assumptions

It is virtually impossible to avoid the making of assumptions in developing production models. The analyst bases findings on data that are standard or nonstandard, depending on what is being examined, and on the basis of these assumptions designs a model for an altered facility. In this step, the impact of data assumptions, that is, the degree to which it is critical to be accurate in the assumptions, is investigated. The more critical a data factor, the more necessary it is to be as accurate as possible in the prognostication of results based on that data. By this, we mean that major critical areas in the process, the ones that can make or break the entire process in terms of cost and efficiency, are the ones for which it is necessary to be certain of the data. It is not critically important to have absolute accuracy in data that is concerned with minor cost factors, or with minor steps that could as easily be done by hand as by machine, but the impact of changing a whole process step from one approach to another is of great importance. As an example, if the question is whether or not to change from two-wheeled to four-wheeled dollies in moving materials, the accuracy of the data concerning the cost and efficiency of each approach is of minor interest, compared with a decision to switch from gas-fired to electrical steel furnaces, an irreversible and incredibly expensive change in methodology. A miscalculation due to faulty data in the first case is an inconvenience; in the second case, it is a complete disaster. The more expensive and influential the change, then, the more critical it is to have accurate data upon which to make the decision.

E. Recognize the Redesign Model's Detrimental Idiosyncrasies and Look for Ways to Improve It

This is as important an element in the analysis of the problem as recognizing the original shortcomings of the present system. When constructed, the redesign model represents a best guess as to how the productivity and economic efficiency of the plant can be maximized. But inherent in that model are flaws, which may be found by applying the same analysis to the model as was applied to the original facility. If the redesign is to be implemented, it needs to be as efficient as possible to ensure that the firm is not trading one set of problems for another. Minor imperfections must be rooted out and eliminated where possible. Changes in the proposed redesign should be analyzed and fitted against present conditions. Eventually the law of diminishing returns sets in, there being an infinite amount of change that can take place. Like an artist, the design engineer and analyst must decide when it is time to stop redesigning, that is, when the savings from the change are less than the expense of the change itself. At that point, the new design is close to maximum efficiency.

F. Determine the Long-Run Implications of the Problem

This gives a clearer picture of just how important solving the problem is, and often results in the discovery of other problems not seen before. By determining the long-run implications as well as the more obvious short-run implications, a stronger case for making the change can also be seen. It must be remembered that a facility is itself a long-term project, and in dealing only with short-run, present problems, the analyst denies the long-run nature of the process. What is quite acceptable now may not be in five, ten, or fifteen years. By altering the structure now, money saved later and efficiency gains achieved now may result in even greater benefits later on that contribute to the life of the facility and the profitability of running it. The long-run aspect of the problem must be included in any analysis, or a false picture of conditions will inevitably arise.

G. Examine the Layout Problem as a Systems Problem

Layout particularly is a problem in systems analysis, since we can easily see how each subsystem constitutes a series of cause-and-effect relationships with the other subsystems in the process. By defining layout as successive layers of subsystems, it is possible to completely describe what is happening and to then manipulate elements into more efficient patterns. The systems approach also ensures that no major element in the process is neglected, such as communications or feedback.

H. Weigh the Qualitative Factors

Not all important factors in the construction of a redesign are quantitative. There are also nonmeasurable qualitative factors to consider, such as the impact on employees, the degree of disruption caused by the redesign, the impact on suppliers and other related industries, and how the redesign will be viewed by others. For example, a modern, space-efficient design may look wonderfully efficient on paper, but if it packs workers into tight spaces without sufficient psychological territory to satisfy their needs, all the machinery upgrades and layout rearrangements in the world are not going to increase output efficiency. The use of sharp angles in the architecture, the placement of windows where employees can or cannot see out of them, the type of environmental controls installed, and other factors that are not easily quantified can have a tremendous effect on the viability of a given design.

I. Select the Proper Tools for Analysis

Match the type of analysis undertaken and the tools utilized to the problem as it really is. At times, this means changing one's paradigm about what is really going on. There are many ways to analyze the performance and nature of an industrial system, and although all of them yield information, it is not all always useful information. The tools used in analysis should reflect the nature of the problem or the questions being asked. The importance goes beyond simply avoiding wasted effort in gathering unnecessary data. The wrong type of statistical tool may be misleading rather than useless. As it has been said (more than once) there are three kinds of lies: lies, white lies, and statistics. Since we have a tendency to work within our personal paradigms, we risk choosing tools that will give us the answers that agree with our paradigm rather than those that reflect reality. Care should be taken in the type of analysis used, whether it be direct observation, financial and statistical analysis, machinery design analysis, or time and motion studies.

J. Seek Wise Counsel from Each Management Area of Concern

This cannot be reiterated often enough. Not only for reasons of avoiding resistance, but also to utilize all available resources, the management involved with any part of a production process that is under consideration for change should be thoroughly informed and brought into the planning process. Numerous peculiarities exist in any production system that are not readily obvious. Those who have experience with the process are most likely to be aware of them, and most likely to have solutions for them as well. Beyond this, the cooperation of those who must carry out and work with the changed system is imperative to the successful completion of the project.

K. Be Flexible; Something Better Could Result

Any loyalty to a single point of view is a self-imposed limitation on available solutions. Flexibility bespeaks a willingness to accept useful change and a desire to find it. Like anyone else, facility designers and analysts may fall into the trap of creating a pet model from which they are not willing to vary. In such cases, and it happens to all creative people at some point, a degree of personal pride is attached to the creation, and any attempt to alter that creation is seen as an attempt to adulterate it. Inspiration and improvement can come from anywhere. When it comes, it should be accepted. All people, from the president of the firm to the custodial crew in the plant, are viable sources of ideas and suggestions. This is not to say that all ideas should be accepted out of hand or implemented. It is simply a matter of being willing to objectively evaluate those ideas and see what works and what does not. The danger lies in forgetting what the purpose of the exercise is in the first place. The idea is to create an efficient, viable production facility, not to build a monument to anyone's creative genius. Indeed, when facility planners open themselves to the ideas of others, the result may well be a very laudable example of that planner's creativity.

IV. CURRENT TRENDS IN REDESIGN: THE JIT FACILITY

Property values are currently rising at an alarming rate. It seems the only cheap land to be found is in the country, yet the most important site selection criterion for facilities is the favorable labor climate available in the cities. Some companies have resolved this problem by using the "just-in-time" or JIT facility concept, which is basically the downsizing of a facility by eliminating almost all inventory storage. It is possible to use a smaller facility and still maintain the volume necessary to make a profit by incorporating some unique points. These include the following:

1. Smaller stockroom space
2. Smaller warehousing space
3. Separate shipping and receiving areas
4. Administrative offices located at the production site
5. Computerized material-handling equipment
6. Smaller queuing areas for inventory-in-process

7. Extra vertical space in the facility used for storage or administration without adding another floor

Though these are all significant characteristics of the JIT facility, other signs will become more apparent as this method of production is fine-tuned by its increased use in the future.

The driving force behind these features is the JIT production schedule that runs the facility. Everything is scheduled, from lunch periods to the arrival of raw materials. The JIT facility is perfectly timed, allowing inventory to be shipped to the customer without being stored, for it was made "just in time" to meet customer demand. Raw material is not stored, because it is not called for until the customer needs the product.

The Japenese have developed a JIT facility that produces houses in their country. The customer who comes into the sales office on Monday is shown designs for homes on a computer. When the customer chooses a house, the computer issues the orders to start the process. At this point, orders for materials to begin production sequencing are placed by the computer. Because of modular building techniques, the house is delivered onto the customer's site and erected, ready to be inhabited by the end of the week. Now that is an efficient JIT facility process! Obviously, timing is the key to the elimination of stockrooms, queuing areas, and warehousing of inventory. Inventory flows through the process and does not need to be stored. Administration is handled right on the spot, saving time and, therefore, money.

Although JIT works well in manufacturing facilities, service facilities can also apply the methodology. Consider a doctor's office that is well scheduled. Waiting rooms may be much smaller and examination rooms fewer, since patients are properly timed. The key here is a master scheduler who knows ahead of time how long the particular problem you have will take to examine and allots time accordingly. Retail stores, which are a type of service facility, are able to receive smaller lots of merchandise more often, usually by daily delivery, in order to keep the inventory moving and looking fresh. The key again is a master scheduler, this time in the form of an inventory control system, that can correctly predict customer demand to the day.

In an existing facility, a JIT system would open up more productive space, since receiving, queuing, and warehousing space would be drastically cut. The plant would then have the ability to expand its production volume, which would increase the chance for greater profits. Before such a redesign can occur, however, one must be certain that the JIT operations system is in place, ready to function, prior to the necessary changes in the physical structure. Pairing the JIT system to a JIT facility can be very profitable, but a JIT facility without a JIT system would be a disaster. Interested readers should contact their local American Production and Inventory Control Society (APICS) for help at 1-800-444-2742.

QUESTIONS

1. How does the redesign of an old facility differ from the design of a new facility? How are they similar?
2. The corporate inertia encountered in a redesign project is different from that encountered with new facility design. Discuss those differences. In what ways would the two types be similar?
3. In redesigning a facility, there is an opportunity to take into account human factors previously unaddressed. What might some of those be?
4. How can redesign exacerbate rather than relieve problems? Offer some examples.
5. Name and define five causes of resistance to facility redesign.
6. List five methods of dealing with resistance to facility redesign.
7. Give some examples of how tact is used to overcome resistance to facility redesign.
8. What are the 11 steps of the redesign process?
9. How is a systems approach useful in redesigning layouts?
10. Are redesign problems primarily technical ones or human ones? Explain.
11. How would a JIT facility differ from a non-JIT facility?
12. What are the advantages and disadvantages of maintaining a JIT facility?
13. Give some examples of JIT service facilities and of JIT manufacturing facilities.
14. Which U.S. businesses would most benefit from JIT production techniques? Which businesses would be unable to use JIT methods?
15. Explain why JIT facilities might make JIT managers jittery.

A Software Appendix

A facility planner should have the ability to produce some rough-cut sketches of the proposed facility. An architect is needed for the detail drawings for actual construction. What the facility planner provides is an approximation of the facility that serves as a rough idea of what the finished facility will look like.

The rough-cut sketches shown in the case study (Chapter 6) were done by using AutoSketch©, a scaled-down version of AutoCad, which works in two dimensions. This is an easy to use CAD program that draws facility exteriors well.

Another program that we found useful for workstation designs was In-a-Vision©. The program works with a "windowing" concept that allows predone workstations to be easily merged into other workstation designs.

Currently, in laboratory classes we are using a CAD program called Roomer 2. This CAD program has the ability to give three dimensions to all drawings. It also has over 100 predrawn pieces of furniture in the program, which makes it a very quick process to develop rough-cut sketches of the proposed interiors of a facility. The following pages show you some of what Roomer 2 is capable of doing.

All of these programs are very inexpensive and work well on PCs. Please read the following brochures:

1. Roomer2 (a CAD program)
2. Means Construction Estimator
3. Storm (Quantitative Modeling for Decision Support)

What's New in ROOMER2!

We have just finished ROOMER2 Designer and greatly improved ROOMER2 Computer Layouts. These products are described on the next two pages.

If you are new to ROOMER2, the workings of the package are explained beginning on page 3. Other products which can be used with ROOMER2 are described beginning on page 5. On page 7 you will find the hardware requirements for running ROOMER2, and an order form.

Reviews

ROOMER2 may be the only product specifically created to lay out and furnish a room.
Chicago Tribune

There are few programs that are both an absolute delight to use and extremely useful or functional. ROOMER2 is one of those rare programs.
Database

ROOMER2 is a charming software program for floor plan and interior design. If you need inexpensive software for conceptual layout of small facilities, then give ROOMER2 a try. It has a lot of features that allow you to do extensive 3D modeling for a modest investment.
Architectural & Engineering Systems

For those who need to make depictions of a room and its contents or floor plans, ROOMER2 by Hufnagel Software provides an excellent and unique alternative to pencil and paper or a traditional computer-aided design program.
Computer Graphics Today

I'm extremely impressed with the professional look that ROOMER2 can give our party consultation services. Once one becomes proficient with it, and with the proper equipment, it can be a real time saver. Try it. You'll like it!
Rental Age

ROOMER2 Designer

ROOMER2 Designer adds over 50 pieces of residential furniture to ROOMER2. Designer also provides a utility for exporting ROOMER2 views to paint programs where they can be colored or shaded.

ROOMER2 Designer has camelback couches, wing chairs, highboys, dressers, beds, tables, plants and pictures. Designer also includes a variety of window, door and curtain treatments which can be incorporated into your designs. Using standard ROOMER2 features you can adjust the dimensions of the furniture to suit your particular needs, and stretch or adjust the proportions of curtains to fit any window.

Some of the Designer furniture was used in creating the view above and one of the views on the last page of this brochure.

An R2GRAB program, which is included with Designer, allows you to convert ROOMER2 views to .PCX format files. Programs such as PC Paintbrush can load these files, and then add color or shading to views generated with ROOMER2. The picture above was shaded using such a paint program, and then printed on an HP LaserJet printer. Views which you color with a paint program can be displayed on your computer screen, and can also be converted into color slides or prints.

ROOMER2 Designer gives you the ability to not only experiment with furniture arrangements, but to experiment with style and color as well.

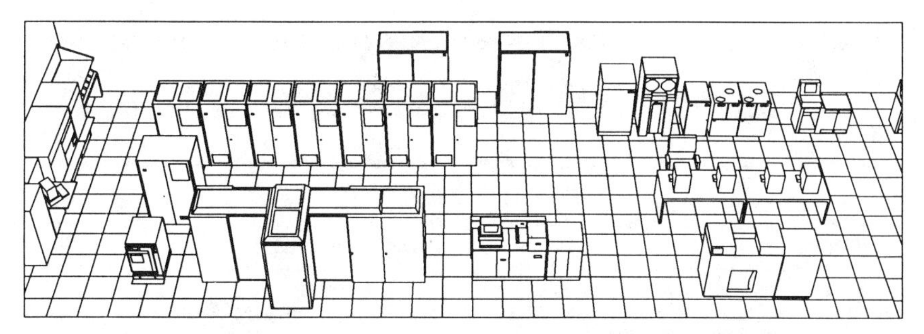

Computer Layouts

The Computer Layouts package allows you to quickly and easily model computer rooms using ROOMER2. Computer Layouts now includes over 80 computer components, and a variety of prebuilt rooms and floor grids. The medium size computer room above and the related plan below were created in a less than 2 hours using ROOMER2 and Computer Layouts. With recent improvements, now even large facilities can be planned.

The prebuilt rooms and grids which are included can be used to create sample systems, or you can use the capabilities of ROOMER2 to create the actual room into which a computer system must fit.

The supplied computer components are loosely based on IBM hardware. Different versions of these components are provided to handle a variety of needs. Non-IBM systems can be approximated.

The Computer Layouts Manual allows computer veterans to get off to a flying start, and is keyed to the standard ROOMER2 Manual so that you can quickly learn more about ROOMER2 if you need to.

Included Components

2440	3287	3726	6262
3082	3350	3745	9309-1
3083	3350F	3746	9309-2
3087	3370A	3746B	38901
3087-1	3370B	3800	38902
3089	3375	3803	38903
3090	3375B1	3820	38904
3090A	3380	3827	38905
3090B	3380J	3835	3890X
3090Unit	3411	3880	A01
3092	3420	3990	A02
3097-1	3422	4234	AS400
3097-2	3430	4245	B10
3174	3480A	4245S	B20
3203	3480B	4248	CSLTAB
3203S	3704	4341	CSU3000
3211	3705	4361	DASD
3262	3705E	4381	DNFLO
3274	3720	5381	J02
3278	3725	5425	TABPC

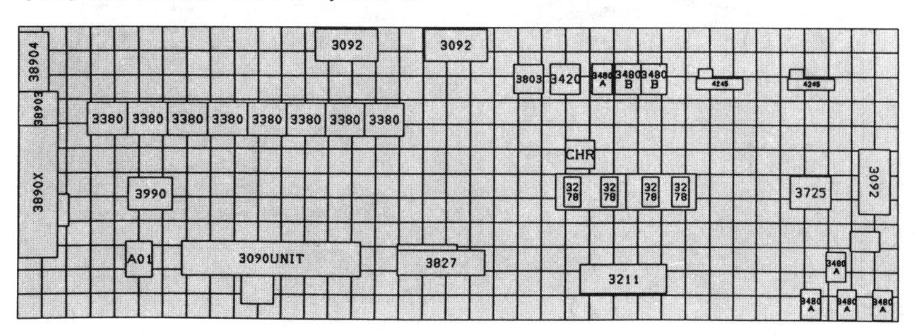

2

How ROOMER2 Works

ROOMER2 is a software package that allows you to create floor plans and room arrangements with your IBM style personal computer. You can also create 3D views of your rooms and arrangements and lay out your final results for maximum impact. ROOMER2 is easy to learn and fun to use. With ROOMER2 you can design offices, homes, kitchens, factory floors, open work areas, show rooms, computer rooms, educational facilities, theater sets; indeed almost any interior space.

ROOMER2 takes a simple approach to space design. A common way to create furniture arrangements is to draw a floor plan on a piece of paper and then move cardboard cutouts of the furniture around on it. This essentially is what ROOMER2 allows you to do with a microcomputer. Of course it's a lot easier to keep everything organized and to scale on a machine, but the best part is that as furniture is moved around on the screen, ROOMER2 builds a three dimensional model of the furniture arrangement in the computer's memory. Once an arrangement has been created the program can produce perspective drawings of the furnished room. The views produced by ROOMER2 give an accurate idea of what a real furniture arrangement will eventually look like.

The sample below shows what ROOMER2 can do, and the descriptions on the next few pages will give you a good idea of how ROOMER2 works.

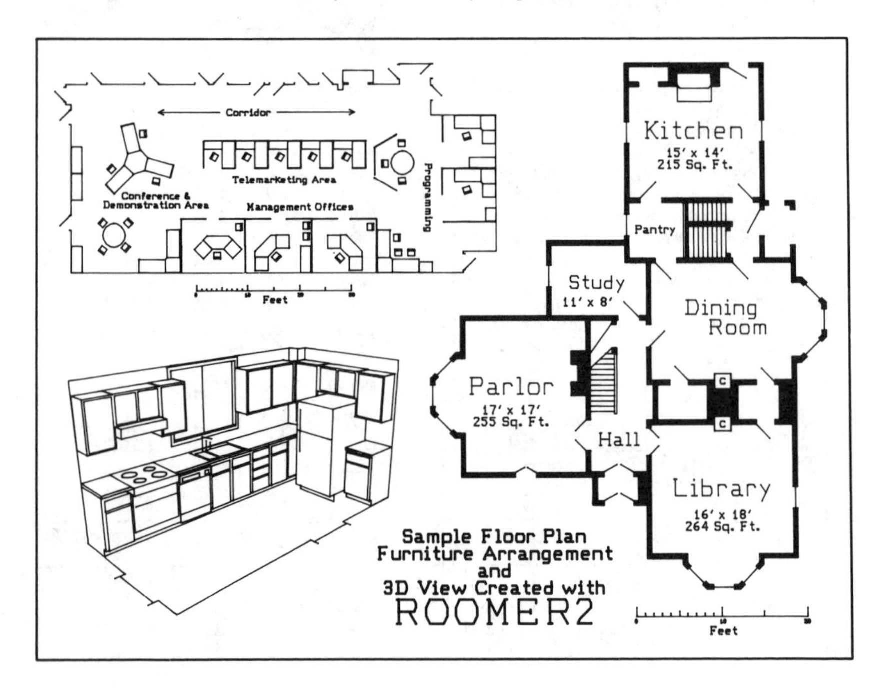

Sample Floor Plan
Furniture Arrangement
and
3D View Created with
ROOMER2

3

General

ROOMER2 consists of several programs which work together to produce a final result. These programs are selected from a Main Menu which is displayed when ROOMER2 starts. The most used programs are briefly described below. All of the art in this pamphlet was produced by ROOMER2.

Making Rooms

The Make Room program produces rooms and floor plans through use of a simple data command language. For instance, the command :WALL 15'6" makes a wall, and :WINDOW 28",72" makes a window. Data commands rapidly create a three dimensional model of rooms and floor plans in computer memory. These can be easily altered to adapt to changes.

```
:SET BASEBOARD 8",CEILING 8'5",TRIM 4"
:WALL 3'4"
:WINDOW 2'5",5'9.5",20.5",4
:WALL 3'9":WINDOW:WALL 3'3"
:TURN RIGHT:WALL 8'7.5"
:DOOR 2'8",6'8",RI,45:WALL 6"
:LINE
:TURN RIGHT:WALL 15"
:TURN LEFT:WALL 6.5"
:TURN RIGHT:WALL 11':DOOR:WALL 5"
:TURN *
:LINE
:WALL 3'8.5":TURN RIGHT:WALL 5"
:TURN LEFT:WALL 4'10"
:TURN LEFT:WALL 5"
:TURN RIGHT:WALL *
```


Data Commands and the Room which they Generate

Arranging Rooms

The Furnish program fills a room with furniture. ROOMER2 comes with a library of over 60 pieces of furniture which can be altered to any size desired. Custom furniture can be created by combining existing pieces. For instance, boxes and lines can be combined to create a refrigerator; or tables, chairs and microcomputers can be combined to make a workstation. Once a piece of furniture is loaded from the library it can be moved about, rotated or placed on top of another piece of furniture.

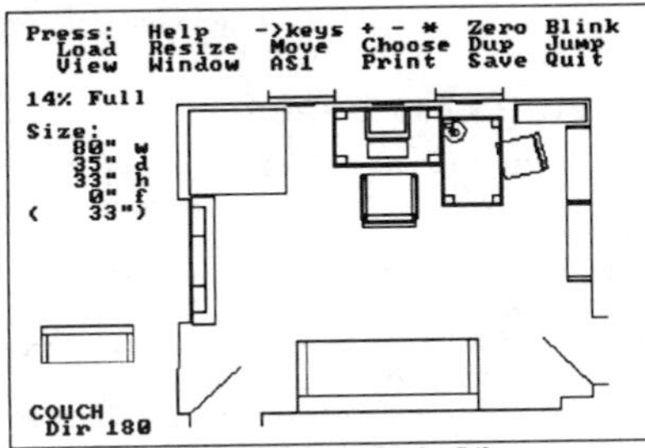

What the Screen Looks Like During Furnishing

Creating Views

The View program produces 3D views of rooms and furniture arrangements. ROOMER2 creates full views or pictures such as a camera might take. Views are created by specifying the point at which the "camera" is located and then pointing it in the desired direction. Wide angle, telephoto and birds-eye views, as well as elevations are possible.

Camera Style View of a Completed Arrangement

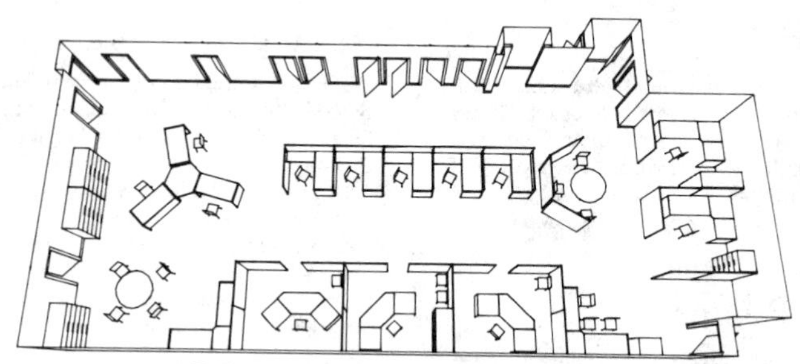

A Full View of the Office Arrangement on Page 3

Laying Out Pages

The Layout Page program assembles floor plans, arrangements and views on an output page. Text can be added and plans can be sized so that they print to scale. Estimates of room and floor plan dimensions, and areas calculated by ROOMER2 can also be added to the page. Finished pages are printed on a dot matrix or laser printer, or plotted. Pages can also be exported to word processors such as WordPerfect 5.0, and to most CAD programs via DXF and HPGL files. The sample on page 3 gives a good idea of the capabilities of the Layout Page Program.

ROOMER2 Plus

ROOMER2 Plus is an option which adds a Report program and more furniture to ROOMER2. The Report program looks up each piece of furniture in an arrangement in price catalogs to obtain pricing and descriptive information. You can use different price catalogs to report on the same arrangement to get comparative price reports.

The 50 pieces of office furniture included with ROOMER2 Plus include pieces for designing modular furniture systems, a variety of chairs, and machines and accessories which are useful in designing office space.

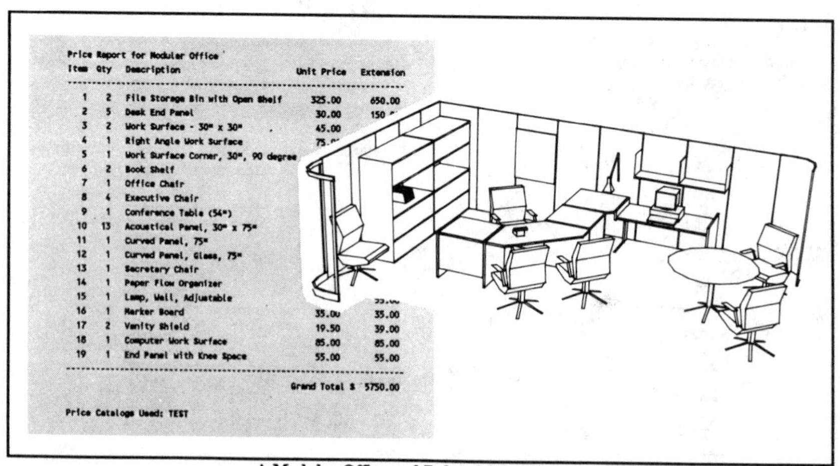

Price Report for Modular Office

Item	Qty	Description	Unit Price	Extension
1	2	File Storage Bin with Open Shelf	325.00	650.00
2	5	Desk End Panel	30.00	150 [illegible]
3	2	Work Surface - 30" x 30"	45.00	
4	1	Right Angle Work Surface	75.[illegible]	
5	1	Work Surface Corner, 30", 90 degree		
6	2	Book Shelf		
7	1	Office Chair		
8	4	Executive Chair		
9	1	Conference Table (54")		
10	13	Acoustical Panel, 30" x 75"		
11	1	Curved Panel, 75"		
12	1	Curved Panel, Glass, 75"		
13	1	Secretary Chair		
14	1	Paper Flow Organizer		
15	1	Lamp, Wall, Adjustable		[illegible]
16	1	Marker Board	35.00	35.00
17	2	Vanity Shield	19.50	39.00
18	1	Computer Work Surface	85.00	85.00
19	1	End Panel with Knee Space	55.00	55.00
			Grand Total $	5750.00

Price Catalogs Used: TEST

A Modular Office and Related Price Report

ROOMER2 Outside

Experienced users of ROOMER2 can use ROOMER2 Outside to make simple building exteriors, lot plans and landscapes.

ROOMER2 Outside comes with a wide variety of house parts, doors and windows which can be resized and combined to create nearly any traditional structure.

18 species of trees and bushes are also included. These look a little like diamonds on sticks, but entire landscapes of trees can be made to "grow" with a single command. Instructions for enhancing landscape views by hand are included in the ROOMER2 Outside manual. This is quite easy to do and you can quickly draw quite presentable results - ROOMER2 Outside does the hard work of getting the perspective right.

The simple landscape below took less than an hour to create. The trees are Red Maples and Blue Spruces. Much more complicated structures and scenes can be created with ROOMER2 Outside.

ROOMER2 was developed for interior design, and designing a landscape involves little more than moving trees, houses and other pieces of "furniture" around on a lot. ROOMER2 Outside has some limitations; some non-traditional buildings cannot be created, and hills, slopes and other terrain features cannot be modeled.

ROOMER2 Twirl

ROOMER2 Twirl allows you to animate the viewing of ROOMER2 furniture, furniture arrangements and rooms. If you plan to build a lot of your own furniture, Twirl will let you quickly check the new pieces for correctness. You can also use Twirl to show off your arrangements, but the animation is only satisfactory for small amounts of data, and hidden lines are not removed. Pieces look a little like the picture below as they twirl.

ROOMER2 Create

ROOMER2 owners will find the furniture creation capabilities of ROOMER2 to be excellent for most purposes. The Create module is for advanced users of ROOMER2 and allows creation of unique pieces of furniture. A Convert program allows drawings to be extruded and turned into ROOMER2 furniture. The dinosaur above and most of the more elaborate pieces of furniture depicted in this brochure were created with the Convert program. A Build program, also included with the Create module, allows you to create a new piece of furniture by drawing lines and planes in space. Pieces such as the trees below can be created with Build.

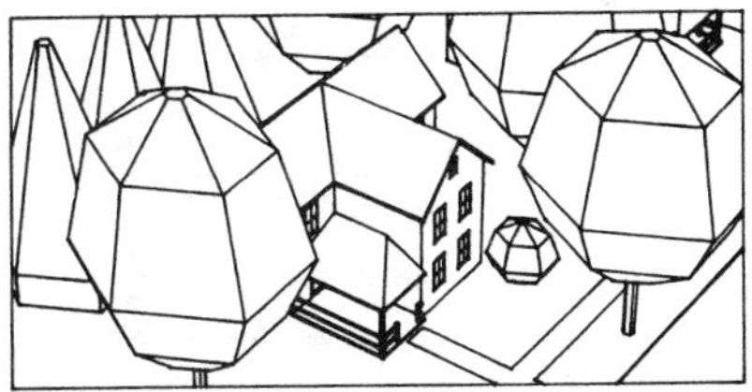

Entire Landscapes can be Made to Grow with ROOMER2 Outside

ROOMER2 Stereo Views

With ROOMER2 Stereo Views you can create and view true 3D images. An inexpensive, but effective stereopticon viewer is included. This allows you to view two slightly different images at the same time, and produces a true 3D effect.

ROOMER2 Stereo Views comes with complete instructions for creating your own "stereo pairs". The process is very simple, and consists of creating two slightly different views with the View program.

ROOMER2 Stereo Views is an interesting toy for most people, but in some instances the ability to see an arrangement in three dimensions is truly useful.

Required Hardware

ROOMER2 requires an IBM PC/XT/AT/PS2 or compatible microcomputer with at least 256 KB of memory and DOS 2.0 or later. Systems without a hard disk require two diskette drives. ROOMER2 will make use of any memory, up to 640KB, present on your system.

ROOMER2 also requires a graphics board and will run on IBM CGA/EGA/VGA graphics boards and equivalents. To use the Hercules Graphics board with ROOMER2 you must purchase separately priced Hercules Support software. ROOMER2 chiefly uses medium resolution CGA graphics.

ROOMER2 needs some sort of output device. If you have a dot matrix printer, or a laser printer similar to the HP LaserJet, or almost any plotter, ROOMER2 will be able to produce excellent results.

ROOMER2 comes with a 179 page manual which includes a detailed tutorial, a reference section and a furniture catalog.

ROOMER2 is not copy protected.

ROOMER2 Video

The ROOMER2 Video is an unpolished but effective 90 minute VHS video tape introduction to using ROOMER2. Hank Hufnagel shows how to use ROOMER2 to produce floor plans, furniture arrangements, three dimensional views and ROOMER2 output on both printers and plotters. This tape has been well received by first time CAD users, by organizations which want to train people in the use of ROOMER2, and by folks who want to get off to the quickest start possible in using ROOMER2. Mr. Hufnagel's style has been compared (unfavorably) to Clint Eastwood's.

Money Back Guarantee

Hufnagel Software offers a 30 day unconditional guarantee on ROOMER2 (less shipping charges). There is a $15 restocking charge if the disk packet has been opened or the manual damaged. This guarantee is valid only for products bought directly from Hufnagel Software.

Order Form

Name________________________

Company______________________

Address______________________

City__________ State___ Zip___

Phone__________________

Check one: __Payment Enclosed

__Purchase order enclosed

Charge my: __Visa __MasterCard

Card no.__________________

Expires_____

Please send me: __ 5¼" disks __ 3½" disks

__	ROOMER2	$295
__	ROOMER2 Plus	$ 45
__	ROOMER2 Designer	$ 45
__	ROOMER2 Computer Layouts	$ 45
__	ROOMER2 Outside	$ 45
__	ROOMER2 Twirl	$ 45
__	ROOMER2 Create	$ 45
__	ROOMER2 Video	$ 25
__	ROOMER2 Stereo Views	$ 15
__	Hercules Support	$ 45

Total Amount (in PA add 6%) ____

Send to: **Hufnagel Software, P.O. Box 747, Clarion, PA 16214**

or Phone: **(814) 226-5600** Visa and MasterCard accepted

References

ADAMS, EVERETT E. *Production and Operations Management.* Englewood Cliffs, N.J.: Prentice-Hall, 1986.

ALCORN, PAUL A. *Social Issues in Technology.* Englewood Cliffs, N.J.: Prentice-Hall, 1986.

BARNES, RALPH M. *Motion and Time Study*, 4th ed. New York: Wiley, 1958.

The Comprehensive Land Development Regulation for Houston County. Perry, Georgia: Houston County Board of Commissioners, May 4, 1976.

COX, BILLY, AND HORSLEY, F. WILLIAM. *Square Foot Estimating.* Kingston, Mass.: Robert Snow Means, 1983.

FOGARTY, DONALD W. *Production and Inventory Management.* Cincinnati: South-Western, 1983.

FRANCIS, RICHARD L. *Facility Layout and Location.* Englewood Cliffs, N.J.: Prentice-Hall, 1974.

FRIEDMAN, EDITH J. *Encyclopedia of Real Estate Appraising.* Englewood Cliffs, N.J.: Prentice-Hall, 1968.

FULLER, BUCKMINSTER R. *Operations Manual Space Ship Earth.* New York: Dutton, 1963.

GAITHER, NORMAN. *Production and Operations Management.* New York: Dryden Press, 1987.

HARWOOD, BRUCE. *Real Estate: An Introduction to the Profession*. Englewood Cliffs, N.J.: Prentice-Hall, 1986.

HENESSEY, JAMES. *Nomadic Furniture*. New York: Random House, 1974.

HOPEMAN, RICHARD J. *Production and Operations Management*. Columbus, Ohio: Charles E. Merrill, 1980.

HORSLEY, F. WILLIAM. *Means Scheduling Manual*, 2nd ed. Kingston, Mass.: Robert Snow Means, 1984.

JAMES, ROBERT W. *Facilities Planning Lab Manual*. Chicago: DeVRY, Inc., 1987.

JAMES, ROBERT W. *Facilities Planning Lesson Plans*. Chicago: DeVRY, Inc., 1986.

LAUFER, ARTHUR C. *Production and Operations Management*. Cincinnati: South-Western, 1984.

LEW, ALAN E. *Means Interior Estimating*. Kingston, Mass.: Robert Snow Means, 1987.

MAHONEY, WILLIAM D., AND CLEVELAND, ALLAN B., EDS. *Means Man-Hour Standards for Construction*, rev. 2nd ed. Kingston, Mass.: Robert Snow Means, 1988.

MAHONEY, WILLIAM D., AND GRANT, ROGER J., EDS. *Means Forms for Building Construction Professionals*. Kingston, Mass.: Robert Snow Means, 1986.

MAHONEY, WILLIAM D., AND HORSLEY, F. WILLIAM. *Means Graphic Construction Standards*. Kingston, Mass.: Robert Snow Means, 1986.

MAHONEY, WILLIAM D., AND LEHIGH, DWAYNE R., EDS. *Means Concrete Cost Data 1989*, 7th annual ed. Kingston, Mass.: Robert Snow Means, 1989.

MENIPAZ, EHUD. *Essentials of Production and Operations Management*. Englewood Cliffs, N.J.: Prentice-Hall, 1984.

MOLNAR, JOHN. *Facilities Management Handbook*. New York: Van Nostrand Reinhold, 1983.

MUNDEL, MARVIN E. *Improving Productivity and Effectiveness*. Englewood Cliffs, N.J.: Prentice-Hall, 1983.

PUTNAM, ROBERT E. *Construction Blueprint Reading*. Reston, Va.: Reston, 1985.

ROSCOE, EDWIN S. *Organization for Production*. Homewood, Ill.: Richard D. Irwin.

SCHMENNER, ROGER W. *Making Business Location Decisions*. Englewood Cliffs, N.J.: Prentice-Hall, 1982.

SCHONBERGER, RICHARD J. *Operations Management*. Plano, Texas: Business Publications, Inc., 1985.

SMIT, KORNELIS, ED. *Means Illustrated Construction Dictionary*. Kingston, Mass.: Robert Snow Means, 1985.

WILLIAMS, J. CLIFTON. *Management and Organization*. Cincinnati: South-Western, 1985.

Glossary

Accelerated depreciation Depreciation methods based on the concept that an asset loses more of its usefulness in the early years of its useful life than in the later years.

Achievement motivation theory A theory developed by David McClelland based on the idea that motivation is dependent on three specific needs: achievement, power, and affiliation.

Activities Actions that are required to complete a task.

Activity analysis The analysis of activities by defining each action taken within a process in order to determine its content.

Adaptability The ability of an individual or system to change in response to environmental or internal stimuli in order to reestablish a state of balance.

Ad hoc approach An approach to developing new facilities by which individuals within the firm are given the task of facilities planning in addition to their present duties.

Adjustable rate mortgage A mortgage loan on which the interest rate rises and falls with changes in prevailing rates.

Agility The ability to move with quick and easy grace.

Agreed risk An insurance policy in which the assumption of risk on the part of the insurance company is on the basis of mutally agreed-upon specific factors.

Air rights The right of the purchaser to the use of the space above the land in question and nothing more.

Alienation clause A clause in a mortgage or promissory note that allows the lender to call the entire loan balance due if the property in question is sold or transferred to a third party.

All-risk insurance The most inclusive form of policy in which, within the stated limits of the policy, the insurance company will accept all risks for negative events without restriction, limitation, or exclusion.

Ambient temperature The natural temperature of a given environment, unaltered by artificial means.

Anthropometry Literally, measurement of the body. In facilities design, the determinations of average bodily measurements and the application of those data to the design of work environments.

Appreciation The increase in the value of real property over time.

Assembly and gang process charts In time and motion study, a large flow diagram that combines a number of processes into a single overview, showing how assemblies take place and giving their individual steps.

Assembly plant A plant dedicated to assembling final goods or services from component parts that have been created elsewhere.

Automation A production methodology in which some or all of the productive processes are performed by machinery controlled through electronic or mechanical servomechanistic devices rather than human operators.

Backstop positions The construction of scenarios representing worst-case situations in which the firm's best plans fail; helps the firm determine how to handle such eventualities.

Balancing the line An analysis technique designed to ensure a steady flow of work in progress through the production process.

Batch operations systems Systems in which operations that take place are periodic, identical runs of specific items.

Blanket mortgage A mortgage that is secured by more than one piece of property.

Blended-rate loan A mortgage in which the present interest rate on an existing loan is combined or "blended" with current rates.

Bracketed return on investment A combination of the worst-case minimum return concept with a best-case maximum expected return to give a range of possible returns that are bracketed by the two extremes. This form of ROI determination takes into account the risk involved in long-range projects.

Brazing processes A joining process similar to welding that uses some medium as a solder to form the join.

Break-even analysis A cost analysis technique that determines the level of production that a firm must have in order to cover all costs connected with the production of the item in question. Below the break-even point, the firm loses money; above it, it begins to show a profit.

Budgets Financial plans.

By-products Secondary products that result from the process of creating a firm's primary product or service, which represent to the firm either an opportunity for additional profit through the sale of such by-products in the market, or a cost for those requiring disposal.

Capital-intensive industries Industries that rely chiefly on the input of capital items, such as automatic machinery and equipment, to achieve their productive goals.

Carryback financing A note accepted by a seller in lieu of cash.

Cash flow projections Projected source and use of funds reports that identify when moneys are expected to be generated by sales from the facility's output, and when and for what those funds will be used. It represents a detailed budget to determine when (and if) shortages of funds will develop so that they may be planned for.

Casting A major forming process in which the material to be formed is liquified and poured or cast into a mold and then allowed to cool to conform to the desired shape of the part or product.

Certain occurrences In insurance, those events that are considered virtually unavoidable. In facilities planning, these include construction delays and changes, weather delays, labor force changes, moving expenses, income losses, transportation delays, and cost of new equipment. Such occurrences are usually uninsurable, though a degree of the occurrence can sometimes be insured against.

Characteristics In production design, those qualities of a product or parts of a process that are considered significant to the achievement of the item's purpose.

Class A knobs Control mechanisms that are to be turned or spun under conditions where position is unimportant.

Class B knobs Control mechanisms that are moved less than one full turn under conditions where position is unimportant.

Class C knobs Control mechanisms requiring less than one turn under conditions where position is important.

Clerical processes The record-keeping function of the facility, and all of the various forms of paperwork that are required for the facility to operate effectively.

Closing The elimination of a facility or facilities no longer considered to be useful or economically viable.

Common law Also referred to as English common law; a body of laws growing up over a long period of time beginning in the Middle Ages and consisting of court decisions based on custom and reason as applied to everyday disputes and events.

Communication and information services The mechanisms within the facility that are in place to gather, store, and provide information for whom and wherever it is needed.

Community business climate The overall business conditions of a given community, including availability of resources, attitudes of the community toward business, cooperative nature of firms in a given area, the general level of prosperity, and the ease with which operation and relocation can take place.

Community property The joint (concurrent) ownership of property by husband and wife simply by virtue of their state of legal marriage.

Concurrent ownership Estates in which the interest in ownership rests with two or more people at the same time.

Construction loan Also known as *interim loan*; a loan in which the borrowed funds are distributed periodically rather than in a lump sum.

Contingency planning A planning technique that seeks to create a planned scenario for a wide range of possible events so that if and when any of these events occur, the mechanism for dealing with them will have already been put in place.

Continuous operations systems An approach to processing in which the operations taking place are repetitive and unvarying, where the product being produced and the demand for that product are relatively constant.

Contract for deed Also known as a *land contract* or *installment contract*, it is a method of selling property by which possession of the property passes to the purchaser, yet the seller retains actual title to the property until the loan agreement is fully executed.

Controls Methods of manipulating the activities of the workstation to perform the various steps in the production process.

Corporate image The way in which a firm is viewed by the public, the business community, and those in political control.

Cost of compliance The cost to the firm of complying with government regulations.

County commissioners The governing body of a county.

Declining-balance depreciation An accelerated method of depreciation in which the value of the asset is reduced at a rate equal to twice the percentage of a straight-line depreciation.

Departmentalization A management technique by which a firm can organize itself into functional subunits, or departments, based on any of a number of formats including function (production, accounting, finance, marketing, and so on), geographic location (northeast, southeast, north central, and so on), or customer (retail, wholesale, industrial, governmental, and the like).

Downsizing The reduction of facilities due to increased technological efficiency or a reduced market for the product.

Drilling A shaping process by which an opening of some specific diameter and depth is created in the material being worked.

Dynamic balance In systemics, the concept that the final purpose and goal of any system is to achieve and maintain a state of balance in which the goal of the system is extant. The balance is dynamic in that, although many elements in the system may be in a state of flux (change), the final result of the sum total of these dynamic actions is that balance is obtained and maintained.

Easements The right of a person to use property owned by another.

Economic use survey A governmentally funded survey indicating how best to support the continued commercial growth and the best-use scenarios that will accomplish these ends.

Economies of scale The concept that, as additional units of input are added to a process, the resulting output rises at a more than proportional rate due to the increased efficiency afforded by those units operating in synergistic cooperation.

Efficiency The creation of goods and services with a minimum of time, money, and effort.

Elemental standard data systems Synthetic time-study methods that determine standard or normal time from previously gathered data.

Eminent domain The legal right of government to seize or authorize the seizure of private property for public use, providing that "just compensation" be provided to the private owner of that property.

Encroachment An intrusion on one's property; a trespassing.

Encumbrance Any restriction that may exist on a property and its ownership that would limit that ownership in some way.

Environmental impact study A study required by the Environmental Protection Agency (EPA), which investigates the impact (effect) of industrial development on the surrounding environment.

Environmental Protection Agency A federal regulating agency whose duties center in protecting the quality and condition of the general environment.

Equity sharing A cooperative approach to financing in which the individual providing the financing for the purchase receives part of the ownership of the property involved.

Ergonomics The study of the manner in which physical work is carried out.

Ergonomic factors Factors that affect the physical dynamics of how work is carried out and the effectiveness of those methods.

Escheat The reversion of ownership of property to a government when there is no legal heir.

Estate at will Also known as *tenancy at will*, a lease arrangement in which the termination of the lease may occur "at will" by either party.

Estate for years A lease estate in which property is conveyed to the leaseholder for a specific number of years, called the *term of the lease*.

Executive branch agencies Governmental regulatory agencies created by executive proclamation rather than by legislative action.

Expansion A strategy by which the firm chooses to increase the productive efficiency and capacity of its operation through the enlargement of present productive capacity.

Expected return on investment (ROI) A measure of profitability determining the amount of profit received on an investment of funds within a given time frame. There are four primary types of return on investment: short term, long term, minimum term, and bracketed return.

Extrusion A process in which a softened material is forced through an opening that is shaped to match the desired final shape of the item being formed.

Facility A building, plant, or other physical structure in which productive activities are carried out and work is accomplished.

Facility building design life The expected useful life in years of a proposed facility building design.

Facility process inputs All elements that affect the facility process, including machinery and equipment, physical plant, labor and capital, customers and suppliers (the market), and the environment (political, economic, and social, as well as physical), which includes technology, the state of international relations, general conditions of life, and attitudes of the public.

Fatigue Physiologically, a reduction in the ability of an organism or parts of an organism to function properly due to extended periods of physical or mental exertion.

Feedback The communication element that represents information returned to the initiator of an action comparing the desired goal of the system with what actually took place; allows further operations to be controlled to adjust for error.

Fee simple The transfer of land from one party to another as a straight exchange for a fee. There are three possible forms of this type of ownership: the simple absolute estate, the life estate, and the qualified fee estate.

Fee simple absolute estate A form of ownership in which the owner of the land has the absolute right to transfer the ownership of the land or real property involved.

Finishing processes A manufacturing process in which assembled goods are readied for final shipment.

Fixed costs Costs that do not change in response to a change in the level of production.

Fixed deductibles The simplest approach to shared insurance risk in which a fixed amount of loss is determined, below which the insurance company will not accept risk.

Fixed-position layout A type of facility layout where the primary deciding factor of design is the fact that the product being constructed cannot be easily moved.

Flow diagram A combination of therblig analysis and diagram of a work area used to completely describe and manipulate the content of job process for any single station.

Forging A process in which the raw material begins as a blank and is hammered into shape under pressure and at temperatures that are high enough to soften the material without causing liquefaction.

Formal approach An approach to developing new facilities in which the firm invests money, manpower, and official status to a development unit for the express purpose of creating a facilities program.

Forming processes Processes that shape the work in progress into a more usable form.

Freehold estate Ownership interests in real property, whether land or other physical property, that will last for an indeterminable period of time. Among the freehold estates are fee simple estates, life estates, and qualified fee estates.

Gantt, Henry L. A scientific management researcher who developed the Gantt production scheduling and coordination charts still in use today.

General-purpose plant strategy A strategy whereby the firm develops facilities utilizing general-purpose machinery capable of producing a wide range of products as opposed to specific purpose or dedicated plants capable of producing only a single or narrow range of products.

Geographical zone Dividing the market up into geographical zones or regions and treating each as a separate market.

Geographic heterogeneity A recent trend in industry in which there is a more uniform distribution of business activities over the entire country, as opposed to a centering of given industries in specific areas of the country.

Gilbreth, Frank and Lillian Scientific management practitioners who were the first to seriously undertake the study of time and motion of workers, using advanced techniques such as motion picture film and micromotion study.

Graduated payment mortgage A fixed-interest-rate loan with a monthly payment that is initially low and increases through time.

Grinding A general-purpose technique designed to remove material from the surface of work in progress in order to shape, smooth, or polish the work as needed.

Habituation To make familiar. The process of becoming habit.

Hard metric conversion A process of metrication by which all machinery and designs are totally converted to the metric system.

Hawthorne effect A psychological phenomenon by which group dynamics can be shown to have a greater effect on productivity than working conditions.

Hawthorne experiments A series of experiments in industrial engineering designed to study the effects of illumination of the workplace on productivity. The results showed what is known as the Hawthorne effect.

Hierarchy of needs A concept developed by Abraham Maslow involving a set of physiological, social, and psychological needs that motivate individuals to behave as they do. In order of relative strength as motivators, the needs are physiological, safety, belonging, self-esteem, and self-actualization.

Independent agencies Governmental regulatory agencies with the power to create law by regulatory decree.

Industrial dynamics A design methodology centered in the acceptance of the philosophy that all productive processes are evolving and thus dynamic in nature, and in the importance of the practical development of a framework that allows for changes to be made in positive and cost-efficient ways.

Inertia In facilities design, a resistance of the firm to the initiation of facilities redesign or expansion projects.

Interest The rent paid for the use of someone else's money. Mathematically, it is $I = Prt$, where I = interest, P = principal, r = the interest rate, and t = time.

Intermittent operations systems Systems in which operations take place on a noncontinuous and nonrepetitive basis using general-purpose rather than special-purpose machinery and equipment.

Internal rate of return A financial analysis method that sets the value of the future income flow equal to the initial capital outlay and finds the discount rate that yields that parity.

Internal testing The testing of production materials, processes, and products carried out within the firm to determine standards and ensure product quality.

Interstate Commerce Commission A federal regulatory agency whose duty it is to ensure that fair and competitive practices are maintained in interstate commerce. It is designed to protect against activities that are in restraint of trade.

Investment credits A tax advantage offered by the federal government to companies that are involved in the construction of facilities.

JIT (just in time) facility A production facility that maintains a minimum of both raw materials and finished goods, relying on materials flow rather than warehousing to control inventory costs.

Job The tasks, or units of work, that are combined in sequence to create a desired end result.

Job design The formulation of the necessary relationships and elements of a job into a system of orderly steps to create a desired end result.

Joining and assembling processes A classification of manufacturing activity designed to combine a number of individually manufactured pieces into a single item.

Joint ownership A condition of ownership in which, upon the death of any owner, the surviving coowners have the right to equally divide the share of ownership of the deceased owner among themselves.

Labor force The quantity and quality of labor available to a firm in a given geographic location.

Labor-intensive industries Industries that rely chiefly on the input of labor to achieve their productive goals.

Labor pool The availability and quality of the labor supply in a given location.

Land banking The process of saving land for future use through the timely purchase of property when it is available.

Land covenants The general class of agreements representing contractual restrictions imposed in the sale of land at the time of that sale.

Lathe A machine designed to turn cylinders.

Law of diminishing returns The concept that, as additional units of input are added to a process, at some point the resulting output will rise at a less than proportional rate, finally ceasing to rise and eventually falling. This is due to the fact that any system has an upper limit on its synergistic capacity, and no further gains can be made by further loading.

Learning curve A graphic representation of the rate at which an individual is able to absorb new information over time.

Leasehold estate A nonfreehold estate existing by virtue of a lease arrangement between the owner of the property (lessor) and the individual or firm seeking to occupy and have use of the property (lessee). The main types of leasehold estates are estate for years, periodic estate, estate at will, and tenancy at sufferance.

Lease purchase A lease agreement that offers lessees the option to purchase the property if they so desire, all or part of the lease fees paid being applied to the purchase of the property.

Leasing A long-term rental agreement.

Leveraging A practice by which borrowed funds are used to produce income that represents a higher percentage return on invested funds than the interest rate representing the cost of capital. Through this practice, income not otherwise realized is developed without the necessity of tying up company funds.

Liens A legal claim against a piece of property that has been used as collateral for a debt.

Life estate A form of freehold estate in which the owner of the land retains ownership of the property only for the term of his or her lifetime.

Line of balance (LOB) A combination scheduling device and production control technique that is designed to ensure that deliveries of products are made on time.

Load-distance model A quantitative-analysis method of process layout and design that bases layout decisions on the distance between workstations and the size of the loads to be moved from one station to another.

Long-term return on investment The expected overall return (profit) realized on an investment in the long-term time frame as defined by the firm. It is generally higher than short-term return on investment.

Lost revenue A loss of revenue resulting from downtime and temporary inefficiencies connected with the startup of new or expanded facilities.

Machine model A view of humanity that deals with the individual as no more than a mechanical input into the system which needs to be engineered as an integral part of the production machine as a whole.

Machining processes A classification of productive processes in which the material being processed is cut in some manner, as opposed to being hammered or extruded. Primary machining processes include turning, drilling, milling, shaping and planing, and grinding.

Maintenance A support-function process by which the firm ensures that the equipment and physical plant will be available and capable of performing assigned functions.

Managerial costs The costs of additional administration connected with the development and opening of a new facility, stemming either from the necessity of procuring additional managerial personnel or from the necessity of dividing present managerial efforts between current duties and duties connected with the facility project.

Man–machine integration An approach that views the productive process as an integrated whole involving both the worker and the technology with which that worker is interfacing, matching the capabilities of each to the other.

Man-made machines Productive artifacts.

Manufacturing facility A facility for the production or manufacture of some physical product.

MAPI system A model developed by the Machinery and Allied Products Institute that determines the rate of return based on a comparison of conditions with and without the anticipated asset.

Market area plant strategy Locating near the market for the goods in order to minimize transportation costs, lead time between production and delivery, and the need for inventory storage.

Marketing sales projection Determination of the probable levels of sales (the potential market and the expected penetration of the company into that potential market) of a product in order to determine production levels, future revenues, and cost factors connected with actually producing goods in the new facility.

Maslow, Abraham A psychologist who developed the hierarchy-of-needs theory of motivation.

Master land use plan A comprehensive plan within a state or community designed to predetermine directions in which the area's growth will be allowed to move.

Merging The combining of two or more facilities into one in order to reduce operating costs and increase efficiency of operation.

Metrication The conversion of measurement methodology to the metric system of measurement.

Milling A grinding process in which cutting is done either by a rotating blade or set of blades, or by an abrasive stone.

Minimum-term return on investment The minimally acceptable return on investment that a firm is willing to receive and still carry out a project. It represents a profitability floor below which the firm is not willing to go.

Mortgage A promissory note in which the loaned funds are secured by a lien on real property, usually real estate.

Mortgage brokers Institutions that act as true middlemen to bring borrowers and lenders together, receiving a finder's fee in the form of points, which are paid at the time the mortgage or loan is contracted.

Mortgage company A firm that makes mortgage loans to individuals and to other firms, and sells them to investors.

Motility Capability to exhibit spontaneous motion.

Motion-study tools Methods of analysis designed to aid in the study of motion and the effects of motion on productivity.

Motion time data systems A time-study approach that relies on previously created data to reach conclusions, indicating the most efficient way of completing a process or movement as well as the expected time, according to the data.

Motivation-hygiene theory An approach to motivation, proposed by Frederick Herzberg in 1959, which divided factors of job motivation into two categories, the first having to do with job content and experience, called *satisfiers*, and a second having to do with job context and environment, called *dissatisfiers*.

Multiplant strategies Strategies that entail expansion by increasing the number of facilities that a firm operates rather than expanding an existing one or moving to a larger facility.

Multitasking model A plant design model centering on the way in which tasks are simultaneously accomplished, and the relationships among them, in terms of creating the final product.

Nature of numerical control A method of controlling the operations of machinery and equipment in which the numerically controlled machines operate according to data stored on punch tape, magnetic tape, or in computer programs, and fed to the machine that actually performs the operations.

Neuromuscular factors In facilities design, considerations relating to the interaction of the nervous system and the musculature of the human body that impact on the efficiency of workers in the productive environment.

Noise Any disruptive or distractive physical phenomenon that reduces the ability of the individual to respond to other external phenomena. Usually associated with extraneous sound as defined situationally.

Observed time study A form of time study in which the operator is actually observed and timed, either in person or by use of motion pictures or video tape.

Office facilities Facilities designed to support the carrying out of nonmanufacturing activities and "white collar" jobs.

Off-site expansion A facilities expansion strategy in which a firm's productive capacity is increased through the development of facilities at some site noncontiguous to current productive facilities for the purpose of either increasing productive capacity or modernizing some or all of the productive process.

On-site expansion A facilities expansion strategy in which a firm's productive capacity is increased through an enlargement and/or improvement of present facilities.

Operational analysis The most detailed level of job analysis, in which each action is broken down into micromotions to be analyzed and changed to maximize efficiency.

Operations system Any of three primary operations systems used in the manufacture of physical goods, including continuous flow, batch, and intermittent.

Opportunity costs The cost incurred by forgoing the opportunity to make a profit by choosing instead to undertake an alternative opportunity that yields a higher profit. The profit not realized by denial of the second-best alternative is considered a cost.

Option The right to purchase or lease a given parcel of property, at some time in the future, at a set price.

Organizational Safety and Health Administration (OSHA) A federal regulatory agency whose duties include the maintenance of a reasonable level of health and safety for workers in the workplace.

Package mortgage A mortgage secured by both real estate and personal property.

Parallel search and engineering projects A method of reducing cost and time spent searching for facilities sites by having both engineering and search projects proceed simultaneously, so that when the site is found, most of the engineering is already complete and actual site development can take place without delay.

Payment cap A maximum level of payment that can be required on an adjustable rate mortgage. Payment caps are often given in terms of percentage increases in a given time period.

Percentage deductible A form of insurance deductible that allows the insurer and the insured to share loss on a predetermined and specified percentage basis.

Periodic estate An estate for years in which the period of the lease is unspecified. The lease renews itself automatically, unless one or the other of the parties entering into the agreement chooses to terminate the arrangement.

Physical production All of those activities actually involved in the production of the good or service in question.

Physiological response factors In facilities design, the study of physiological factors that affect the capacity of, and manner in which, individuals respond to various stimuli.

Physiological work measurement A process design technique that investigates the physiological implication of a work design, such as oxygen consumption, heart rate, the effects of environmental stress, and fatigue.

Planning board A municipal regulatory agency whose duties include the responsibility of ensuring the orderly and efficient development of its community.

Plant output capacity The quantity of final goods and services that a plant can produce and ship in a given period of time.

Present value The dollar value at the present time of a future payment or series of payments which have been adjusted for the time value of money. Present value is given by the formula:

$$PV = FV_t \left[\frac{1}{(1 + r)^t} \right]$$

where PV is the present value, FV_t is the future value in year t, t is the number of years into the future that the money is realized, and r is the opportunity rate or discount rate used.

Primary mortgage market The market where lenders originate loans.

Probable occurrences In insurance, events that have a high degree of probability of occurrence, although their occurrence is not inevitable. In facilities planning, such occurrences include employee, customer, and construction accidents, land-title problems, and storm damage. Because of the degree of uncertainty involved with such occurrences, they are insurable, the cost being commensurate with the degree of risk involved and the probability of occurrence.

Process All of the activities and internal elements necessary to convert productive inputs into a viable finished product or service. In production, a series of related steps leading to a given goal, usually a physical product or subunit thereof.

Process analysis The study of the flow of material through a production station to determine the different processes that take place at that station and to redefine the movements to increase production or reduce required time.

Process layout A facility layout in which the deciding factor in the development of the layout is the process to be performed.

Process plant strategy A strategy by which the firm isolates the production of the component parts of a product in different facilities, finally bringing them together at a separate facility for final assembly.

Production design A reflection of the product content, it is a top-down systemic design process whereby the individual elements of the production system are defined and then their interrelationships are determined to create a model of the production system in its entirety.

Productive output The goods or services produced by a firm, the company image, and the returned support of the environment and community within which the firm is operating. Note that this is a far broader concept than economic output.

Product layout A facility layout in which the deciding factor in the development of the layout is the product to be produced. The approach is often used where large or continuous runs of the same products or product groups are to be performed in the facility.

Product life cycle The expected sales life of a given product in the marketplace.

Product plant strategies A strategy by which a firm creates separate facilities for each product or product group that the firm produces. This is particularly useful if the products are fundamentally different in nature.

Profitability index A financial analysis approach that calculates the expected profitability of the project through the use of the formula:

$$PI = (\text{Original cost} + \text{Net present value})/(\text{Original cost})$$

Project utility The value of benefits and satisfaction received by the firm from a given project.

Qualified estate A form of freehold estate whose status is determined by certain restrictions or qualifications.

Quality circle A method of problem solving and decision making that brings together individuals from all levels of the company and many different areas of expertise to collectively arrive at solutions.

Quality control A continuous and preventative process designed to ensure that the product produced is identical to the product the firm wishes to produce.

Quality of life The overall economic and social well-being experienced by the population of some geographic area, in terms of economic well-being, health, prosperity, and available opportunities and services.

Reciprocity In systemics, the concept that the results of the operation of a system is directly dependent on what is introduced into the system. That is, what you get out of a system is equal to what you put into that system. The inputs and outputs are reciprocal in nature.

Reductionism The dividing of a system into its constituent parts and then further dividing those subdivisions into constituent parts, continuing this process to some acceptable level.

Regulatory laws Federal state and local laws designed to attain and maintain a degree of equitability in the development of business properties, avoiding conflict with other private and public use of land and at the same time ensuring that the rights of the producers of goods and services are protected so that they may carry out their business activities efficiently and safely.

Relocation A strategy by which the firm chooses to move its entire productive operation from one location to another.

Remodeling The redesign or rebuilding of present facilities to create greater efficiency and take advantage of changes in technology, replace depreciated assets, and better utilize existing available space.

Remotics A method of performing work similar to that of robotics but in which remotely located machines perform operations according to instructions received directly from human operators through the use of telecommunication devices.

Rent based on sales An arrangement by which the rental fee paid is dependent on the size of the sales volume connected with the facility.

Renting A short-term agreement by which the party known as the lessee agrees to take possession of a property in return for periodic compensation. The time period for this type of rental agreement is usually unspecified.

Replacement cost Costs connected with replacing machinery and equipment from an existing facility with new equipment in that expanded facility or at a new facility.

Reproduction costs Moving costs connected with reproducing conditions, specialized layouts within a facility, and other operational mechanisms in the new plant as they existed in the old.

Retail facility A facility designed to support the sale of goods and services to the public through the provision of space, support, and accessibility of the public to the goods or services in question.

Reverse mortgage A mortgage in which the lender makes monthly payments to the borrower, who later repays in a lump sum.

Risk The probability of negative events' occurring during a project that result in increased costs of operation to the firm.

Robotics The use of computer-control machines that can be programmed to perform a series of complex operations in a manner similar to that done by humans.

Salvage value The value of a fully depreciated asset. An accounting concept, this value may or may not coincide with the price that the item will command in the open market.

Savings and loan associations Institutions for the purpose of saving monies and making loans. Unlike commercial banks, savings and loans have limited portfolios of loans centered in long-term loans such as mortgages.

Secondary mortgage market A market in which primary lenders sell mortgages to investors, thus freeing up capital for the further generation of mortgage loans.

Securities Exchange Commission A federal regulatory agency whose job it is to maintain a reasonable level of competitiveness in the exchange of stocks and other securities on open markets.

Service facility A facility designed for either the performance of or support of the performance of salable services.

Shaping and planing Machining processes, the purposes of which are to cut specific shapes into the surface of work in progress.

Shared appreciation mortgage A mortgage in which the borrower receives a lower interest rate on the loan in return for assigning a portion of the property's appreciation through time to the lender.

Shared risk An insurance policy in which the coverage is generally limited in terms of dollar liability on the part of the insurer, rather than the events covered.

Short-term return on investment The expected return (profit) realized on an investment in the short-term time frame as defined by the firm.

Simple interest rate Also referred to as the *nominal interest rate*, it is the stated interest rate of a loan and represents the amount of interest paid on a loan at maturity. The amount that will be due at the end of the loan period is the combination of the borrowed amount (principal) and the interest (I).

Site evaluation The determination of site suitability in terms of profitability to see if the profits received from the expansion would be potentially high enough to warrant the project.

Site-survey library Investigations by a firm provide potential expansion sites, and the information developed from these surveys is kept in a site library data bank to minimize the need for additional research to find a suitable site.

Sociotechnical approach An approach to facilities planning that takes into account the fact that both technological and social factors affect the efficiency of facility design and implementation.

Sociotechnical systems Any system containing elements of both human interaction and technological interface, where the actions of both human relations and man–machine integration become important factors (subsystems) of that system.

Soft metric conversion A process of metrication by which machinery and tolerances are modified to present both English and metric scales so that they are simultaneously available.

Stamping and forming A process in which the medium (metal, plastic, and the like) is hammered quickly, under pressure, to force adherence to the desired shape.

Standardization A definition of the important characteristics of a product in quantitatively descriptive terms.

Start-up costs The extraordinary and nonrecurring costs connected with opening a new facility, exclusive of normal production costs and operational overhead.

Statute law A body of law consisting of officially enacted statutes existing by virtue of legislative actions taken by governmental bodies within a society.

Statutory estate An estate in which the property is owned through certain statutory provisions not found in normal common law.

Straight-line depreciation A method of depreciating the value of an asset based on the assumption that the usefulness of the asset decreases uniformly over the life of the asset, thus decreasing by an equal amount in each year of the asset's useful life.

Subordination A process in which the lender agrees to subordinate all rights under the mortgage, that is, to take a lesser priority of payment among debtors.

Subprocess sequence design The design of individual steps in the productive process (subprocesses) in terms of their sequence of occurrence.

Subsurface rights Also known as *mineral rights*, an encumbrance that restricts the owner of real estate in the disposal of mineral deposits or use of subsurface space connected with the property.

Sum of year's digits depreciation An accelerated method of depreciation in which the depreciation of an item is determined by the formula:

$$\text{Depreciation} = \frac{\text{Remaining useful life}}{\text{Sum of year's digits}} * (\text{Cost} - \text{Salvage value})$$

Synergy The whole is greater than the sum of the parts. That is, in any system, there is some quality of the system that exists only because the system is extant.

Synthetic time study A form of time study in which the work is standardized through an analysis of machinery, and estimated times are based on experience with similar operations. No actual observation takes place.

System A collection of interrelated elements that act cooperatively to perform some function or achieve some goal. Systems operate synergistically, on the basis of reciprocal action, to achieve and maintain a state of balance defined by the goal of the system.

System approach An approach to the study of real-world phenomena that depends on a view of those phenomena as systems, having the characteristics of systems and following the behavioral patterns of systems.

Tactility Sensitivity of touch sensations.

Taylor, Frederick W. The father of scientific management, who used scientific methodology to study and improve work efficiency.

Technical processes The state of the art in producing a product and the whole body of technical expertise that goes into developing and carrying out the productive process.

Technology Ideas, methodology, and man-made productive artifacts that assist in carrying out the productive process.

Tenancy at sufferance A lease arrangement by which the tenancy of the lessee is at the sufferance of the lessor and can be terminated at any time.

Tenancy by the entirety An agreement between spouses allowing for the right of survivorship, thus facilitating the legal transfer of property upon the death of a spouse and avoiding any disagreement as to inheritance of the property.

Tenancy in common A concurrent ownership in a property that is undivided and proportional.

Theory X One of two management theories developed by Douglas McGregor; it states that employees are basically lazy and unmotivated and must be forced to perform through the use of threats and cajoling.

Theory Y One of two management theories developed by Douglas McGregor; it states that employees are basically motivated and want to achieve and are therefore more motivated by feelings of accomplishment and opportunities to excel than they are by threats and monetary reward.

Theory Z A management theory based on what has become known as the Japanese school of management, which views the worker as a member of the corporate family, motivated by feelings of loyalty and belongingness to a particular firm rather than by personal gain. It is a form of managerial paternalism.

Therbligs Classifications of simple hand movements and other worker activities into individual elements, originally developed by the Gilbreths.

Time-reduction curve A graphic representation of the rate at which workers vary their level of productivity depending on the length of time they have been at a job.

Time study That part of operations standards development which investigates and analyzes operations in an attempt to determine standard time, or the performance that the average worker, operating at normal speed and efficiency, is expected to maintain.

Tolerance In the production process, the minimal acceptable degree of accuracy inherent in the physical characteristics of a part or production.

Total cost Fixed cost plus variable cost.

Total revenues Mathematically, $TR = (p)(q)$, where TR = total revenue, p = price per unit, and q = quantity of units sold.

Transportation hub An urban area serving as the center of transportation and communications activities for a geographic region, which connects that region with other transportation hubs.

Turning A shaping process in which a cylindrical piece is placed in a machine that rotates it about the axis of the cylinder and a blade is applied to the surface and material is cut away, creating a cylindrical surface of desired diameter and shape.

Unadjusted rate of return on average investment An approach to the determination of a simple rate of return on investment in which both initial investment costs and ongoing costs are taken into account. Mathematically, it is equal to (average net income)/(average investment).

Uncertain occurrences Relatively rare events, whose probability of occurrence is extremely small. These occurrences include changes in climatic conditions, union and regulation problems, war, and changes in the market. Because of the extremely small probability of occurrence, these events are insurable, although many companies choose to deal with them through contingency planning instead.

Uniform commercial code (UCC) A code of law accepted in all states with the exception of Louisiana, which allows for consistency in commercial practices from one location to another.

Usury The practice of charging a higher interest rate than is allowable under the law.

Variable costs Costs that are directly related to the production of goods and therefore change, or vary, with the level of production.

Variable deductible A hybrid form of fixed-risk insurance in which the amount of the deduction varies with conditions, circumstances and time.

Venture analysis Investigation of potential production facilities projects in terms of costs, legal responsibilities, and project impact on the community and the environment.

Vibration Any repetitive physical movement or agitation, usually of a relatively high frequency.

Warehouse facilities Facilities designed to store and protect raw materials and goods until such time as they are transferred for processing or shipped for sale.

Welding A joining process in which two pieces to be joined are heated along the joining surfaces until they are hot enough to melt together or hot enough to be hammered together.

Wholesale facilities Facilities designed to accomplish or support the sale of goods and services to retailers or other middlemen wholesalers.

Work The exertion of energy to achieve some goal.

Work content The actual physical steps and duties involved in a specific body of work.

Work motivation The collection of physical and psychological factors that encourage workers to carry out the duties of their respective jobs.

Work sampling An analysis technique that samples activities over a long period of time and at random intervals, to determine the activities taking place, and the percentage of time devoted to each.

Wraparound mortgage A mortgage that includes existing mortgages and is considered subordinate to those mortgages.

Zoning commission A regulatory board in a given municipality whose duties include the determination of the types and allowable locations of industry for their community, before construction can begin.

Zoning laws Laws representing the most common land-use control mechanism at the local level, which divides land into zones and controls the uses to which that land can be legally put, regulating both the type of use allowable and the intensity of use allowable for the land within a given zone.

Zoning variance A suspension of zoning regulations for a specific piece of property when compliance would create an undue hardship on the landowner.

Problem Appendix

CHAPTER 2

1. A businessperson was analyzing some information about the business the facility was producing. This person observed the following data.

Inputs	1	2	3	4	5	6	7	8	9	10
Outputs	2	6	14	30	42	50	58	64	68	60

   At what point does the relationship between inputs and outputs begin to decrease?
2. Using the input/output information from Problem 1, graph the data.
3. On the graph for Problem 2, indicate where the output is still increasing but at a decreasing rate. Also, indicate where the output begins to fall.
4. How many outputs does a business need to break even given the following figures from an accountant?

   Selling price = $1500
   Variable price = $1000
   Fixed costs = $9750

5. Using the break-even information from Problem 4, graph the break-even point for the facility.
6. Given the following information about the operating expenses of a facility, determine the break-even points, using a graph (the expenses, or inputs, are expressed in thousands; outputs are expressed in units).

Output	10	20	30	40	50	60	70	80
Input	$20	$32	$45	$52	$60	$80	$105	$140

Selling price = $1,500/unit

7. Using the information from Problem 6, determine where the most profitable output level would be.

CHAPTER 11

1. Using a learning rate of 80 percent, what would be the amount of time needed to work the eighth unit, if 160 hours are needed to work the first unit?
2. Using a learning rate of 80 percent, how long would it take to produce the fourth unit, if it takes 80 hours to produce the first?
3. Using a learning rate of 90 percent, how long would it take to produce the fourth unit, if it takes 20 hours to produce the first?
4. Using a learning rate of 90 percent, how long would it take to produce the fifteenth unit, if the first unit takes 1,000 hours?
5. Using a learning rate of 90 percent, how long would it take to produce five units, if the first unit takes 580 hours?

CHAPTER 13

1. What is the rate of return per year of an $15,000 investment that would yield you $2,800 per year?
2. Using the five-year straight-line depreciation method, what is the depreciated value (book value) of an item purchased for $10,000 at the end of the third year? The salvage value of the item is $800.
3. Using the sum of the digits methods of depreciation, refigure Problem 2.
4. Using the double declining balance method of depreciation, refigure Problem 2.
5. If at the end of five years the value of money is worth $10,000 because it was invested at an 8 percent earnings rate. What was the original investment?
6. What is the internal rate of return for a project that costs $50,000 and yields $12,000 per year for five years?
7. A business broker locates for you a business to purchase. By using $50,000 from your retirement account, you can purchase the business. Your retirement account earns 6 percent. This business nets $3,800 in income each year. Should you buy the business?
8. What is the value of a $500 bill if you hid it under your mattress for one year when the inflation rate was 8 percent?
9. If you were promised a $500 bill a year from now, what would you take now in cash, if the inflation rate is 6 percent per year?
10. What is the present value of money, if after four years you would receive $10,000. The current inflation rate is 8 percent.

11. What is the company's internal rate of return, when your actual cash flow per year is $8,000 for eight years? Your initial cash investment into the business was $50,000.
12. What is your internal rate of return, when your actual cash flow per year is $7,500 for six years with an initial cash investment of $95,000?
13. What is the after-tax return for the project, when the net gain from the project is $48,000 and the net investment is $510,000?
14. What is the profitability index of a high-tech machine, if the original cost was $110,000 and the future value of the machine in five years is $185,000 when the internal rate of return for this business is 8 percent?
15. Which facility site plan is more profitable? Site A with an original cost of $1,100,000 and a future value of $2,500,000 with a projected internal rate of return for this facility of 8 percent? Or site B, with an original cost of $800,000 and a future value of $1,200,000 with a projected internal rate of return for this facility of 6 percent? Both facilities will be used for five years.

CHAPTER 14

Use the load distance table below to answer the following questions.

Department	*1*	*2*	*3*	*4*
1. Cleaning		120		
2. Welding			80	
3. Polishing				60
4. Painting				

The distance between departments has a value of 2.

1. Using the preceding load distance model, solve the following facility arrangement for the load/distance factor, if the facility is arranged in this departmental order: 1→2→3→4.
2. Using the preceding load distance model, solve the following facility arrangement for the load/distance factor, if the facility is arranged in this departmental order: 1→3→4→2.
3. Assume that three products have facility arrangement load factors as follows:

   Product A = 230
   Product B = 489
   Product C = 980

   What is the combined facility load factor if the products are manufactured using the facility with this ratio of production: 45:35:20?
4. What is the combined facility load factor for Problem 3 if the ratio changes to 20:35:45?
5. If the combined maximum load factor for the facility is 500, could we produce the products in Problem 3 in this plant using the last two ratios?

CHAPTER 15

Using the following information table, solve the following work standard problems.

Sales
(000s)

Jan	*Feb*	*Mar*	*Apr*	*May*	*Jun*	*Jul*	*Aug*	*Sep*	*Oct*	*Nov*	*Dec*	*Total*
$30	$28	$26	$31	$30	$24	$18	$28	$60	$35	$36	$30	$376

1. Using the following established work standard for productivity of $40 per employee work hour, estimate the annual number of employee work hours needed to operate this facility.
2. Using the same work standard, figure out how many employee work hours are needed every month.
3. Using a full-time employee work standard of 172 hours per month and the monthly hours needed from Problem 2, figure out how many full-time employees are needed in the slowest month (July).
4. Using a part-time employee work standard of 86 hours per month, figure out how many part-time employees are needed in the months of January, July, and September when you already have three full-timers on the payroll.
5. If the work standard were raised to $50 per hour, what would the new total required hours be for the facility? How much of an increase in productivity would this be?

Solutions to Problems

CHAPTER 2

1.

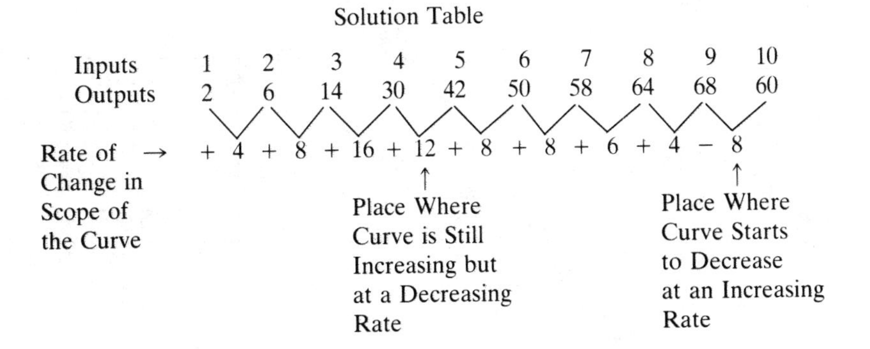

2. and 3.

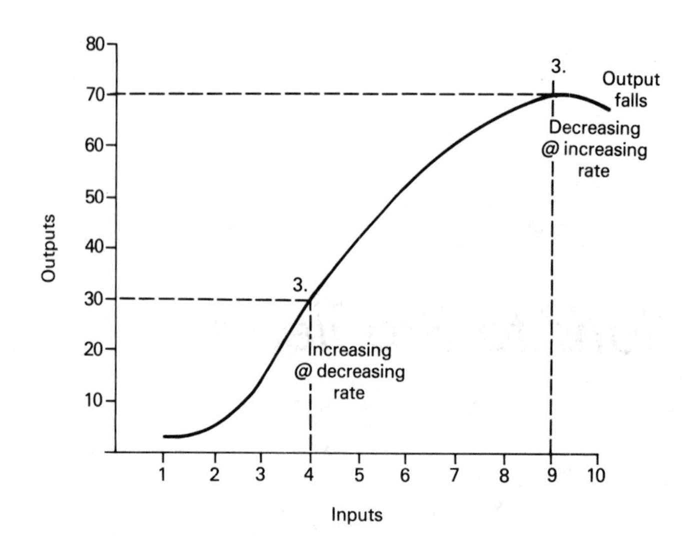

4.

Breakeven Point = Fixed Cost/Selling Price − Variable Costs

Breakeven Point = \$9,750/\$1,500 − \$1000

Breakeven Point = \$9,750/\$500 = 19.5 units (20 units)

5.

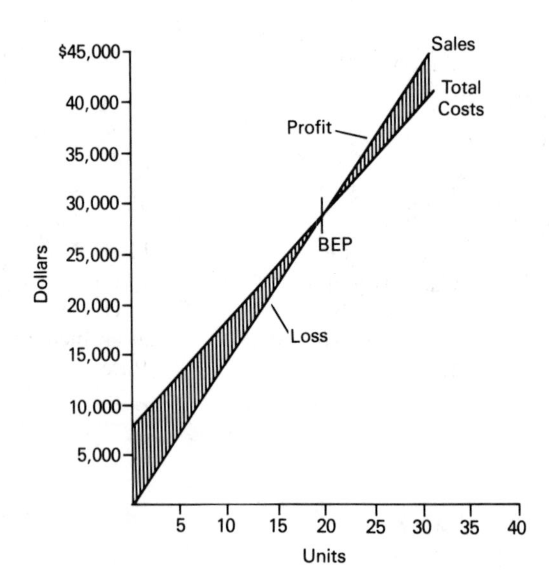

Solution Table

Units	Sales ($)	Total Costs ($)
5	7,500	14,750
10	15,000	19,750
15	22,500	24,750
20	20,000	29,750
25	37,500	34,750
30	45,000	39,750

$$\text{Total costs} = \text{FC} + \text{VC (units)}$$

$$\text{Total costs (5 units)} = \$9{,}750 + \$1{,}000(5) = \$14{,}750$$

$$\text{Total costs (10 units)} = \$9{,}750 + \$1{,}000(10) = \$19{,}750$$

$$\text{Sales} = \text{(Units) (Selling price)}$$

$$\text{Sales} = (5)(\$1{,}500) = \$7{,}500$$

$$\text{Sales} = (10)(\$1{,}500) = \$15{,}000$$

6. and 7.

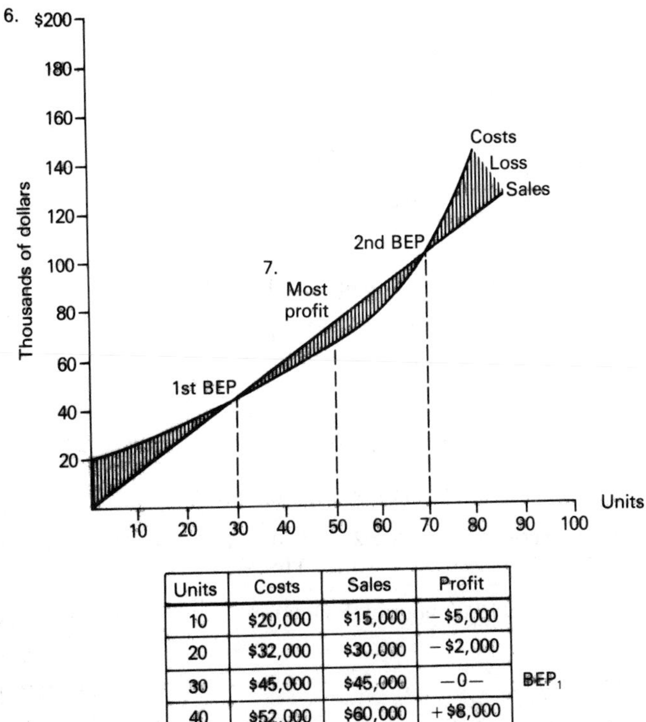

Units	Costs	Sales	Profit	
10	\$20,000	\$15,000	– \$5,000	
20	\$32,000	\$30,000	– \$2,000	
30	\$45,000	\$45,000	–0–	BEP_1
40	\$52,000	\$60,000	+ \$8,000	
50	\$60,000	\$75,000	+ \$15,000	Most Profit
60	\$80,000	\$90,000	+ \$10,000	
70	\$105,000	\$105,000	–0–	BEP_2
80	\$140,000	\$120,000	– \$20,000	

Solution Table

See Problem #5 for Formulas to Solve Solutions Table

CHAPTER 11

1.

T_N rate 80% = 1st unit 1.00 × 100% = 1.00

T_2 rate 80% = 2nd unit 1.00 × 80% = 0.80

T_4 rate 80% = 4th unit 0.80 × 80% = 0.64

T_8 rate 80% = 8th unit 0.64 × 80% = 0.512

$(T_1)(T_N \text{ rate})$

(160 hr)(0.512) = 81.92 hours

2.

T_N rate (see #1)

T_4 rate = 0.64

$(T_1)(T_N \text{ rate})$

(80 hr)(0.64) = 51.2 hours

3.

T_N rate 90% = 1 unit 1.00 × 100% = 1.00

T_2 rate 90% = 2 units 1.00 × 90% = 0.90

T_4 rate 90% = 4 units 0.90 × 90% = 0.81

$(T_1)(T_N \text{ rate})$

(20 hr)(0.81) = 16.2 hours

4. Because not in exponent sequence 2, 4, 8, 16, 32, etc., must use the learning curve formula $T_N = T_1 N^b$ where $b = -0.1520$ for 90%

$$T_{15} = (1{,}000 \text{ hours})(15 \text{ units})^{-0.1520} = 662.6 \text{ hours}$$

5.

$$T_N = T_{1N}{}^b$$

T_1 = (580 hours)(1st unit)$^{-0.1520}$ = (580)(1) = 580 hours

T_2 = (580 hours)(2nd unit)$^{-0.1520}$ = (580)(.90) = 522 hours

T_3 = (580 hours)(3rd unit)$^{-0.1520}$ = (580)(.85) = 493 hours

T_4 = (580 hours)(4th unit)$^{-0.1520}$ = (580)(.81) = 470 hours

T_5 = (580 hours)(5th unit)$^{-0.1520}$ = (580)(.78) = 452 hours

2,517 hours total to make 5 units

CHAPTER 13

1.

$$\text{Rate of return} = \frac{\text{Net income}}{\text{Total investment}} \times 100\% =$$

$$\text{Rate of return} = \frac{\$2{,}800}{\$15{,}000} \times 100\% = 18.67\%$$

2.

(a) $10,000 cost
−800 salvage
$ 9,200 depreciation

(b) $\frac{9{,}200}{5 \text{ years}} = \$1{,}840/\text{year} \times 3 \text{ year} = \$5{,}520$

(c) $10,000 cost
−5,520 depreciation
$4,480 book value

3. $$\text{Depreciation}_{3\text{ years}} = \frac{\text{Remaining useful life}}{\text{Sum of digits}}(\text{Cost} - \text{Salvage})$$

$$\text{Depreciation}_{1\text{ year}} = \frac{5}{1+2+3+4+5}(\$10{,}000 - \$800) = \$3{,}067$$

$$\text{Depreciation}_{2\text{ years}} = \frac{4}{1+2+3+4+5}(\$10{,}000 - \$800) = \$2{,}453$$

$$\text{Depreciation}_{3\text{ years}} = \frac{3}{1+2+3+4+5}(\$10{,}000 - \$800) = \$1{,}840$$

$7,360 total depreciation

$10,000 cost
−7,360 depr
$ 2,640 book value (depreciated value)

4.

(a) $10,000 cost
−$800
$ 9,200

(b) $\left(\frac{1 \text{ year}}{5 \text{ years}} = 20\%\right) \times 2 = 40\%$ double declining factor

(c) ($9,200) 40% = $3,680 year 1
($9,200 − $3,680) 40% = $2,208 year 2
($9,200 − $5,888) 40% = $1,325 year 3
$7,213 total depreciation

(*Note: $2,208 + $3,680 = $5,888)

$10,000 cost
−7,213 depreciation
$ 2,787 book value

5.

$$PV = FV_t \left[\frac{1}{(1 + r)^t} \right]$$

$$PV_5 = \$10{,}000 \left[\frac{1}{(1 + 0.08)^5} \right] = \$6{,}806$$

6.

$$IRR = \sqrt[t]{\frac{\text{Actual cash flow (money} \times \text{years)}}{\text{Initial outlay}}} - 1\ (100\%)$$

$$IRR = \sqrt[5]{\frac{(\$12{,}000 \times 5\text{ yrs})}{\$50{,}000}} - 1\ (100\%) = 1.037 - 1.00\ (100\%) = 3.7\%$$

7.

$$\text{Rate of return} = \frac{\text{Net income}}{\text{Total investment}}$$

$$\text{Rate of return} = \frac{\$3{,}800}{\$50{,}000} = 7.6\%$$

Buy business, +1.6% better return

8.

For investment $PV = FV \left[\frac{1}{(1 + r)^t} \right]$; for *not* investing $PV = FV \left[\frac{1}{(1 - r)^t} \right]$

$$\frac{1}{FV} = \frac{\left[\frac{1}{(1 - r)^t} \right]}{PV} = \frac{1}{FV} = \frac{\left[\frac{1}{(1 - .08)^1} \right]}{\$500} = 0.0021739 \qquad FV = \frac{1}{0.0021739} = \$460.00$$

or $PV = FV\left[\frac{1}{(1 - r)^t} \right]$; $\$500 = FV\left[\frac{1}{(1 - .08)^t} \right]$; $\$500 = FV[1.086] = \$460.00$

9.

$$PV = FV \left[\frac{1}{(1 + r)^t} \right]$$

$$PV = \$500 \left[\frac{1}{(1 + 0.06)^1} \right] = \$471.70$$

10.

$$PV = FV_t \left[\frac{1}{(1 + r)^t} \right]$$

$$PV = \$10{,}000 \left[\frac{1}{(1 + 0.08)^4} \right] = \$7{,}350$$

11.

$$IO = \frac{ACF}{(1 + IRR)^t}$$

$$IRR = \left(\sqrt[t]{\frac{ACF}{IO}} - 1\right)(100\%) = \left(\sqrt[8]{\frac{\$8{,}000 \times 8 \text{ yr}}{\$50{,}000}} - 1\right)(100\%) = +3.13\%$$

12.

$$IO = \frac{ACF}{(1 + IRR)^t}$$

$$IRR = \left(\sqrt[t]{\frac{ACF}{IO}} - 1\right)(100\%) = \left(\sqrt[6]{\frac{\$7{,}500 \times 6 \text{ yr}}{\$95{,}000}} - 1\right)(100\%)$$

$$= (+0.8829 - 1.0000)(100\%)$$

$$= -11.70\% \text{ (loss)}$$

13.

$$\text{After-tax return} = \frac{\text{Net gain after tax}}{\text{Net investment}}(100\%)$$

$$ATR = \left(\frac{\$48{,}000}{\$510{,}000}\right)100\% = 9.41\%$$

14.

$$\text{Profitability index} = \frac{\text{Original cost} + \text{Net present value}}{\text{Original cost}}$$

$$PV = FV_t\left[\frac{1}{(1 + r)^t}\right]$$

$$PV = \$185{,}000\left[\frac{1}{(1.08)^5}\right] = \$125{,}908$$

$$PI = \frac{\$110{,}000 + \$125{,}908}{\$110{,}000} = 2.145$$

15.

	Site A	*Site B*
PV =	$\$2{,}500{,}000\left[\frac{1}{(1 + 0.08)^5}\right]$	$\$1{,}200{,}000\left[\frac{1}{(1 + 0.06)^5}\right]$
PV =	\$1,701,458	\$896,710
PI =	$\frac{\$1{,}100{,}000 + \$1{,}701{,}458}{\$1{,}100{,}000}$	$\frac{\$800{,}000 + \$896{,}710}{\$800{,}000}$
PI =	2.547 (more profitable)	2.121

CHAPTER 14

1.

$$\text{Load factor} = \sum_{i=1}^{n}\sum_{j=1}^{n} L_{ij}D_{ij} = \sum \text{(Load)(Distance)}$$

In 1 | 2 | 3 | 4 → Out

Work flow table
Departmental order

Hint: 1 to 2 + 2 to 3 + 3 to 4

LF = (120)(2) + (80)(2) + (60)(2) =

LF = 240 + 160 + 120 = 520

2.

$$\text{Load factor} = \sum_{i=1}^{n}\sum_{j=1}^{n} L_{ij}D_{ij} = \sum \text{(Load)(Distance)}$$

In 1 | 3 | 4 | 2 → Out

Work flow table
Departmental order

Hint: 1 to 2 + 2 to 3 + 3 to 4

LF = (120)(2)(2)(2) + (80)(2)(2) + (60)(2) =

LF = 960 + 320 + 120 = 1400

3.

230 × 45% = 103.5

489 × 35% = 171.2

980 × 20% = 196.0

470.7 load factor

4.

230 × 20% = 46

489 × 35% = 171.2

980 × 45% = 441.0

658.2 load factor

5. Problem 3 with a load factor of 470.7: Product combination could be built. Problem 4 with a load factor of 658.2: Product combination could *not* be built since the combined load factor of the factory cannot exceed 500.

CHAPTER 15

1.

$$\frac{\$376{,}000 \text{ sales}}{\$40 \text{ sales/hour}} = 9{,}400$$ total employee work hours needed to operate facility

2.

$$\text{Jan} = \frac{\$30{,}000}{\$40} = 750 \text{ hr} \qquad \text{Jul} = \frac{\$18{,}000}{\$40} = 450 \text{ hr}$$

$$\text{Feb} = \frac{\$28{,}000}{\$40} = 700 \qquad \text{Aug} = \frac{\$28{,}000}{\$40} = 700$$

$$\text{Mar} = \frac{\$26{,}000}{\$40} = 650 \qquad \text{Sep} = \frac{\$60{,}000}{\$40} = 1500$$

$$\text{Apr} = \frac{\$31{,}000}{\$40} = 775 \qquad \text{Oct} = \frac{\$35{,}000}{\$40} = 875$$

$$\text{May} = \frac{\$30{,}000}{\$40} = 750 \qquad \text{Nov} = \frac{\$36{,}000}{\$40} = 900$$

$$\text{Jun} = \frac{\$24{,}000}{\$40} = 600 \qquad \text{Dec} = \frac{\$30{,}000}{\$40} = 750$$

Total 9,400 payroll hours needed to operate facility. See Problem 1.

3.

$$\text{July} = \frac{450 \text{ hours}}{172 \text{ hours/full-timer}} = 2.61 \text{ or } 3 \text{ full-timers needed}$$

4.

$$\text{January} = 750 \text{ hr} - (3 \text{ full-timers} \times 172 \text{ hr}) = 234 \text{ part-time hours required}$$

$$= \frac{234 \text{ part-time hours}}{86 \text{ hours/part-timer}}$$

$$= 2.72 \text{ or } 3 \text{ part-timers}$$

$$\text{July} = 450 \text{ hr} - (3 \times 172 \text{ hr})$$

$$= -66 \text{ part-time hours (send a full-timer on vacation)}$$

$$\text{September} = 1{,}500 \text{ hr} - (3 \times 172 \text{ hr}) = 984 \text{ part-time hours required}$$

$$= \frac{984 \text{ part-time hours}}{86 \text{ hours/part-timer}}$$

$$= 11.4 \text{ or } 11 \text{ part-timers}$$

5.

$$\frac{\$376{,}000 \text{ sales}}{\$50 \text{ sales/hr}} = 7{,}520 \text{ hours total for year}$$

$$\frac{9{,}400 \text{ old hours}}{7{,}520 \text{ new hours}} = 1.25$$

or a 25% increase in facility productivity

Index